Principles of
Computer Science

Principles of Computer Science

Editor

Donald R. Franceschetti, PhD

The University of Memphis

SALEM PRESS

A Division of EBSCO Information Services, Inc.

Ipswich, Massachusetts

GREY HOUSE PUBLISHING

Publisher's Cataloging-In-Publication Data
(Prepared by The Donohue Group, Inc.)

Names: Franceschetti, Donald R., 1947- editor.
Title: Principles of computer science / editor, Donald R. Franceschetti, PhD, the University of Memphis.
Description: [First edition]. | Ipswich, Massachusetts : Salem Press, a division of EBSCO Information Services, Inc. ; [Amenia, New York] : Grey House Publishing, [2016] | Series: Principles of | Includes bibliographical references and index.
Identifiers: ISBN 978-1-68217-139-4 (hardcover)
Subjects: LCSH: Computer science.
Classification: LCC QA76 .P75 2016 | DDC 004—dc23

PRINTED IN THE UNITED STATES OF AMERICA

CONTENTS

Publisher's Note

Salem Press is pleased to add *Principles of Computer Science* as the fourth title in a new *Principles of* series that includes *Chemistry, Physics, Astronomy,* and *Computer Science.* This new resource introduces students and researchers to the fundamentals of computer science using easy-to-understand language, giving readers a solid start and deeper understanding and appreciation of this complex subject.

The 117 entries range from 3D printing to Workplace Monitoring and are arranged in an A to Z order, making it easy to find the topic of interest. Entries include the following:

- Related fields of study to illustrate the connections between the various branches of computer science including computer engineering, software engineering, biotechnology, security, robotics, gaming, and programming languages;
- A brief, concrete summary of the topic and how the entry is organized;
- Principal terms that are fundamental to the discussion and to understanding the concepts presented;
- Illustrations that clarify difficult concepts via models, diagrams, and charts of such key topics as wide area networks (WAN), electronic circuits, and quantum computing;
- Photographs of significant contributors to the field of computer science;
- Sample problems that further demonstrate the concept presented;
- Further reading lists that relate to the entry.

This reference work begins with a comprehensive introduction to the field, written by editor Donald R. Franceschetti, PhD, starting with the development of the first computers, an explanation of analog versus digital computing, neural networks, the contributions of Alan Turing, and a discussion of how advances from vacuum tubes to digital processors have fundamentally changed the way we live.

The book's backmatter is another valuable resource and includes:

- A timeline of the developments that related to and led up to the first modern computer, starting in 2400 BCE with the invention of the abacus in Babylonia and ending with the 1949 *Popular Mechanics* prediction that computers of the future might weigh no more than 1.5 tons;
- A time of the development of microprocessors from the Intel 4004 in 1971 to the IBM POWER8 in 2014;
- Nobel Notes that explain the significance of the prizes in physics to the study of the science and
- List of important figures in computer science and their key accomplishment;
- Glossary;
- General bibliography; and
- Subject index.

Salem Press and Grey House Publishing extend their appreciation to all involved in the development and production of this work. The entries have been written by experts in the field. Their names and affiliations follow the Editor's Introduction.

Principles of Computer Science, as well as all Salem Press reference books, is available in print and as an e-book. Please visit www.salempress.com for more information.

Editor's Introduction

What is Computer Science?

Computer science generally refers to the body of knowledge that allows humans to use mechanical aids to do mathematical calculations very efficiently. While the term computer originally referred to a human who was able to do simple arithmetic quickly and accurately, the name has been transferred to programmable automata, which are able to do the same thousands to billions of times faster.

Analogue or Digital

There are two basic kinds of computers: analogue and digital. In an analog process, the system of interest is modeled by a system obeying the same dynamical laws. Thus a system of mechanical vibrations might be modeled by a system of capacitors, inductors, and resistors. In a digital computer, the quantity of interest is generally encoded in some way and one obtains coded representation of the solution. Some of the advances are fairly mundane and are actually possible to achieve with low-tech adding machines, if one has enough time. Others are a bit more esoteric, requiring some abstract mathematics. Almost all computing machines, with the notable exceptions of the abacus and slide rule, have come into existence since the beginning of the twentieth century and rest on the fundamental discoveries made about electromagnetism since 1800, as well as more recent insights gained into the process of computation.

Digital Computation Requires a Code

Digital computation begins with the selection of a code so that information can be stored and sent over significant distances. It can be difficult for those living in the twenty-first century to appreciate the full impact of the development of telegraphy, by which information (letters and numbers) could be transmitted from one place to another as a series of dots and dashes, using a pair of telegraph wires strung between two cities along with the necessary relay equipment. For the first time, the results of a baseball game or a battle or the day's stock trading could be known almost instantly. It's not surprising that so many innovators in the area of information exchange began their careers as telegraph operators.

The earliest innovations in telegraphy greatly increased the number of messages that could be sent simultaneously. It was eventually learned that by using more than two conductors, several symbols could be sent at the same time. Of course, it should be noted that the original Morse code used to tap out messages with a series of dots and dashes was highly inefficient and error prone. Suppose seven wires carried voltage simultaneously. With seven voltages, one can encode $2^7 = 128$ symbol choices, a calculation that accounts for the digits of the decimal system plus upper and lower case letters plus several punctuation marks. The telegraph thus becomes a teletype machine capable of transmitting information a hundred times faster.

Once the process of telegraphy was in place, it was a simple matter to switch from the network of telegraph stations to radio. Once amplitude modulation was discovered voice could be transmitted over the air or over wires with ease. Television eventually followed. Many of the pioneers of electronic technology were self-educated men—men like Thomas Edison, and Michael Faraday before him—who could read well enough but functioned outside the establishment of colleges and academic degrees. Even today there is some controversy over the value of traditional education as many pioneers of the computer sciences dropped their pursuit of traditional degrees, such as Bill Gates and Steve Jobs, while means that the question still remains open.

Still, formal education came to be an important aspect of the inexorable march toward the computer age. In the United Kingdom, while social standing may have prevent many from attending a university enrollment, entire institutions were set up to teach those who had not the pedigree to be accepted at Oxford or Cambridge, notably the Royal Institution, a school operating under Royal charter and featuring both Sir Humphrey Davy and Michael Faraday (who both declined to be knighted) as lecturers. Things were somewhat more democratic in the United States, with public universities in almost every state that were supported by agriculture and expected to make a contribution in return. Private institutions set up chapters of Phi Beta Kappa to encourage traditional study of

Latin and Greek, while Institutes of Technology like the Massachusetts Institute of Technology offered degrees in the Sciences and Engineering.

One of the important areas of computer science is the selection and security of codes. An in-depth treatment of the topic is provided by Claude Shannon and Warren Weaver in two papers originally published in *Scientific American* and in the *Bell System Technical Journal*. The very existence of a widely respected scientific journal published not by a non-profit organization, but rather as a house organ of Bell System shows how the system of scientific publication had to adapt to the emergence of a very large corporation that of necessity has control of a major portion of the work to be done.

Shannon begins by defining the entropy of a text in terms of the probabilities of the symbols. Scrambling the symbols in a text corresponds to an increase in entropy and the probabilities of certain errors. Shannon and others have made very effective use of the entropy concept, which was originally proposed by Ludwig Boltzmann in the context of the molecular theory of heat. Among other things, the creation of information results in the generation of heat and so one of the major technical issues in the design of highly powerful computers is the need to transport the heat away as it is generated.

The Development of the Digital Computer
In 1900, the great German mathematician David Hilbert gave a talk at the Second International Conference of Mathematicians in Paris, France, during which he described twenty-three mathematical problems that he expected to be solved in the twentieth century. His tenth problem dealt with polynomial equations with integer coefficients (Diophantine equations). Hilbert asked if it could be determined in a finite number of steps whether the equation could be solved in rational numbers. The matter in question was not to find the solution, but simply to find out if a solution could be found.

One of the world's true polymaths, Alan Turing, took on the challenge of solving Hilbert's problem–the so-called "decision problem" or *Entscheidungsproblem*. To solve this problem, Turing looked closely at the process of solution. He begins by looking at a solution as a practical problem, in other words, as the sequence of steps that a mathematician or a young child might go through solving

an arithmetic problem by working out the solution one digit at a time. It is a truism that the solution to practical problems is sometimes found by "thinking outside the box" and considering mechanisms that might seem to have nothing to do with the problem at hand. In the process of solving Hilbert's decision problem, Turing conceived of a machine that could read and write one digit at a time, based on instructions that could be written on the same tape that would carry the solution—known today as the Turing Machine.

It has been suggested by a number of Turing biographers that since Alan enjoyed working with actual typewriters, it was in fact his practical bent that led him to conceptualize the Turing machine and then the universal Turing machine. The set of instructions constitutes the computer program. The numbers written on the tape at the start of the tape constitute the *input* or *data* and the numbers written on the tape after the computer is done are the *output*. Numbers which might be written by such a machine are called computable numbers. Turing published his first paper "On Computable Numbers with an application to the *Entscheidungsproblem*" in 1937. This paper marks the theoretical development of the digital computer.

Turing received his PhD in mathematics in 1938 and by 1939 was at work as a cryptographer in the British Foreign Office. It was during this period of his life—which became the subject of the Academy Award-winning movie, *The Imitation Game*—that he was able to apply his efforts to produce the "enigma" coding machine capable of breaking the German code and ultimately, saving numerous lives.

It turns out that one can write a program capable of accepting the description of another Turing machine as data and then emulating it. Such programs are called universal Turing machines. The modern programmable digital computer is, from an abstract point of view, just such a machine. As one of the first individuals to take the possibility of artificial intelligence seriously, Turing proposed a test, known today as the Turing test, to tell whether a machine could be considered intelligent.

Background of the Turing Machine
The digital computer has undergone many changes since Turing first developed it to deal with the urgencies of the Second World War. Today's digital

computer as a practical device is the result of developments in a large number of fields.

The nineteenth century had been a trying time for the natural sciences. While the physics of 1900 left little room for new discoveries, the discovery of subatomic particles like the electron and the nuclear structure of the atom called for a fundamental reexamination of assumptions not challenged since the time of Isaac Newton. Einstein had opened the possibility that the geometry of space was not that proposed by Euclid over two thousand years ago. Philosophically-minded scientists turned to mathematics for sure and certain knowledge. Of all areas of mathematics, the most noncontroversial was the theory of arithmetic, in which there was only one item of unfinished business: the principle of mathematical induction, which seemed to be an unnecessary part of the structure of mathematics. This principle stated that if N was an integer, and if P(N) was a true statement about N that implied the truth of the statement P(N+1), then P(N) was true for all N. A number of philosophers of mathematics went to work trying to show that mathematics could be constructed in a self- consistent way without the troublesome postulate.

By far the most thoroughgoing approach to eliminating the principle of mathematical induction could be found the three-volume *Principia Mathematica,* published in 1911 (and still in print!) by Bertrand Russell and Alfred North Whitehead. Although a fallacy inherent in *Principia Mathematica* was soon discovered, the book remains a valuable exposition of symbolic logic. In 1931 Kurt Godel ,who would later become a very close friend of Albert Einstein published a paper "On formally undecidable systems…" that put an end to the matter by showing that in any system of axioms for arithmetic that allowed multiplication, there would necessarily be propositions which could not be decided within the system.

Neural Networks
In 1943, however, Warren S. McCulloch and Walter Pitts published "A logical calculus of the ideas immanent in nervous activity" showing that for each formula in the notation of *Principia Mathematica,* an equivalent network of all or none neurons could be found. Work on neural networks has continued to this day.

Von Neumann's Contributions
Von Neumann was a very unusual physicist, one of a group of immigrants known collectively as "Martians" because their Hungarian speech resembled no known European language; other Martians included Edward Teller and Leo Szilard. Unlike the majority of academics, he had adequate financial means and was able to attend classes at several universities. He did not shy away from positions of political and social influence, even accepting President Eisenhower's appointment as Commissioner of Atomic Energy. While his PhD degree was given for a thesis on the "Mathematical Foundations of Quantum Mechanics," he was also interested in practical electronics. As a member of the Institute for Advanced Studies at Princeton New Jersey, he oversaw the construction of one of the first vacuum tube computers. In the last years of his life, Dr. John von Neumann prepared to give the Silliman lectures at Yale University, his topic: The Computer and the Brain. While he was unable to deliver the Silliman lectures for health reasons, the manuscript was eventually published,

The First Electronic Computers
Prepared with a basic understanding of neural networks as devices that model human thought, we turn now to the problems of electronic devices that can undertake computation. The essential requirement is for a two-state (or greater) device. The Edison effect, observed by Thomas Edison in 1883, provided the basis for the first such devices. Edison had noted the emission of electrons from a heated metal electrode when maintained in vacuum, but he merely noted the effect in his laboratory notebook. Eventually, however, the British engineer, Sir John Ambrose Fleming, used a heated cathode in the first vacuum tube diode, also called a rectifier since it only allowed electron flow from the heated cathode. An American inventor, Lee de Forest, added a third element, or grid, that allowed a small voltage to control a much larger current.

The Vacuum Tube Era
The first electronic computers employed vacuum tube circuitry and were not always dependable. Vacuum tubes contained a heated element, or filament, that was essential to its proper function. Once the filament burned out, the tube would not conduct when it should. There were two possible approaches

to solving the problem. One could design a test calculation that would use every tube in the computer and for which the result is known. If one ran the test calculation before and after a new calculation was run and obtained the same results, one could be relatively sure that no tubes had burned out during the calculation. An alternative suggested by von Neumann was to build redundancy into the circuit so that the results obtained could be trusted even if a few components proved unreliable.

The Transistor Era: Computers become Profitable

While there were fortunes to be made making business machines, the computer industry offered new opportunities to creative individuals. As semiconductor devices replaced vacuum tubes, it became clear that a few creative individuals could compete effectively against older and well-established corporations. While there was still a need for large mainframe computers, the personal computer had the virtues of both versatility and low cost. The leaders in the field became the hobbyists, many of them still in high school, who established Microsoft and Apple Computer. Bill Gates, founder of Microsoft, would become one of the world's richest men. Steve Jobs, cofounder of Apple Computer, brought the standalone microcomputer to new levels of competitiveness. Facebook's evolution created an increasing appetite for social media. New mechanisms of marketing were found. The internet, first started in order to connect a few physics research laboratories, became a "place" where all manner of goods and services could be bought and sold.

The Future

Clearly computer science is in the midst of a revolution of its own making. It is not possible to predict where or when things will "settle down," or if they ever actually will. Nonetheless, a few outcomes are already clear: Information will become more and more available and there will be an increased need for professionals in the great many fields impacted by the advances described here. If this volume helps the reader to see even a bit more clearly through the maze of opportunity, it has been well worth the effort to bring it together.

—*Donald Franceschetti, PhD*

Bodanis, David. *Electric Universe: The Shocking True Story of Electricity.* New York: Crown Publishers, 2005. Print.

Penrose, Roger. *The Emperor's New Mind: Concerning Computers, Minds, and the Laws of Physics.* Oxford: Oxford University Press, 1989. Print.

Shannon, Claude E, and Warren Weaver. *The Mathematical Theory of Communication.* Urbana: University of Illinois Press, 1999. Print.

Yandell, Ben. The Honors Class: Hilbert's Problems and Their Solvers. Natick, Mass: A.K. Peters, 2002. Print.

Von, Neumann J, and Ray Kurzweil. *The Computer & the Brain.* New Haven, Conn.; London : Yale University Press, 2012. Print.

Contributors

Andrew Farrell, MLIS

Andrew Hoelscher, MEng

Daniel Horowitz

Donald R. Franceschetti, PhD

John Vines

Joseph Dewey

Joy Crelin

Kenrick Vezina, MS

Marianne M. Madsen, MS

Maura Valentino, MSLIS

Melvin O

Micah L. Issitt

Richard M. Renneboog, MSc

Scott Zimmer

Teresa E. Schmidt

Trudy Mercadal, PhD

3-D PRINTING

FIELDS OF STUDY

Computer Science; Digital Media

ABSTRACT

Additive manufacturing (AM), or 3-D printing, comprises several automated processes for building three-dimensional objects from layers of plastic, paper, glass, or metal. AM creates strong, light 3-D objects quickly and efficiently.

PRINICIPAL TERMS

- **binder jetting:** the use of a liquid binding agent to fuse layers of powder together.
- **directed energy deposition:** a process that deposits wire or powdered material onto an object and then melts it using a laser, electron beam, or plasma arc.
- **material extrusion:** a process in which heated filament is extruded through a nozzle and deposited in layers, usually around a removable support.
- **material jetting:** a process in which drops of liquid photopolymer are deposited through a printer head and heated to form a dry, stable solid.
- **powder bed fusion:** the use of a laser to heat layers of powdered material in a movable powder bed.
- **sheet lamination:** a process in which thin layered sheets of material are adhered or fused together and then extra material is removed with cutting implements or lasers.
- **vat photopolymerization:** a process in which a laser hardens layers of light-sensitive material in a vat.

ADDITIVE MANUFACTURING

3-D printing, also called additive manufacturing (AM), builds three-dimensional objects by adding successive layers of material onto a platform. AM differs from traditional, or subtractive, manufacturing, also called machining. In machining, material is removed from a starting sample until the desired structure remains. Most AM processes use less raw material and are therefore less wasteful than machining.

The first AM process was developed in the 1980s, using liquid resin hardened by ultraviolet (UV) light. By the 2000s, several different AM processes had been developed. Most of these processes use liquid, powder, or extrusion techniques. Combined with complex computer modeling and robotics, AM could launch a new era in manufacturing. Soon even complex mechanical objects could be created by AM.

SOFTWARE AND MODELING

3-D printing begins with a computer-aided design (CAD) drawing or 3-D scan of an object. These drawings or scans are usually saved in a digital file format known as STL, originally short for "stereolithography" but since given other meanings, such as "surface tessellation language." STL files "tessellate" the object—that is, cover its surface in a repeated pattern of shapes. Though any shape can be used, STL files use a series of non-overlapping triangles to model the curves and angles of a 3-D object. Errors in the file may need repair. "Slices" of the STL file determine the number and thickness of the layers of material needed.

LIQUID 3-D PRINTING

The earliest AM technique was stereolithography (SLA), patented in 1986 by Chuck Hull. SLA uses liquid resin or polymer hardened by UV light to create a 3-D object. A basic SLA printer consists of an elevator platform suspended in a tank filled with light-sensitive liquid polymer. A UV laser hardens a thin layer of resin. The platform is lowered, and the laser hardens the next layer, fusing it to the first. This process is repeated until the object is complete. The

1

FDM Process

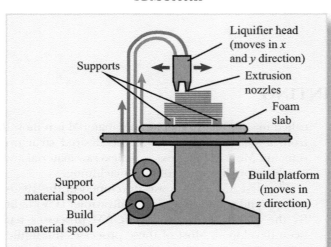

SLA Process

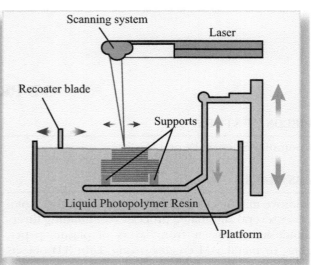

SLS Process

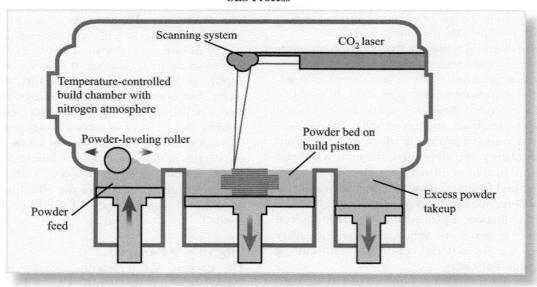

This presents a comparison of the three common 3-D printing processes: SLA (in which liquid polymer resin is solidified by a laser and support material is removed after completion), SLS (in which powder is fused by a CO_2 laser and unfused powder acts as support), and FDM (in which liquid modeling material is extruded through extrusion nozzles and solidifies quickly, and a build material and a support material can be used in tandem, with the support material being removed after completion). EBSCO illustration.

object is then cleaned and cured by UV. This AM technique is also called vat photopolymerization because it takes place within a vat of liquid resin. Various types of SLA printing processes have been given alternate names, such as "photofabrication" and "photo-solidification."

POWDER-BASED 3-D PRINTING
In the 1980s, engineers at the University of Texas created an alternate process that uses powdered solids instead of liquid. Selective layer sintering (SLS), or powder bed fusion, heats powdered glass, metal, ceramic, or plastic in a powder bed until the material is "sintered." To sinter something is to cause its particles to fuse through heat or pressure without liquefying it. A laser is used to selectively sinter thin layers of the powder, with the unfused powder underneath giving structural support. The platform is lowered and the powder compacted as the laser passes over the object again.

EXTRUSION PRINTING
Material extrusion printing heats plastic or polymer filament and extrudes it through nozzles to deposit a layer of material on a platform. One example of this process is called fused deposition modeling (FDM). As the material cools, the platform is lowered and another layer is added atop the last layer. Creating extruded models often requires the use of a structural support to prevent the object from collapsing. Extrusion printing is the most affordable and commonly available 3-D printing process.

EMERGING AND ALTERNATIVE METHODS
Several other 3-D printing methods are also emerging. In material jetting, an inkjet printer head deposits liquefied plastic or other light-sensitive material onto a surface, which is then hardened with UV light. Another inkjet printing technique is binder jetting, which uses an inkjet printer head to deposit drops of glue-like liquid into a powdered medium. The liquid then soaks into and solidifies the medium. In directed energy deposition (DED), metal wire or powder is deposited in thin layers over a support before being melted with a laser or other heat source. Sheet lamination fuses together thin sheets of paper, metal, or plastic with adhesive. The resulting object is then cut with a laser or other cutting tool to refine the shape. This method is less costly but also less accurate than others.

THE FUTURE OF 3-D PRINTING
While AM techniques have been in use since the 1980s, engineers believe that the technology has not yet reached its full potential. Its primary use has been in rapid prototyping, in which a 3-D printer is used to quickly create a 3-D model that can be used to guide production. In many cases, 3-D printing can create objects that are stronger, lighter, and more customizable than objects made through machining. Printed parts are already being used for planes, race cars, medical implants, and dental crowns, among other items. Because AM wastes far less material than subtractive manufacturing, it is of interest for conservation, waste management, and cost reduction. The technology could also democratize manufacturing, as small-scale 3-D printers allow individuals and small businesses to create products that traditionally require industrial manufacturing facilities. However, intellectual property disputes could also occur more often as AM use becomes more widespread.

—*Micah L. Issitt*

BIBLIOGRAPHY
"About Additive Manufacturing." *Additive Manufacturing Research Group.* Loughborough U, 2015. Web. 6 Jan. 2016.

Hutchinson, Lee. "Home 3D Printers Take Us on a Maddening Journey into Another Dimension." *Ars Technica.* Condé Nast, 27 Aug. 2013. Web. 6 Jan. 2016.

"Knowledge Base: Technologies in 3D Printing." *DesignTech.* DesignTech Systems, n.d. Web. 6 Jan. 2016.

Matulka, Rebecca. "How 3D Printers Work." *Energy. gov.* Dept. of Energy, 19 June 2014. Web. 6 Jan. 2016.

"The Printed World." *Economist.* Economist Newspaper, 10 Feb. 2011. Web. 6 Jan. 2016.

"3D Printing Processes: The Free Beginner's Guide." *3D Printing Industry.* 3D Printing Industry, 2015. Web. 6 Jan. 2016.

AGILE ROBOTICS

FIELDS OF STUDY

Robotics

ABSTRACT

Movement poses a challenge for robot design. Wheels are relatively easy to use but are severely limited in their ability to navigate rough terrain. Agile robotics seeks to mimic animals' biomechanical design to achieve dexterity and expand robots' usefulness in various environments.

PRINICIPAL TERMS

- **anthropomorphic:** resembling a human in shape or behavior; from the Greek words *anthropos* (human) and *morphe* (form).
- **autonomous:** able to operate independently, without external control.
- **biomechanics:** the study of how living things move and the laws governing their movement.
- **dexterity:** finesse; skill at performing delicate or precise tasks.
- **dynamic balance:** the ability to maintain balance while in motion.
- **humanoid:** resembling a human.

Robots That Can Walk

Developing robots that can match humans' and other animals' ability to navigate and manipulate their environment is a serious challenge for scientists and engineers. Wheels offer a relatively simple solution for many robot designs. However, they have severe limitations. A wheeled robot cannot navigate simple stairs, to say nothing of ladders, uneven terrain, or the aftermath of an earthquake. In such scenarios, legs are much more useful. Likewise, tools such as simple pincers are useful for gripping objects, but they do not approach the sophistication and adaptability of a human hand with opposable thumbs. The cross-disciplinary subfield devoted to creating robots that can match the dexterity of living things is known as "agile robotics."

Inspired by Biology

Agile robotics often takes inspiration from nature. Biomechanics is particularly useful in this respect, combining physics, biology, and chemistry to describe how the structures that make up living things work. For example, biomechanics would describe a running human in terms of how the human body—muscles, bones, circulation—interacts with forces such as gravity and momentum. Analyzing the activities of living beings in these terms allows roboticists to attempt to recreate these processes. This, in turn, often reveals new insights into biomechanics. Evolution has been shaping life for millions of years through a process of high-stakes trial-and-error. Although evolution's "goals" are not necessarily those of scientists and engineers, they often align remarkably well.

Boston Dynamics, a robotics company based in Cambridge, Massachusetts, has developed a prototype robot known as the Cheetah. This robot mimics the four-legged form of its namesake in an attempt to recreate its famous speed. The Cheetah has achieved a land speed of twenty-nine miles per hour—slower than a real cheetah, but faster than any other legged robot to date. Boston Dynamics has another four-legged robot, the LS3, which looks like a sturdy mule and was designed to carry heavy supplies over rough terrain inaccessible to wheeled transport. (The LS3 was designed for military use, but the project was shelved in December 2015 because it was too noisy.) Researchers at the Massachusetts Institute of Technology (MIT) have built a soft robotic fish. There are robots in varying stages of development

Agile robots are designed to have dexterity, flexibility, and a wider range of motions to allow for more responsiveness to their surroundings. By Manfred Werner - Tsui, CC-BY-SA-3.0 (http://creativecommons.org/licenses/by-sa/3.0/), via Wikimedia Commons.

that mimic snakes' slithering motion or caterpillars' soft-bodied flexibility, to better access cramped spaces.

In nature, such designs help creatures succeed in their niches. Cheetahs are effective hunters because of their extreme speed. Caterpillars' flexibility and strength allow them to climb through a complex world of leaves and branches. Those same traits could be incredibly useful in a disaster situation. A small, autonomous robot that moved like a caterpillar could maneuver through rubble to locate survivors without the need for a human to steer it.

HUMANOID ROBOTS IN A HUMAN WORLD
Humans do not always compare favorably to other animals when it comes to physical challenges. Primates are often much better climbers. Bears are much stronger, cheetahs much faster. Why design anthropomorphic robots if the human body is, in physical terms, relatively unimpressive?

NASA has developed two different robots, Robonauts 1 and 2, that look much like a person in a space suit. This is no accident. The Robonaut is designed to fulfill the same roles as a flesh-and-blood astronaut, particularly for jobs that are too dangerous or dull for humans. Its most remarkable feature is its hands. They are close enough in design and ability to human hands that it can use tools designed for human hands without special modifications.

Consider the weakness of wheels in dealing with stairs. Stairs are a very common feature in the houses and communities that humans have built for themselves. A robot meant to integrate into human society could get around much more easily if it shared a similar body plan. Another reason to create humanoid robots is psychological. Robots that appear more human will be more accepted in health care, customer service, or other jobs that traditionally require human interaction.

Perhaps the hardest part of designing robots that can copy humans' ability to walk on two legs is achieving dynamic balance. To walk on two legs, one must adjust one's balance in real time in response to each step taken. For four-legged robots, this is less of an issue. However, a two-legged robot needs sophisticated sensors and processing power to detect and respond quickly to its own shifting mass. Without this, bipedal robots tend to walk slowly and awkwardly, if they can remain upright at all.

THE FUTURE OF AGILE ROBOTICS

As scientists and engineers work out the major challenges of agile robotics, the array of tasks that can be given to robots will increase markedly. Instead of being limited to tires, treads, or tracks, robots will navigate their environments with the coordination and agility of living beings. They will prove invaluable not just in daily human environments but also in more specialized situations, such as cramped-space disaster relief or expeditions into rugged terrain.

—*Kenrick Vezina, MS*

BIBLIOGRAPHY

Bibby, Joe. "Robonaut: Home." *Robonaut.* NASA, 31 May 2013. Web. 21 Jan. 2016.

Gibbs, Samuel. "Google's Massive Humanoid Robot Can Now Walk and Move without Wires." *Guardian.* Guardian News and Media, 21 Jan. 2015. Web. 21 Jan. 2016.

Murphy, Michael P., and Metin Sitti. "Waalbot: Agile Climbing with Synthetic Fibrillar Dry Adhesives." *2009 IEEE International Conference on Robotics and Automation.* Piscataway: IEEE, 2009. *IEEE Xplore.* Web. 21 Jan. 2016.

Sabbatini, Renato M. E. "Imitation of Life: A History of the First Robots." *Brain & Mind* 9 (1999): n. pag. Web. 21 Jan. 2016.

Schwartz, John. "In the Lab: Robots That Slink and Squirm." *New York Times.* New York Times, 27 Mar. 2007. Web. 21 Jan. 2016.

Wieber, Pierre-Brice, Russ Tedrake, and Scott Kuindersma. "Modeling and Control of Legged Robots." *Handbook of Robotics.* Ed. Bruno Siciliano and Oussama Khatib. 2nd ed. N.p.: Springer, n.d. (forthcoming). *Scott Kuindersma—Harvard University.* Web. 6 Jan. 2016

ALGOL

FIELDS OF STUDY

Programming Languages

ABSTRACT

The ALGOL programming language was developed in 1958 as a program for the display of algorithms. It was elegant and included several design features that have since become staple features of advanced programming languages. ALGOL programs and procedures employ a head-body format, and procedures can be nested within other procedures. ALGOL was the first programming language to make use of start-end delimiters for processes within procedures, a feature now common in advanced object-oriented programming languages. The language allowed recursion and iterative procedures, dynamic array structures, and user-defined data types. Despite its elegance and advanced features, however, ALGOL never became widely used and is now one of the oldest and least used of programming languages.

PRINICIPAL TERMS

- **algorithm:** a set of step-by-step instructions for performing computations.
- **character:** a unit of information that represents a single letter, number, punctuation mark, blank space, or other symbol used in written language.
- **function:** instructions read by a computer's processor to execute specific events or operations.
- **object-oriented programming:** a type of programming in which the source code is organized into objects, which are elements with a unique identity that have a defined set of attributes and behaviors.

- **main loop:** the overarching process being carried out by a computer program, which may then invoke subprocesses.
- **programming languages:** sets of terms and rules of syntax used by computer programmers to create instructions for computers to follow. This code is then compiled into binary instructions for a computer to execute.
- **syntax:** rules that describe how to correctly structure the symbols that comprise a language.

HISTORY AND DEVELOPMENT OF ALGOL

The name ALGOL is a contraction from ALGOrithmic Language. It was developed by a committee of American and European computer scientists at ETH, in Zurich, Switzerland, as a language for the display of algorithms. The detailed specifications for the language were first reported in 1960, as ALGOL-58, and there have been several revisions and variations of the language since its inception. These include the updated ALGOL-60, ALGOL-N, ALGOL-68, ALGOL-W and Burroughs Extended ALGOL. The language was developed from the FORTRAN language, which appeared in 1956. Both ALGOL and FORTRAN strongly influenced the development of later languages such as BASIC, PL/1 and PL/C, Euler and Pascal. Many ALGOL programmers agree that the development of ALGOL-68 made the language much more complex and less elegant than the previous version ALGOL-60, ultimately leading to its fall into disfavor with computer users. It is now one of the oldest and least used of all programming languages.

PROGRAM CHARACTERISTICS

The syntax of ALGOL is rather logical, using natural-language reserved keywords such as comment, begin and end and the ":" (colon) character to identify standard arithmetical operators. The ALGOL program format utilizes a two-part "block" structure in its main loop that has since become a familiar feature of most modern computer languages such as C/C++, Java and many others. The first block, called the "head," consists of a series of type declarations similar to those used in FORTRAN type declaration statements. The declarations specify the class of the entities used in the program, such as integers, real

numbers, arrays, and so on. For example, the head section of an ALGOL-60 program may be

```
comment An ALGOL-60 sorting program
procedure Sort (Y, Z)
    value Z;
    integer Z; real array Y;
```

As can be seen, comment lines for documentation are identified by the keyword comment, and procedures are identified by the keyword procedure followed by the name of the procedure. The second block, which is the main part of the program, consists of a sequence of statements that are to be executed. It is initiated by the keyword begin and terminated by the keyword end. In fact this is the common structure of ALGOL procedures, and the program format supports procedures within procedures. For example, the following body section of a sorting procedure contains a second procedure within it, as

```
begin
    real x;
    integer i, j;
    for i := 2 until Z do begin
        x := Y[i];
        for j := i-1 step -1 until 1 do
            if x >= A[j] then begin
                A[j+1] := x; goto Found
            end else
                A]j+1] := A[j];
        A[1] := x;
    Found:
        end
    end
end Sort
```

Combining these two code segments produces an ALGOL-60 program that sorts a series of numbers. Examination of this body code segment reveals that the begin statement is followed by the head of a nested procedure, which is in turn followed by the body statements of the secondary procedure before the terminating end statements. The ability to support nested procedures foreshadowed the concept of a function in the context of object-oriented programming. It is a vitally important aspect of computer programming, and has proven the robustness of the form in many languages since the development of ALGOL.

SAMPLE PROBLEM

Comment the following section of code:

```
begin
      real x;
      integer i, j;
      for i := 2 until Z do begin
            x := Y[i];
```

Answer:

```
begin
comment identify the variable types as real
numbers and integers
      real x;
      integer i, j;
comment an iterative routine to as-
sign elements in the array
for i := 2 until Z do begin
            comment assigns the value of x
            to the array element
            x := Y[i];
```

THE ALGOL LEGACY

While the ALGOL programming language was overshadowed by the FORTRAN programming language and did not become popular, it featured a number of innovative aspects that have since become staples of essentially all major programming languages. In particular, the nested procedure structure foreshadowed the object-oriented programming style. It was the first language to make use of start-end identifiers as block delimiters, a convention that has carried over in object-oriented languages using function delimiters such as the { and } brace characters. ALGOL procedures included conditional statements (if...then and if...else), and iterative statements (for... until). Functions could call themselves in recursive computations (as in the statement i = i + j), and ALGOL enabled dynamic arrays in which the subscript range of the array elements was defined by the value of a variable, as in the array element i_j. ALGOL also used reserved keywords that could not be used as identifiers by a programmer, but it also allowed user-defined data types.

—*Richard M. Renneboog M.Sc.*

BIBLIOGRAPHY

Malcolme-Lawes, D.J. (1969) *Programming – ALGOL* London, UK: Pergamon Press. Print.

Organick, E.I., Forsythe, A.I. and Plummer, R.P. (1978) *Programming Language Structures* New York, NY: Academic Press. Print.

O'Regan, Gerard (2012) *A Brief History of Computing* 2nd ed., New York, NY: Springer-Verlag. Print.

Wexelblat, Richard L. (1981) *History of Programming L:anguages* New York, NY: Academic Press. Print.

Rutishauer, Heinz (1967) "Description of ALGOL-60" Chapter in Bauer, F.L., Householder, I.S., Olver, F.W.J., Rutishauer, H., Samelson, K. and Stiefel, E., eds. *Handbook for Automatic Computation* Berlin, GER: Springer-Verlag. Print.

ALGORITHMS

FIELDS OF STUDY

Computer Science; Operating Systems; Software Engineering

ABSTRACT

An algorithm is set of precise, computable instructions that, when executed in the correct order, will provide a solution to a certain problem. Algorithms are widely used in mathematics and engineering, and understanding the design of algorithms is fundamental to computer science.

PRINICIPAL TERMS

- **deterministic algorithm:** an algorithm that when given a particular input will always produce the same output.
- **distributed algorithm:** an algorithm designed to run across multiple processing centers and so is capable of directing a concentrated action between several computer systems.
- **drakon chart:** a flowchart used to model algorithms and programmed in the hybrid DRAKON computer language.
- **function:** instructions read by a computer's processor to execute specific events or operations.
- **recursive:** describes a method for problem solving that involves solving multiple smaller instances of the central problem.
- **state:** a technical term for all of the stored information, and the configuration thereof, that a program or circuit can access at a given time.

AN ANCIENT IDEA

The term "algorithm" is derived from the name al-Khwarizmi. Muhammad ibn Musa al-Khwarizmi was a ninth-century Persian mathematician who is credited with introducing the decimal system to the West. He has been celebrated around the world as a pioneer of mathematics and conceptual problem solving.

"Algorithm" has no precise definition. Broadly, it refers to a finite set of instructions, arranged in a specific order and described in a specific language, for solving a particular problem. In other words, an algorithm is like a plan or a map that tells a person or a machine what steps to take in order to complete a given task.

Algorithm Basics

In computer science, an algorithm is a series of instructions that tells a computer to perform a certain function, such as sorting, calculating, or finding data. Each step in the instructions causes the computer to transition from one state to another until it reaches the desired end state.

Any procedure that takes a certain set of inputs (a data list, numbers, information, etc.) and reaches a desired goal (finding a specific datum, sorting the list, etc.) is an algorithm. However, not all algorithms are equal. Algorithms can be evaluated for "elegance," which measures the simplicity of the coding. An elegant algorithm is one that takes a minimum number of steps to complete. Algorithms can also be evaluated in terms of "goodness," which measures the speed with which an algorithm reaches completion.

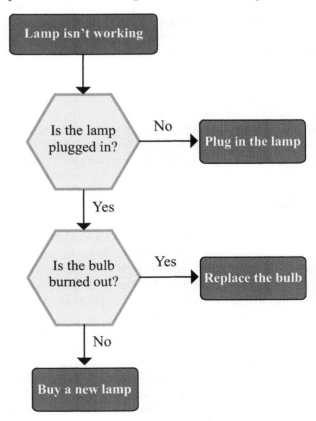

Algorithms are a set of operations or a procedure for solving a problem or processing data. Flowcharts are often used as a visualization of the process, showing the order in which steps are performed. EBSCO illustration.

Algorithms can be described in a number of ways. Flowcharts are often used to visualize and map the steps of an algorithm. The DRAKON computer language, developed in the 1980s, allows users to program algorithms into a computer by creating a flowchart that shows the steps of each algorithm. Such a flowchart is sometimes called a drakon chart.

Algorithms often specify conditional processes that occur only when a certain condition has been met. For instance, an algorithm about eating lunch might begin with the question, "Are you hungry?" If the answer is "yes," the algorithm will instruct the user to eat a sandwich. It will then ask again if the user is hungry. If the answer is still yes, the "eat a sandwich" instruction will be repeated. If the answer is "no," the algorithm will instruct the user to stop eating sandwiches. In this example, the algorithm repeats the "eat a sandwich" step until the condition "not hungry" is reached, at which point the algorithm ends.

TYPES OF ALGORITHMS
Various types of algorithms take different approaches to solving problems. An iterative algorithm is a simple form of algorithm that repeats the same exact steps in the same way until the desired result is obtained. A recursive algorithm attempts to solve a problem by completing smaller instances of the same problem. One example of a recursive algorithm is called a "divide and conquer" algorithm. This type of algorithm addresses a complex problem by solving less complex examples of the same problem and then combining the results to estimate the correct solution.

Algorithms can also be serialized, meaning that the algorithm tells the computer to execute one instruction at a time in a specific order. Other types of algorithms may specify that certain instructions should be executed simultaneously. Distributed algorithms are an example of this type. Different parts of the algorithm are executed in multiple computers or nodes at once and then combined.

An algorithm may have a single, predetermined output, or its output may vary based on factors other than the input. Deterministic algorithms use exact, specific calculations at every step to reach an answer to a problem. A computer running a deterministic algorithm will always proceed through the same sequence of states. Nondeterministic algorithms incorporate random data or "guessing" at some stage of the process. This allows such algorithms to be used

as predictive or modeling tools, investigating problems for which specific data is lacking. In computational biology, for instance, evolutionary algorithms can be used to predict how populations will change over time, given estimations of population levels, breeding rates, and other environmental pressures.

ALGORITHM APPLICATIONS
One of the most famous applications of algorithms is the creation of "search" programs used to find information on the Internet. The Google search engine can search through vast amounts of data and rank millions of search results in a specific order for different users. Sorting large lists of data was one of the earliest problems that computer scientists attempted to solve using algorithms. In the 1960s, the quicksort algorithm was the most successful sorting algorithm. Using a random element from the list as a "pivot," quicksort tells the computer to pick other elements from the list and compare them to the pivot. If the element is less than the pivot, it is placed above it; if it is greater, it is placed below. The process is repeated until each pivot is in its proper place and the data is sorted into a list. Computer scientists are still attempting to find search and sorting algorithms that are more "elegant" or "good" in terms of completing the function quickly and with the least demand on resources.

Searching and sorting are the most famous examples of algorithms. However, these are just two of the thousands of algorithm applications that computer scientists have developed. The study of algorithm design has become a thriving subfield within computer science.

—*Micah L. Issitt*

BIBLIOGRAPHY
"Intro to Algorithms." *Khan Academy*. Khan Acad., 2015. Web. 19 Jan. 2016.
Anthes, Gary. "Back to Basics: Algorithms." *Computerworld*. Computerworld, 24 Mar. 2008. Web. 19 Jan. 2016.
Bell, Tim, et al. "Algorithms." *Computer Science Field Guide*. U of Canterbury, 3 Feb. 2015. Web. 19 Jan. 2016.
Cormen, Thomas H. *Algorithms Unlocked*. Cambridge: MIT P, 2013. Print.
Cormen, Thomas H., et al. *Introduction to Algorithms*. 3rd ed. Cambridge: MIT P, 2009. Print.
Toal, Ray. "Algorithms and Data Structures." *Ray Toal*. Loyola Marymount U, n.d. Web. 19 Jan. 2016.

ANDROID OS

FIELDS OF STUDY

Computer Science; Operating Systems; Mobile Platforms

ABSTRACT

This article briefly discusses the general issues involved with mobile computing and presents a history and analysis of Google's Android operating system. It concludes with a look at Android's future in the growing market for mobile technology.

PRINICIPAL TERMS

- **application program interface (API):** the code that defines how two pieces of software interact, particularly a software application and the operating system on which it runs.
- **immersive mode:** a full-screen mode in which the status and navigation bars are hidden from view when not in use.
- **Material Design:** a comprehensive guide for visual, motion, and interaction design across Google platforms and devices.
- **multitasking:** in the mobile phone environment, allowing different apps to run concurrently, much like the ability to work in multiple open windows on a PC.
- **multi-touch gestures:** touch-screen technology that allows for different gestures to trigger the behavior of installed software.

A FORCE IN MOBILE COMPUTING

Mobile computing is the fastest-growing segment of the tech market. As pricing has become more affordable, developing nations, particularly in Africa, are the largest growing market for smartphones. With smartphones, users shop, gather information, connect via social media such as Twitter and Facebook, and communicate—one of the uses more traditionally associated with phones.

By far the most popular operation system running on mobile phones is Android. It has outpaced Apple's iOS with nearly double the sales. As of 2014, more than a million Android devices were being activated daily. Since its launch in 2008, Android has far and away overtaken the competition.

ANDROID TAKES OFF

Android came about amid a transformative moment in mobile technology. Prior to 2007, slide-out keyboards mimicked the typing experience of desktop PCs. In June of that year, Apple released its first iPhone, forever altering the landscape of mobile phones. Apple focused on multi-touch gestures and touch-screen technology. Nearly concurrent with

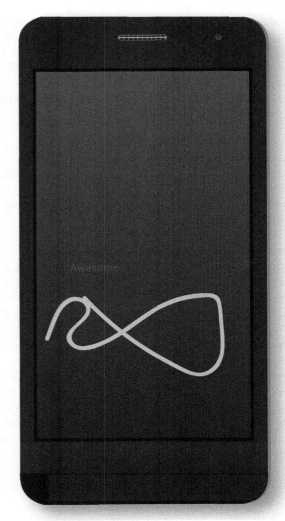

The swype keyboard, originally designed for Android operating systems, was developed to speed up typing capabilities by allowing the user to slide a finger over the keyboard from letter to letter without lifting their finger to choose each character. This standard software in the Android operating system allows for quick texting. EBSCO illustration.

this, Google's Android released its first application program interface (API).

The original API of Google's new operating system (OS) first appeared in October 2008. The Android OS was first installed on the T-Mobile G1, also known as the HTC Dream. This prototype had a very small set of preinstalled apps, and as it had a slide-out QWERTY keyboard, there were no touchscreen capabilities. It did have native multitasking, which Apple's iOS did not yet have. Still, to compete with Apple, Google was forced to replace physical keyboards and access buttons with virtual onscreen controls. The next iteration of Android shipped with the HTC Magic and was accompanied by a virtual keyboard and a more robust app marketplace. Among the other early features that have stood the test of time are the pull-down notification list, home-screen widgets, and strong integration with Google's Gmail service.

One later feature, the full-screen immersive mode, has become quite popular as it reduces distractions. First released with Android 4.4, "KitKat," in 2013, it hides the navigation and status bars while certain apps are in use. It was retained for the release of Android 5.0, "Lollipop," in 2015.

ANDROID CHANGES AND GROWS

Both of Google's operating systems—Android and its cloud-based desktop OS, Chrome— are based on the free open-source OS Linux, created by engineer Linus Torvalds and first released in 1991. Open-source software is created using publicly available source code. The open-source development of Android has allowed manufacturers to produce robust, affordable products that contribute to its widespread popularity in emerging and developing markets. This may be one reason why Android has had more than twice as many new users as its closest rival, Apple's iOS. This strategy has kept costs down and has also helped build Android's app marketplace, which offers more than one million native apps, many free of charge. By 2014 Android made up 54 percent of the global smartphone market.

This open-source development of Android has had one adverse effect: the phenomenon known as "forking," which occurs primarily in China. Forking is when a private company takes the OS and creates their own products apart from native Google services such as e-mail. Google seeks to prevent this loss of

control (and revenue) by not supporting these companies or including their apps in its marketplace. Forked versions of Android made up nearly a quarter of the global market in early 2014.

Google's business model has always focused on a "rapid-iteration, web-style update cycle." By contrast, rivals such as Microsoft and Apple have had a far slower, more deliberate pace due to hardware issues. One benefit of Google's faster approach is the ability to address issues and problems in a timely manner. A drawback is the phenomenon known as "cloud rot." As the cloud-based OS grows older, servers that were once devoted to earlier versions are repurposed. Since changes to the OS initially came every few months, apps that worked a month prior would suddenly lose functionality or become completely unusable. Later Android updates have been released on a timescale of six months or more.

ANDROID'S FUTURE

In 2014 alone, more than one billion devices using Android were activated. One of the biggest concerns about Android's future is the issue of forking. Making the code available to developers at no cost has made Android a desirable and cost-effective alternative to higher-end makers such as Microsoft and Apple, but it has also made Google a target of competitors.

Another consideration for Android's future is its inextricable link to the Chrome OS. Google plans to keep the two separate. Further, Google executives have made it clear that Chromebooks (laptops that run Chrome) and Android devices have distinct purposes. Android's focus has been on touch-screen technology, multi-touch gesturing, and screen resolution, making it a purely mobile OS for phones, tablets, and more recently wearable devices and TVs. Meanwhile, Chrome has developed tools that are more useful in the PC and laptop environment, such as keyboard shortcuts. However, an effort to unify the appearance and functionality of Google's different platforms and devices called Material Design was introduced in 2014. Further, Google has ensured that Android apps can be executed on Chrome through Apps Runtime on Chrome (ARC). Such implementations suggest a slow merging of the Android and Chrome user experiences.

—*Andrew Farrell, MLIS*

12

BIBLIOGRAPHY

Amadeo, Ron. "The History of Android." *Ars Technica*. Condé Nast, 15 June 2014. Web. 2 Jan. 2016.

"Android: A Visual History." *Verge*. Vox Media, 7 Dec. 2011. Web. 2 Jan. 2016.

Bajarin, Tim. "Google Is at a Major Crossroads with Android and Chrome OS." *PCMag*. Ziff Davis, 21 Dec. 2015. Web. 4 Jan. 2016.

Edwards, Jim. "Proof That Android Really Is for the Poor." *Business Insider*. Business Insider, 27 June 2014. Web. 4 Jan. 2016.

Goldsborough, Reid. "Android on the Rise." *Tech Directions* May 2014: 12. *Academic Search Complete*. Web. 2 Jan. 2016.

Manjoo, Farhad. "Planet Android's Shaky Orbit." *New York Times* 28 May 2015: B1. Print.

Newman, Jared. "Android Laptops: The $200 Price Is Right, but the OS May Not Be." *PCWorld*. IDG Consumer & SMB, 26 Apr. 2013. Web. 27 Jan. 2016.

Newman, Jared. "With Android Lollipop, Mobile Multitasking Takes a Great Leap Forward." *Fast Company*. Mansueto Ventures, 6 Nov. 2014. Web. 27 Jan. 2016.

APPLICATION

FIELDS OF STUDY

Applications; Software Engineering

ABSTRACT

In the field of information technology, an application is a piece of software created to perform a task, such as word processing, web browsing, or chess playing. Each application is designed to run on a particular platform, which is a type of system software that is installed on desktop computers, laptops, or mobile devices such as tablet computers or smartphones.

PRINICIPAL TERMS

- **app:** an abbreviation for "application," a program designed to perform a particular task on a computer or mobile device.
- **application suite:** a set of programs designed to work closely together, such as an office suite that includes a word processor, spreadsheet, presentation creator, and database application.
- **platform:** the specific hardware or software infrastructure that underlies a computer system; often refers to an operating system, such as Windows, Mac OS, or Linux.
- **system software:** the basic software that manages the computer's resources for use by hardware and other software.
- **utility program:** a type of system software that performs one or more routine functions, such as disk partitioning and maintenance, software installation and removal, or virus protection.
- **web application:** an application that is downloaded either wholly or in part from the Internet each time it is used.

APPLICATIONS IN CONTEXT

Applications are software programs that perform particular tasks, such as word processing or web browsing. They are designed to run on one or more specific platforms. The term "platform" can refer to any basic computer infrastructure, including the hardware itself and the operating system (OS) that manages it. An OS is a type of system software that manages a device's hardware and other software resources. Application designers may create different versions of an application to run on different platforms. A cross-platform application is one that can be run on more than one platform.

In the context of mobile devices such as tablets and smartphones, the term "application" is typically shortened to app. Since the introduction of the iPhone in 2007, apps have taken center stage in the realm of consumer electronics. Previously, consumers tended to be attracted more to a device's hardware or OS features. A consumer might have liked a certain phone for its solid design or fast processor, or they might have preferred the graphical interface of Microsoft Windows and Mac OS to the command-line interface of Linux. These features have since become much less of a concern for the average consumer. Instead, consumers tend to be more interested in finding a device that supports the specific apps they wish to use.

EVOLUTION OF APPLICATIONS

Over the years, apps have become more and more specialized. Even basic utility programs that were once included with OSs are now available for purchase as separate apps. In some cases, these apps are a more advanced version of the utility software that comes with the OS. For example, an OS may come with free antivirus software, but a user may choose to purchase a different program that offers better protection.

Some software companies offer application suites of interoperating programs. Adobe Creative Cloud is a cloud-based graphic design suite that includes popular design and editing programs such as Photoshop and InDesign. Microsoft Office is an office suite consisting of a word processor (Word), a spreadsheet program (Excel), and other applications commonly used in office settings. These programs are expensive and can take up large amounts of storage space on a user's computer. Before broadband Internet access became widely available, application suites were distributed on portable media such as floppy disks, CD-ROMs, or DVDs, because downloading them over a dial-up connection would have taken too long.

As high-speed Internet access has become much more common, application developers have taken a different approach. Instead of investing in bulky application suites, users often have the option of using web applications. These applications run partly or entirely on remote servers, avoiding the need to install them on the computer's hard drive.

TYPES OF APPLICATIONS

Many different types of software fall under the broad heading of applications. A large segment of the application market is focused on office and productivity software. This category includes e-mail applications, word processors, spreadsheet software, presentation software, and database management systems. In an office environment, it is critical that users be able to create documents using these applications and share them with others. This often means that a business or organization will select a particular application suite and then require all employees to use it.

Other types of applications include games, audio-video editing and production software, and even software that helps programmers write new software. Due the complexity of software engineering, programmers have developed many applications to help them produce more polished, bug-free programs. Software developers may use multiple applications to code a single program. They might use a word processor or text editor to write the source code and explanatory comments, a debugging tool to check the code for errors, and a compiler to convert the code into machine language that a computer can execute. There is even a type of application that can emulate a virtual computer running inside another computer. These applications are often used by web-hosting companies. Instead of having to set up a new physical server for each customer that signs up, they can create another virtual server for the user to access.

SECURITY IMPLICATIONS

Applications must have certain privileges in order to use the resources of the computer they are running on. As a result, they can sometimes be a point of weakness for attackers to exploit. A clever attacker can take over a vulnerable application and then use its privileges to make the computer behave in ways it should not. For example, the attacker could send spam e-mails, host illegally shared files, or even launch additional attacks against other computers on the same network.

CAREERS IN APPLICATIONS

Applications are the focus of a variety of career options for those interested in working with software. Apart from the obvious role of computer programmer, there are several other paths one might take. One option is quality assurance. Quality assurance staff are responsible for testing software under development to make sure it performs as it should. Technical support is another option. Technical support specialists assist users with operating the software

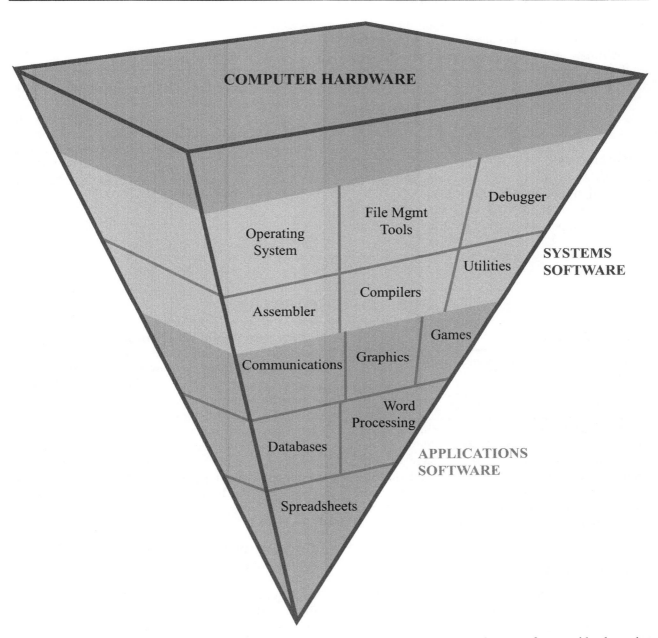

The variety and quantity of application software available is massive compared to the limited array of system software and hardware that support them. EBSCO illustration.

and fixing errors it might cause. Yet another path is technical writing. Technical writers create software user manuals and training materials. Finally, some applications are so complex that using them can be a career in itself.

—Scott Zimmer, JD

BIBLIOGRAPHY

Bell, Tom. *Programming: A Primer; Coding for Beginners.* London: Imperial Coll. P, 2016. Print.

Calude, Cristian S., ed. *The Human Face of Computing.* London: Imperial Coll. P, 2016. Print.

Dey, Pradip, and Manas Ghosh. *Computer Fundamentals and Programming in C.* 2nd ed. New Delhi: Oxford UP, 2013. Print.

Goriunova, Olga, ed. *Fun and Software: Exploring Pleasure, Paradox, and Pain in Computing.* New York: Bloomsbury, 2014. Print.

Neapolitan, Richard E. *Foundations of Algorithms.* 5th ed. Burlington: Jones, 2015. Print.

Talbot, James, and Justin McLean. *Learning Android Application Programming: A Hands- On Guide to Building Android Applications.* Upper Saddle River: Addison, 2014. Print.

APPLIED LINGUISTICS

FIELDS OF STUDY

Computer Science; Programming Language; Software Engineering

ABSTRACT

Applied linguistics is a linguistics subfield that focuses on solving the linguistics problems of daily life. Areas of interest include language learning, language preservation, and automated linguistic tools. Applied and computational linguistics overlap in the design of algorithms and computer programs for educational and commercial applications.

PRINICIPAL TERMS

- **computational linguistics:** a branch of linguistics that uses computer science to analyze and model language and speech.
- **lexicon:** the total vocabulary of a person, language, or field of study.
- **morphology:** a branch of linguistics that studies the forms of words.
- **semantics:** a branch of linguistics that studies the meanings of words and phrases.
- **syntax:** a branch of linguistics that studies how words and phrases are arranged in sentences to create meaning.

LINGUISTICS AND TECHNOLOGY

Applied linguistics seeks to address real-world language issues. While its emergence predates computer science, modern applied linguistics overlaps with computational linguistics. In particular, computer science tools are useful for modeling linguistic concepts and creating automated translation tools. Applied and computational linguistics research has produced many Internet-based language learning programs, as well as programs that allow machines to recognize and respond to linguistic cues.

DEVELOPMENT OF THE FIELD

Applied linguistics was first recognized as a distinct field of linguistics research in the 1950s. However, the application of linguistics knowledge is a far older tradition within academic linguistics. Some of the first applied linguistics departments and organizations in the United States focused on guiding language policy in government initiatives.

Another early area of concern was the application of linguistics research to language learning, both for native language speakers and for those learning a second language. In the 1960s, linguistics researchers began working on ways to use computer technology in linguistics education and research. Computer-assisted language learning (CALL), first developed

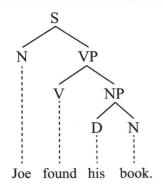

sentence: Joe found his book.

S = sentence
N = noun
VP = verb phrase
V = verb
NP = noun phrase
D = determiner

Joe found his book.

A parse tree, also known as a "syntax tree," shows the grammatical structure of a sentence. In the example sentence (S), "Joe found his book," there are two nouns (N), "Joe" and "book." Based on the verb "found," we can determine that "his book" is part of the predicate, and the subject—the noun doing the finding— is "Joe." Determiners (D) are used to distinguish the specific noun about which the sentence is talking. These parse trees are used as a basis for developing computer programs that can identify grammatical structure and draw meaning from a sentence in the same way human brains do. EBSCO illustration.

in the 1960s, studies ways to use computers to assist in language acquisition. In the twenty-first century, tools and techniques developed for CALL programs are regularly used to design software-based language learning programs.

Computational linguistics focuses on both applied and theoretical research. It grew out of military efforts in the 1950s to use computers to translate foreign languages, particularly Russian. Computers were widely used in linguistics research by the late 1980s. In the 1990s and 2000s, web-based linguistic applications began to become common.

Linguistics research focuses generally on "natural languages," which develop "naturally" through human use and refinement. They are distinct from formal languages, such as computer programming languages, and constructed languages, such as Esperanto. Developing better learning materials, both traditional text-based materials and modern software-based materials, is a primary focus of applied linguistics.

LINGUISTIC ANALYSIS

One major area of linguistics research is linguistic analysis, which seeks to dissect natural languages into their component parts to better understand how languages function and change. Each language can be analyzed in terms of its lexicon. This is the total vocabulary available in a certain language or subset of a language. The term can also refer to an individual's total vocabulary. To understand a language, its specific lexicon must be broken into smaller units. Linguists may study the morphology of words (the specific pattern of characters and phonemes that make up each word). They may also study a language's semantics (how words and phrases relate to what they represent). Beyond semantics is syntax, which looks at how words and phrases are arranged to form sentences and other higher-order units of language. In the digital age, linguistic analysis increasingly relies on software that can compile and evaluate properties of language.

COMPUTER SCIENCE APPLICATIONS

Automated translation is among the best-known examples of applied computational linguistics. Web-based services such as Babelfish and Google Translate use algorithms to analyze user-provided words and sentences and translate the semantic concepts into equivalent expressions in other languages. Computer scientists and linguists are working to refine such software to provide more accurate translations. Applied linguistics has also resulted in the development of software systems that allow users to "speak" to automated systems. Apple's iOS assistant Siri is an example of a program that can analyze the syntax and semantics of a spoken prompt and respond in kind.

Computer science and linguistics have created innovative solutions to a wide range of linguistic problems. For instance, the Enable Talk glove is a Bluetooth-enabled glove that, when worn, can translate sign language into spoken language. This enables sign-language users to communicate with individuals who do not understand sign language. Various similar therapeutic applications exist for individuals with speech, hearing, or other linguistic challenges. Another real-world issue being addressed through applied linguistics is the preservation, documentation, and analysis of the world's "endangered languages." These languages are in danger of being lost because of a lack of native speakers or a low rate of transmission to new generations.

Applied computational linguistics also seeks to create tools that can search through text and written data. Computer scientists and linguistics specialists have developed software that can use programmed linguistic rules and structural cues to locate keywords, sentences, and other elements within text. Basic keyword searches are one type of computational linguistic search tool. Computer scientists are working on new search tools that will allow more complex searches and meta-searches within text data. Given the wealth of textual data online, better text search and analysis tools could have a major impact on the future of education.

—*Micah L. Issitt*

BIBLIOGRAPHY

Davies, Alan. *An Introduction to Applied Linguistics: From Practice to Theory.* 2nd ed. Edinburgh: Edinburgh UP, 2007. Print.

Jurafasky, Daniel, and James H. Martin. *Speech and Language Processing.* 2nd ed. Upper Saddle River: Prentice Hall, 2008. Print.

Lardinois, Frederic. "Ukrainian Students Develop Gloves That Translate Sign Language into Speech." *TechCrunch.* AOL, 9 July 2012. Web. 19 Jan 2016.

Ramasubbu, Suren. "How Technology Can Help Language Learning." *Huffington Post.* TheHuffingtonPost.com, 3 June 2015. Web. 19 Jan. 2016.

Simpson, James, ed. *The Routledge Handbook of Applied Linguistics.* New York: Routledge, 2011. Print.

Tomlinson, Brian, ed. *Applied Linguistics and Materials Development.* New York: Bloomsbury, 2013. Print.

ARCHITECTURE SOFTWARE

FIELDS OF STUDY

Applications; Graphic Design

ABSTRACT

Architecture software is a category of specialized computer software intended for use by architects. Through the use of such software, architects may design, model, and modify plans for homes or other buildings digitally rather than on paper. As the type and scale of the structures being designed vary greatly, many types of architecture software are available to address the differing requirements of individual projects.

PRINICIPAL TERMS

- **building information modeling (BIM):** the creation of a model of a building or facility that accounts for its function, physical attributes, cost, and other characteristics.
- **modeling:** the process of creating a 2-D or 3-D representation of the structure being designed.
- **postproduction:** the period after a model has been designed and an image has been rendered, when the architect may manipulate the created image by adding effects or making other aesthetic changes.
- **raster:** a means of storing, displaying, and editing image data using individual pixels.
- **rendering:** the process of transforming one or more models into a single image.
- **vector:** a means of storing, displaying, and editing image data using defined points and lines.

THE INTRODUCTION OF SOFTWARE TO ARCHITECTURE

During the late twentieth and early twenty-first centuries, architecture became one of many fields to be dramatically transformed through the use of computer technology. For centuries, architects designed single structures such as homes as well as larger, multibuilding facilities by hand. They drafted plans on paper and created 3-D scale models out of materials such as cardboard, wood, and foam. These processes relied heavily on physical components. Therefore, it

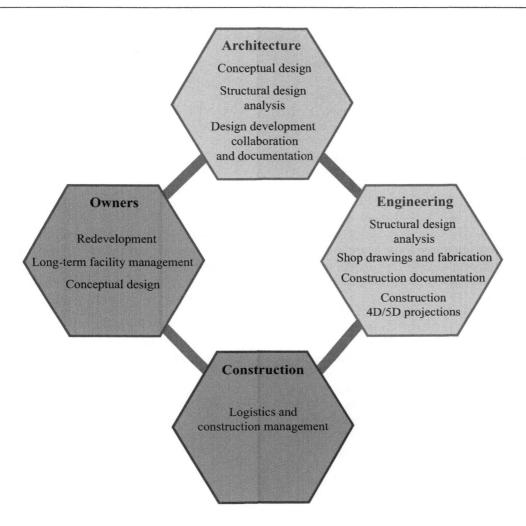

Architecture software is designed to assist in one or more of the many tasks required for building information modeling (BIM). Depending on one's role and how one relates to the building process, certain tasks
may be more important. Determining the proper architecture software first requires one to know what BIM
tasks one hopes to take on. EBSCO illustration.

was relatively difficult to collaborate with fellow architects or other professionals who were not located
nearby. Making changes to one's work, particularly to
models that were already assembled, could be time
consuming and difficult. The advent of architecture
software brought advances in areas such as drafting,
modeling, and modifying models and floorplans as
well as collaborating with other professionals.

Although typically designed for and marketed to
architects, architecture software is similar in many
ways to other types of software used in the broader

field of art and design. Some architecture programs
are essentially specialized versions of computer-aided
design (CAD) software. CAD programs are typically
used in fields such as manufacturing. For example,
one particularly popular architecture program,
AutoCAD Architecture, is sold by the software company Autodesk. Autodesk is known for its CAD and
computer-aided manufacturing (CAM) programs.
Architects began to adopt the use of CAD programs
in the 1980s. By the following decade, such software
was widely used in the field.

FEATURES

Architecture software is created and sold by a number of different technology companies. Therefore, the available programs vary in terms of features, although there are certain common ones. The first and perhaps most essential feature is the ability to draft floor plans, a process that was historically done with paper and pencil. Architecture software also typically allows users to create 3-D models of structures and easily insert elements such as doors and windows. Unlike physical models of proposed structures, digital models can be modified with just a few clicks of a mouse. Architecture software also sometimes includes features that support collaboration among architects and other professionals. These can include the ability to check files in and out, add notes to floor plans, and record measurements.

Specialized software devoted to building information modeling (BIM) has much in common with standard architecture software. However, BIM also allows the user to incorporate information such as expected time frames, costs, and the functions of a particular building or facility into the design. BIM allows the various professionals involved in the design process to evaluate and modify the design prior to moving forward with the project.

RENDERING AND POSTPRODUCTION

In addition to creating floor plans and 3-D models, architects are frequently tasked with creating images of their designs for clients. Transforming a 3-D model into a 2-D image is known as rendering. In postproduction, an architect may use graphic design software to add visual effects to the rendered image. For instance, if the architect has designed a waterfront apartment complex, he or she may use a graphic design program such as Adobe Photoshop to add photorealistic water textures to the image and increase its aesthetic appeal.

When creating floor plans and models, architects typically use vector images. Vector images are based on defined points and lines. This allows the images to remain true to scale no matter how much their overall size is modified. Graphic design software such as Photoshop, on the other hand, is a raster-based program. In a raster-based program, the images are based on individual pixels. As such, architects also sometimes use software capable of converting from vector to raster.

ADVANTAGES AND DISADVANTAGES

The use of specialized software in the field of architecture has a number of advantages and disadvantages. Chief among the advantages is a potential increase in productivity and time savings. Digital tools to draft floor plans and create models have largely replaced more time-consuming physical methods. The abilities to modify designs, test designs through digital modeling, and easily collaborate with fellow professionals on a project are cited as key benefits of architecture software.

However, the use of architecture software is not a substitute for architectural training. Some architects argue that the availability of such software can enable some users to produce substandard work. In addition, ongoing changes in the field require architects to maintain up-to-date knowledge of current technology. This can put architects accustomed to working with older systems at a disadvantage.

—*Joy Crelin*

BIBLIOGRAPHY

Ambrose, Gavin, Paul Harris, and Sally Stone. *The Visual Dictionary of Architecture.* Lausanne: AVA, 2008. Print.

Bergin, Michael S. "History of BIM." *Architecture Research Lab.* Architecture Research Lab, 21 Aug. 2011. Web. 31 Jan. 2016.

"A History of Technology in the Architecture Office." *Architizer.* Architizer, 23 Dec. 2014. Web. 31 Jan. 2016.

Kilkelly, Michael. "Which Architectural Software Should You Be Using?" *ArchDaily.* ArchDaily, 4 May 2015. Web. 31 Jan. 2016.

Marble, Scott, ed. *Digital Workflows in Architecture.* Basel: Birkhäuser, 2012. Print.

Solomon, Nancy B., ed. *Architecture: Celebrating the Past, Designing the Future.* New York: Visual Reference, 2008. Print.

ASCII

FIELDS OF STUDY

Computer Science; Computer Engineering

ABSTRACT

The American Standard Code for Information Interchange (ASCII) is a character encoding system. It enables computers and other electronic communication devices to store, process, transmit, print, and display text and other graphic characters. Initially published in 1963, ASCII formed the basis for several other character encoding systems developed for use with PCs and the Internet.

PRINICIPAL TERMS

- **bit width:** the number of bits used by a computer or other device to store integer values or other data.
- **character:** a unit of information that represents a single letter, number, punctuation mark, blank space, or other symbol used in written language.
- **control characters:** units of information used to control the manner in which computers and other devices process text and other characters.
- **hamming distance:** a measurement of the difference between two characters or control characters that effects character processing, error detection, and error correction.
- **printable characters:** characters that can be written, printed, or displayed in a manner that can be read by a human.

UNDERSTANDING CHARACTER ENCODING

Written language, or text, is composed of a variety of graphic symbols called characters. In many languages, these characters include letters, numbers, punctuation marks, and blank spaces. Such characters are also called printable characters because they can be printed or otherwise displayed in a form that can be read by humans. Another type of character is a control character. Control characters effect the processing of other characters. For example, a control character might instruct a printer to print the next character on a new line. Character encoding is the process of converting characters into a format that

can be used by an electronic device such as a computer or telegraph.

Originally designed for use with Samuel Morse's telegraph system, Morse code was one of the first character encoding schemes adopted for widespread use. Telegraphs transmit information by sending electronic pulses over telegraph wires. Morse code assigns each character to a unique combination of short and long pulses. For example, the letter *A* was assigned to the combination of one short followed by one long pulse, while the letter *T* was assigned to a single long pulse. Using Morse code, a telegraph operator can send messages by transmitting a sequence of pulses. The sequence, or string, of pulses represents the characters that comprise the message text.

Other character encoding systems were created to meet the needs of new types of electronic devices including teleprinters and computers. By the early 1960s, the use of character encoding systems had become widespread. However, no standard character encoding system existed to ensure that systems from different manufacturers could communicate with each other. In fact, by 1963, over sixty different encoding systems were in use. Nine different systems were used by IBM alone. To address this issue, the American Standards Association (ASA) X3.4 Committee developed a standardized character encoding scheme called ASCII.

UNDERSTANDING THE ASCII STANDARD

The ASCII standard is based on English. It encodes 128 characters into integer values from 0 to 127. Thirty-three of the characters are control characters, and ninety-five are printable characters that include the upper- and lowercase letters from *A* to *Z*, the numbers zero to nine, punctuation marks, and a blank space. For example, the letter *A* is encoded as 65 and a comma as 44.

The encoded integers are then converted to bits, the smallest unit of data that can be stored by a computer system. A single bit can have a value of either zero or one. In order to store integers larger than one, additional bits must be used. The number of bits used to store a value is called the bit width. ASCII specifies a bit width of seven. For example, in ASCII, the integer value 65 is stored using seven bits, which can be represented as the bit string 1000001.

ASCII TABLE

Decimal	Hex	Char	Decimal	Hex	Char	Decimal	Hex	Char	Decimal	Hex	Char
0	0	[NULL]	32	20	[SPACE]	64	40	@	96	60	`
1	1	[START OF HEADING]	33	21	!	65	41	A	97	61	a
2	2	[START OF TEXT]	34	22	"	66	42	B	98	62	b
3	3	[END OF TEXT]	35	23	#	67	43	C	99	63	c
4	4	[END OF TRANSMISSION]	36	24	$	68	44	D	100	64	d
5	5	[ENQUIRY]	37	25	%	69	45	E	101	65	e
6	6	[ACKNOWLEDGE]	38	26	&	70	46	F	102	66	f
7	7	[BELL]	39	27	'	71	47	G	103	67	g
8	8	[BACKSPACE]	40	28	(	72	48	H	104	68	h
9	9	[HORIZONTAL TAB]	41	29	)	73	49	I	105	69	i
10	A	[LINE FEED]	42	2A	*	74	4A	J	106	6A	j
11	B	[VERTICAL TAB]	43	2B	+	75	4B	K	107	6B	k
12	C	[FORM FEED]	44	2C	,	76	4C	L	108	6C	l
13	D	[CARRIAGE RETURN]	45	2D	-	77	4D	M	109	6D	m
14	E	[SHIFT OUT]	46	2E	.	78	4E	N	110	6E	n
15	F	[SHIFT IN]	47	2F	/	79	4F	O	111	6F	o
16	10	[DATA LINK ESCAPE]	48	30	0	80	50	P	112	70	p
17	11	[DEVICE CONTROL 1]	49	31	1	81	51	Q	113	71	q
18	12	[DEVICE CONTROL 2]	50	32	2	82	52	R	114	72	r
19	13	[DEVICE CONTROL 3]	51	33	3	83	53	S	115	73	s
20	14	[DEVICE CONTROL 4]	52	34	4	84	54	T	116	74	t
21	15	[NEGATIVE ACKNOWLEDGE]	53	35	5	85	55	U	117	75	u
22	16	[SYNCHRONOUS IDLE]	54	36	6	86	56	V	118	76	v
23	17	[ENG OF TRANS. BLOCK]	55	37	7	87	57	W	119	77	w
24	18	[CANCEL]	56	38	8	88	58	X	120	78	x
25	19	[END OF MEDIUM]	57	39	9	89	59	Y	121	79	y
26	1A	[SUBSTITUTE]	58	3A	:	90	5A	Z	122	7A	z
27	1B	[ESCAPE]	59	3B	;	91	5B	[	123	7B	{
28	1C	[FILE SEPARATOR]	60	3C	<	92	5C	\	124	7C		
29	1D	[GROUP SEPARATOR]	61	3D	=	93	5D	]	125	7D	}
30	1E	[RECORD SEPARATOR]	62	3E	>	94	5E	^	126	7E	~
31	1F	[UNIT SEPARATOR]	63	3F	?	95	5F	_	127	7F	[DEL]

This chart presents the decimal and hexadecimal ASCII codes for common characters on a keyboard. Public domain, via Wikimedia Commons

---

SAMPLE PROBLEM

ASCII defines the integer values for the first eleven lowercase letters of the alphabet as follows:

$a = 97$; $b = 98$; $c = 99$; $d = 100$; $e = 101$; $f = 102$; $g = 103$; $h = 104$; $i = 105$; $j = 106$; $k = 107$

Using this information, translate the word *hijack* to the correct ASCII integer values.

Answer:

The ASCII representation of the word *hijack* can be determined by comparing each character in the word to its defined decimal value as follows:

h i j a c k

h (104) i (105) j (106) a (97) c (99) k (107)

104 105 106 97 99 107

The correct ASCII encoding for the word *hijack* is 104 105 106 97 99 107.

The ASCII seven-bit integer values for specific characters were not randomly assigned. Rather, the integer values of specific characters were selected to maximize the hamming distance between each value. Hamming distance is the number of bits set to different values when comparing two bit strings. For example, the bit strings 0000001 (decimal value 1) and 0000011 (decimal value 3) have a hamming distance of 1 as only the second to last bit differs between the two strings. The bit patterns 0000111 (decimal value 7) and 0000001 (decimal value 1) have a hamming distance of two as the bit in the third to last position also differs between the two strings. ASCII was designed to maximize hamming distance because larger hamming distances enable more efficient data processing as well as improved error detection and handling.

BEYOND ASCII

Following its introduction in 1963, ASCII continued to be refined. It was gradually adopted for use on a wide range of computer systems including the first IBM PC. Other manufacturers soon followed IBM's lead. The ASCII standard was also widely adopted for use on the Internet. However, as the need for more characters to support languages other than English grew, other standards were developed to meet this need. One such standard, Unicode can encode more than 120,000 characters. ASCII remains an important technology, however. Many systems still use ASCII. Character encoding systems such as Unicode incorporate ASCII to promote compatibility with existing systems.

—*Maura Valentino, MSLIS*

BIBLIOGRAPHY

Amer. Standards Assn. *American Standard Code for Information Interchange.* Amer. Standards Assn., 17 June 1963. Digital file.

Anderson, Deborah. "Global Linguistic Diversity for the Internet." *Communications of the ACM* Jan. 2005: 27. PDF file.

Fischer, Eric. *The Evolution of Character Codes, 1874–1968.* N.p.: Fischer, n.d. *Trafficways.org.* Web. 22 Feb. 2016.

Jennings, Tom. "An Annotated History of Some Character Codes." *World Power Systems.* Tom Jennings, 29 Oct. 2004. Web. 16 Feb. 2016.

McConnell, Robert, James Haynes, and Richard Warren. "Understanding ASCII Codes." *NADCOMM.* NADCOMM, 14 May 2011. Web. 16 Feb. 2016.

Silver, H. Ward. "Digital Code Basics." *Qst* 98.8 (2014): 58–59. PDF file.

"Timeline of Computer History: 1963." *Computer History Museum.* Computer History Museum, 1 May 2015. Web. 23 Feb. 2016.

ASSEMBLY LANGUAGE

FIELDS OF STUDY

Programming Languages; System-level Programming; Embedded Systems

ABSTRACT

Assembly language is the most fundamental programming language. It is used to directly manipulate data in working memory, allowing the programmer to account for each clock cycle and use the available computing resources most efficiently. Assembly language programming is very compact, making it ideal for use in embedded systems that are designed to carry out specific functions or are not generally accessible to the end user.

PRINICIPAL TERMS

- **embedded systems:** computer systems that are incorporated into larger devices or systems to monitor performance or to regulate system functions.
- **hardware:** the physical parts that make up a computer. These include the motherboard and processor, as well as input and output devices such as monitors, keyboards, and mice.
- **hexadecimal:** a base-16 number system that uses the digits 0 through 9 and the letters A, B, C, D, E, and F as symbols to represent numbers.
- **imperative programming:** programming that produces code that consists largely of commands issued to the computer, instructing it to perform specific actions.
- **main loop:** the overarching process being carried out by a computer program, which may then invoke subprocesses.
- **process:** the execution of instructions in a computer program.
- **syntax:** rules that describe how to correctly structure the symbols that comprise a language.

CPU ARCHITECTURE AND ADDRESSING

Digital electronic devices function by the manipulation of binary signals, typically the voltages applied to a series of conducting circuits under the control of a central processing unit, or CPU. A binary signal has only two states, either on ("high") or off ("low"). In CPU architecture and other systems within the hardware of a computer or digital device, a parallel set of binary signals is controlled by a "clock" signal that triggers the progression from one set of binary signals to the next in a process. Each series of signals, and all other counting procedures in the device, including the ordering of memory locations, are expressed as a hexadecimal number that corresponds to the pattern of high and low signals in the particular series`. Some locations are reserved for specific purposes and functions, and cannot be altered. These locations are vital components for the functioning of an assembly language program by "accumulating" and "registering" data values that the program uses. The special registers are identified by specific symbols in the syntax of assembly language and their addresses are typically specific to the architecture of the CPU chip, although assembly language itself is a general standard. Assembly language programs are strictly linear, and all operations and logic are contained within the main loop of the program. Assembly language programs are not compiled, but are entered directly into the CPU memory locations for their operation. Assembly language programs are accordingly the most basic examples of imperative programming.

ASSEMBLY LANGUAGE PROGRAM FORMAT
The principal register in CPU memory is the accumulator. It is the standard location for the accumulated value of a combination of instructions that are being carried out. Locations called "R registers" hold

values being used in intermediate processes to generate the value that will go into the accumulator. The third vital register is the "program counter," which holds the location of the next instruction to be executed in code memory. Other locations are termed "data pointers," and contain the location in memory at which a series of stored data is located. Each line of an assembly language program consists typically of three parts: an operation instruction, followed by a memory location and data, and finally a documentation phrase. Though not absolutely required, it is perhaps the most essential for understanding what the purpose of the instruction. It is a statement that is not recognized as an instruction or data by the CPU. For example, the statement

MOV A,30h ; load value from memory address 30h into accumulator

is made readily understandable to a human by the documentation phrase, while the CPU understands only the MOV A,30h statement. Data can be addressed in an Assembly language program in various ways. The statement MOV A,#20h is an example of "immediate addressing," in which the value to be stored in memory immediately follows the "op-code" in code memory. The above statement MOV A,30h is an example of "direct addressing." The statement MOV A,@R0 is an example of "indirect addressing," in which the value to be stored in memory is retrieved from the location in internal RAM specified, or "pointed to," by the value stored in the first R-register (R0). Two commands are used for "external direct" addressing. The first, MOVX A,@DPTR reads a value in external RAM at the location identified by the value stored in the DPTR (data pointer) register and loads it into the accumulator. The second, MOVX @DPTR,A writes the value in the accumulator to the location in external RAM identified by the value stored in the DPTR register. The statement MOVX @R0,A is an example of external indirect addressing and functions in the same way using the value stored in the specified R-register. A final example is termed "code indirect" addressing, and is used to access data stored in the program code memory. Since code memory represents only a fairly small amount of actual memory, code indirect addressing is generally useful only for very small projects.

SAMPLE PROBLEM

Provide documentation phrases for the following statements:

MOV DPTR,#2021h
CLR A
MOVC A,@A+DPTR

Answer:

MOV DPTR,#2021h; sets the value of the DPTR register to 2021h
CLR A; clears the accumulator (sets the value in the accumulator to 00)
MOVC A,@A+DPTR; reads the value stored in code memory location 2021h and loads it into the accumulator

ASSEMBLY LANGUAGE IS EVERYWHERE

Because the coding of Assembly language programs is so compact and functions directly within the CPU of a device, it is the programming language of choice for all manner of embedded systems and other digital control devices that do not require the power or resources of a larger computer. The programs are typically installed directly into the device during manufacture and so are not accessible to the user afterwards. The vast majority of electronically-controlled devices such as household appliances, mp3 players, digital cameras, and any number of other devices, are controlled by the functioning of Assembly language programming.

—*Richard M. Renneboog M.Sc.*

BIBLIOGRAPHY

Cavanagh, Joseph (2013) *X86 Assembly Language and C Fundamentals* Boca Raton, FL: CRC Press. Print.

Duntemann, Jeff (2011) *Assembly Language Programming Step-by-Step: Programming with Linux* 3rd ed., Hoboken, NJ: John Wiley & Sons. Print.

Gilder, Jules H. (1986) *Apple Iic and Iie Assembly Language* New York, NY: Chapman and Hall. Print.

Hyde, Randall (2010) *The Art of Assembly Language* 2nd ed., San Francisco, CA: No Starch Press. Print.

Margush, Timothy S. (2012) *Some Assembly Required. Assembly Language Programming with the AVR Microcontroller* Boca Raton, FL: CRC Press. Print.

Peterson, James L. (1978) *Computer Organization and Assembly Language Programming* New York, NY: Academic Press. Print.

Steiner, Craig (1990) *The 8051/8052 Microcontroller: Architecture, Assembly Language and Hardware Interfacing* Boca Raton, FL: Universal Publishers. Print.

Streib, James T. (2011) *Guide to Assembly Language. A Concise Introduction* New York, NY: Springer. Print.

AUTONOMIC COMPUTING

FIELDS OF STUDY

Computer Science; Embedded Systems; System-Level Programming

ABSTRACT

Autonomic computing is a subfield of computer science that focuses on enabling computers to operate independently of user input. First articulated by IBM in 2001, the concept has particular relevance to fields such as robotics, artificial intelligence (AI), and machine learning.

PRINICIPAL TERMS

- **autonomic components:** self-contained software or hardware modules with an embedded capacity for self-management, connected via input/outputs to other components in the system.
- **bootstrapping:** a self-starting process in a computer system, configured to automatically initiate other processes after the booting process has been initiated.
- **multi-agent system:** a system consisting of multiple separate agents, either software or hardware systems, that can cooperate and organize to solve problems.
- **resource distribution:** the locations of resources available to a computing system through various software or hardware components or networked computer systems.
- **self-star properties:** a list of component and system properties required for a computing system to be classified as an autonomic system.

SELF-MANAGING SYSTEMS

Autonomic computing is a branch of computer science aimed at developing computers capable of some autonomous operation. An autonomic system is one that is, in one or more respects, self-managing. Such systems are sometimes described as "self-*" or "self-star." The asterisk, or "star," represents different properties of autonomic systems (self-organization, self-maintenance). Autonomic computing aims to develop systems that require less outside input, allowing users to focus on other activities.

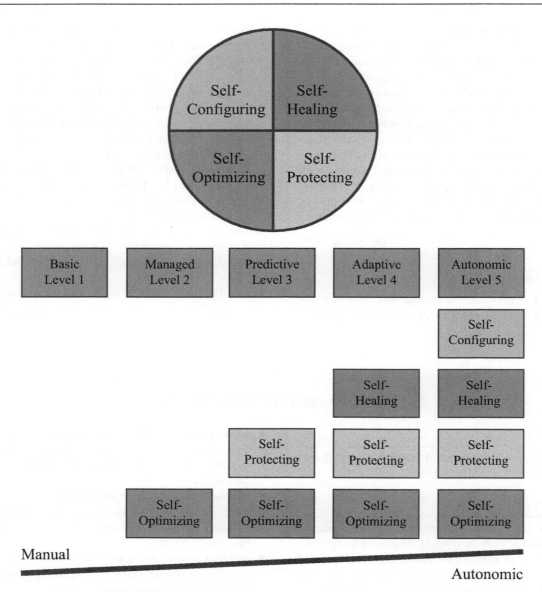

As computer systems have advanced from very basic technologies needing intense IT management toward autonomic systems that can self-manage, there have been four major stepping stones: self-optimizing, self-protecting, self-healing, and self-configuring. Each of these steps toward fully autonomic systems allows for more expansive computing while reducing the skill level required of the end users. EBSCO illustration.

SELF-STAR SYSTEMS

The concept of autonomic computing is based on autonomic systems found in nature. Examples of such systems include the autonomic nervous system of humans and the self-regulation of colonial insects such as bees and ants. In an autonomic system, the behaviors of individual components lead to higher-order self-maintenance properties of the group as a whole.

The properties that a system needs to function autonomically are often called self-star properties. One is self-management, meaning that the system can manage itself without outside input after an initial setup. Computer scientists disagree about what other self-star properties a system must have to be considered autonomic. Proposed properties include:

self-stabilization, the ability to return to a stable state after a change in configuration;

self-healing, the ability to recover from external damage or internal errors;

self-organization, the ability to organize component parts and processes toward a goal;

self-protection, the ability to combat external threats to operation; and

self-optimization, the ability to manage all resources and components to optimize operation.

Autonomic systems may also display self-awareness and self-learning. Self-awareness in a computer system differs from self-awareness in a biological system. In a computer system, self-awareness is better defined as the system's knowledge of its internal components and configuration. Self-learning is the ability to learn from experiences without a user programming new information into the system.

DESIGN OF AUTONOMIC SYSTEMS

An autonomic computer system is typically envisioned as having autonomic components (ACs), which are at least partly self-managing. An example of an AC is bootstrapping. Bootstrapping is the process by which a computer configures and initiates various processes during start-up. After a user turns on the computer, the bootstrapping process is self-managed. It proceeds through a self-diagnostic check and then activates various hardware and software components.

There are two basic models for autonomic computer design: a feedback control system and a multi-agent system. In a feedback control system, changing conditions provide feedback to the system that triggers changes in the system's function. Feedback control is often found in biological systems. In the autonomic nervous system, for example, levels of various neurotransmitters are linked to feedback systems that activate or deactivate ACs. A multi-agent system uses the collective functions of separate components to complete higher-order functions. For instance, groups of computers can be networked such that by performing individual functions, the components can collectively manage the system's functions and resources with reduced need for outside input. Linking multiple processors together changes the system's resource distribution, requiring that it be able to locate computing resources within all connected agents in order to handle tasks effectively.

Semi-autonomic software and hardware systems are commonly used. Peer-to-peer (P2P) systems for social networking and communication generally have some autonomic properties, including self-configuration, self-tuning, and self-organization. These systems can determine a user's particular computer setup, tune themselves to function in various environments, and self-organize in response to changing data or configuration. Most modern computing systems contain ACs but are not considered fully autonomic, as they still require some external management.

THE PROMISE OF AUTONOMIC SYSTEMS

The main goal of autonomic computing is to enable computer systems to perform basic maintenance and optimization tasks on their own. Maximizing the number of automated functions that a computer can handle allows engineers and system administrators to focus on other activities. It also enhances ease of operation, especially for those less adept at data management or system maintenance. For instance, the bootstrapping system and the self-regulatory systems that detect and correct errors have made computing more friendly for the average user.

Autonomic computer systems are particularly important in robotics, artificial intelligence (AI), and machine learning. These fields seek to design machines that can work unaided after initial setup and programming. The science of autonomic computing is still in its infancy, but it has been greatly enhanced by advancements in processing power and dynamic computer networking. For instance, the AI system Amelia, developed by former IT specialist Chetan Dube, not only responds to verbal queries and answers questions but can also learn by listening to human operators answer questions that it cannot answer. To some, systems that can learn and alter their own programming are the ultimate goal of autonomic design.

—*Micah L. Issitt*

BIBLIOGRAPHY

Bajo, Javier, et al., eds. *Highlights of Practical Applications of Agents, Multi-Agent Systems, and Sustainability.* Proc. of the International Workshops of PAAMS 2015, June 3–4, 2015, Salamanca, Spain. Cham: Springer, 2015. Print.

Berns, Andrew, and Sukumar Ghosh. "Dissecting Self-* Properties." *SASO 2009: Third IEEE International Conference on Self-Adaptive and Self-Organizing Systems.* Los Alamitos: IEEE, 2009. 10–19. *Andrew Berns: Homepage.* Web. 20 Jan. 2016.

Follin, Steve. "Preparing for IT Infrastructure Autonomics." *IndustryWeek.* Penton, 19 Nov. 2015. Web. 20 Jan. 2016.

Gibbs, W. Wayt. "Autonomic Computing." *Scientific American.* Nature Amer., 6 May 2002. Web. 20 Jan. 2016.

Lalanda, Philippe, Julie A. McCann, and Ada Diaconescu, eds. *Autonomic Computing: Principles, Design and Implementation.* London: Springer, 2013. Print.

Parashar, Manish, and Salim Hariri, eds. *Autonomic Computing: Concepts, Infrastructure, and Applications.* Boca Raton: CRC, 2007. Print.

AVATARS AND SIMULATION

FIELDS OF STUDY

Digital Media; Graphic Design

ABSTRACT

Avatars and simulation are elements of virtual reality (VR), which attempts to create immersive worlds for computer users to enter. Simulation is the method by which the real world is imitated or approximated by the images and sounds of a computer. An avatar is the personal manifestation of a particular person. Simulation and VR are used for many applications, from entertainment to business.

PRINICIPAL TERMS

- **animation variables (avars):** defined variables used in computer animation to control the movement of an animated figure or object.
- **keyframing:** a part of the computer animation process that shows, usually in the form of a drawing, the position and appearance of an object at the beginning of a sequence and at the end.
- **modeling:** reproducing real-world objects, people, or other elements via computer simulation.
- **render farm:** a cluster of powerful computers that combine their efforts to render graphics for animation applications.
- **virtual reality:** the use of technology to create a simulated world into which a user may be immersed through visual and auditory input.

VIRTUAL WORLDS

Computer simulation and virtual reality (VR) have existed since the early 1960s. While simulation has been used in manufacturing since the 1980s, avatars and virtual worlds have yet to be widely embraced outside gaming and entertainment. VR uses computerized sounds, images, and even vibrations to model some or all of the sensory input that human beings constantly receive from their surroundings every day. Users can define the rules of how a VR world works in ways that are not possible in everyday life. In the real world, people cannot fly, drink fire, or punch through walls. In VR, however, all of these things are possible, because the rules are defined by human coders, and they can be changed or even deleted. This is why users' avatars can appear in these virtual worlds as almost anything one can imagine—a loaf of bread, a sports car, or a penguin, for example. Many users of virtual worlds are drawn to them because of this type of freedom.

Because a VR simulation does not occur in physical space, people can "meet" regardless of how far apart they are in the real world. Thus, in a company that uses a simulated world for conducting its meetings, staff from Hong Kong and New York can both occupy the same VR room via their avatars. Such virtual meeting spaces allow users to convey nonverbal cues as well as speech. This allows for a greater degree of authenticity than in telephone conferencing.

Avatars were around long before social media. As computers have become more powerful and the rendering capabilities more efficient, avatars have improved in detail and diversity. (left) Anandeeta Gurung, CC0, via Wikimedia Commons. (right) Jason Rohrer, public domain, via Wikimedia Commons

MECHANICS OF ANIMATION

The animation of avatars in computer simulations often requires more computing power than a single workstation can provide. Studios that produce animated films use render farms to create the smooth and sophisticated effects audiences expect.

Before the rendering stage, a great deal of effort goes into designing how an animated character or avatar will look, how it will move, and how its textures will behave during that movement. For example, a fur-covered avatar that moves swiftly outdoors in the wind should have a furry or hairy texture, with fibers that appear to blow in the wind. All of this must be designed and coordinated by computer animators. Typically, one of the first steps is keyframing, in which animators decide what the starting and ending positions and appearance of the animated object will be. Then they design the movements between the beginning and end by assigning animation variables (avars) to different points on the object. This stage is called "in-betweening," or "tweening." Once avars are assigned, a computer algorithm can automatically change the avar values in coordination with one another. Alternatively, an animator can change "in-between" graphics by hand. When the program is run, the visual representation of the changing avars will appear as an animation.

In general, the more avars specified, the more detailed and realistic that animation will be in its movements. In an animated film, the main characters often have hundreds of avars associated with them. For instance, the 1995 film *Toy Story* used 712 avars for the cowboy Woody. This ensures that the characters' actions are lifelike, since the audience will focus attention on them most of the time. Coding standards for normal expressions and motions have been developed based on muscle movements. The MPEG-4 international standard includes 86 face parameters and 196 body parameters for animating human and humanoid movements. These parameters are encoded into an animation file and can affect the bit rate (data encoded per second) or size of the file.

EDUCATIONAL APPLICATIONS

Simulation has long been a useful method of training in various occupations. Pilots are trained in flight simulators, and driving simulators are used to prepare for licensing exams. Newer applications have included training teachers for the classroom and improving counseling in the military. VR holds the promise of making such vocational simulations much more realistic. As more computing power is added, simulated environments can include stimuli that better approximate the many distractions and detailed surroundings of the typical driving or flying situation, for instance.

VR IN 3-D

Most instances of VR that people have experienced so far have been two-dimensional (2-D), occurring on a computer or movie screen. While entertaining, such experiences do not really capture the concept of VR. Three-dimensional (3-D) VR headsets such as the Oculus Rift may one day facilitate more lifelike business meetings and product planning. They may also offer richer vocational simulations for military and emergency personnel, among others.

—Scott Zimmer, JD

BIBLIOGRAPHY

Chan, Melanie. *Virtual Reality: Representations in Contemporary Media.* New York: Bloomsbury, 2014. Print.

Gee, James Paul. *Unified Discourse Analysis: Language, Reality, Virtual Worlds, and Video Games.* New York: Routledge, 2015. Print.

Griffiths, Devin C. *Virtual Ascendance: Video Games and the Remaking of Reality.* Lanham: Rowman, 2013. Print.

Hart, Archibald D., and Sylvia Hart Frejd. *The Digital Invasion: How Technology Is Shaping You and Your Relationships*. Grand Rapids: Baker, 2013. Print.

Kizza, Joseph Migga. *Ethical and Social Issues in the Information Age*. 5th ed. London: Springer, 2013. Print.

Lien, Tracey. "Virtual Reality Isn't Just for Video Games." *Los Angeles Times*. Tribune, 8 Jan. 2015. Web. 23 Mar. 2016.

Parisi, Tony. *Learning Virtual Reality: Developing Immersive Experiences and Applications for Desktop, Web, and Mobile*. Sebastopol: O'Reilly, 2015. Print.

B

BASIC

FIELDS OF STUDY

Programming language

ABSTRACT

BASIC is an imperative programming language that employs a linear logic format. Subroutines are integral components of a BASIC program, in contrast to functions in object-oriented language programs. Some aspects of the BASIC language facilitate errors in program code, but are normally easily amended. Modern versions of the BASIC language provide features of the object-oriented programming approach, making it more powerful and versatile. The majority of .NET developers use Visual BASIC.NET exclusively.

PRINICIPAL TERMS

- **algorithm:** a set of step-by-step instructions for performing computations.
- **cathode ray tube (CRT):** a vacuum tube used to create images in devices such as older television and computer monitors.
- **central processing unit (CPU):** electronic circuitry that provides instructions for how a computer handles processes and manages data from applications and programs.
- **imperative programming:** programming that produces code that consists largely of commands issued to the computer, instructing it to perform specific actions.
- **main loop:** the overarching process being carried out by a computer program, which may then invoke subprocesses.
- **programming languages:** sets of terms and rules of syntax used by computer programmers to create instructions for computers to follow. This code is then compiled into binary instructions for a computer to execute.
- **source code:** the set of instructions written in a programming language to create a program.
- **variable:** a symbol representing a quantity with no fixed value.

BASIC BACKGROUND

The Beginners All-Purpose Symbolic Instruction Code, or BASIC, was developed in 1964 by Kemeny and Kurtz and is one of the first programming languages made available to the general public. Prior to its development, programming often required each instruction to be made as a punched card that was then fed into a card reader in sequential order as a 'batch' (hence the term 'batch process'). Preparation of such programs was a specialized skill, primarily of physicists and mathematicians. Programs were run on mainframe computers such as Digital Equipment Corporation's DEC PDP-11 in universities and major businesses that had the financial resources to afford them and required specialized training in languages such as ALGOL or FORTRAN with their complex syntax and terminology. The BASIC language was developed to make computer programming accessible to anyone by employing a simple, linear program structure and familiar 'human' words in its instruction set. The structure of a BASIC program is very similar to the structure of an ASSEMBLY language program, which is often referred to as 'machine language' or 'machine code'. The functioning of the BASIC language can therefore be understood as converting 'human' code into 'machine code'.

In the 1970s, an electronics revolution began with the introduction of the personal or micro computer, the first of which to be released commercially was the MITS Altair 8800. Several other electronics manufacturers released their own versions of personal computers in quick succession, each designed to connect to a TV set or other cathode ray tube (CRT) device and came with a version of BASIC as its standard programming language. The ready availability of such an easy-to-learn programming language resulted in an explosion of user-created programs that could be readily transferred from one machine to another and amended or enhanced at will. These machines used 8-bit central processing units (CPU) designed and manufactured by various companies. Each different CPU chip used a slightly different architecture requiring its own version of ASSEMBLY language, and therefore its own version of BASIC. With only minor alterations, however, a BASIC program written for one type of microcomputer could be made to function on practically any other BASIC-capable microcomputer. With the introduction of the IBM 8088 microcomputer, and the establishment of the Microsoft Corporation's proprietary version of BASIC as the *de facto* standard in those machines, the competition from other manufacturers eventually fell by the wayside. The introduction of the IBM 80286 CPU solidified the place of the IBM standard architecture with the introduction of DOS and Windows operating systems in competition with Apple Corporation's Macintosh graphics-based operating system. Despite these historical changes, however, the BASIC programming language continues to be a valuable and reliable 'workhorse' programming language, having been developed through several iterations. In 2006, for example, it was reported that some 59% of program and application developers working on the .NET Framework for Microsoft Corporation used Visual BASIC.NET exclusively.

BASIC PROGRAM STRUCTURE
BASIC is an imperative programming language that is linear in structure and entirely self-contained, consisting of a logical series of commands. Each BASIC program is essentially an algorithm designed to produce a specific type of output from the data that is provided to it. Within the main loop of the program, or source code, any number of secondary algorithms may be accommodated. Each of these is designed to carry out a specific function or calculation using the values determined or provided for an input variable. First-generation BASIC was capable of carrying out fairly complex mathematical calculations such as complex polynomials and multi-exponent relations, provided that they could be written out in tan executable format. Iterative calculations or operations are carried out using FOR-NEXT commands, and selections are made using IF-THEN-ELSE commands. The user 'communicated' with the BASIC program via the INPUT and PRINT commands. Comments in the program code began with the REM command, allowing the programmer to document the source code without interfering with the program function. The program commands are carried out in sequential order until encountering the END command. Each command is predicated by a line number, followed by the command instruction and the input/output (I/O) variables as appropriate. (Mainframe computers running BASIC normally used printer terminals to display the I/O of the program as hard copy.) More recent versions of BASIC incorporated more advanced features and operations, as well as object classes and structures that are typical of non-linear object-oriented programming languages.

SUBROUTINES VERSUS FUNCTIONS
The linear structure of BASIC programs relies on the use of subroutines that are integrated into the main loop. As the program code is executed line by line, a decision point in the program may direct the execution to a specific section of commands that does not appear in the sequence. That section of code is a secondary algorithm that performs a specific task or operation. When that operation has been completed, execution returns to the main code and continues from where it left off. Different sections of the program code can refer to the same subroutine any number of times. This eliminates the need to repeat the coding throughout the program. In more advanced versions of BASIC, a subroutine can exist outside of the main code, but must be identified by the SUB and END SUB identifiers. A function, on the other hand, is a feature of non-linear "object oriented" programming languages, such as C++ and Java, in which the main program may simply be a short list of functions that are to be executed. Unlike subroutines, functions are not dependent on the main program, and are portable between entirely different programs. The source code of a single

function to be much longer than the source code of the main program.

SAMPLE PROBLEM

Write a simple BASIC program that accepts a value of the temperature in degrees Fahrenheit and outputs the corresponding temperature in degrees Celsius and Kelvin, then prompts for a new input value.

Answer:

```
10    PRINT "Enter temperature in de-
      grees Fahrenheit:"; Tf
20    Tc = ((Tf – 32)*5)/9);
30    Tk = Tc + 273.15;
40    PRINT "Corresponding temperature
      in degrees Celsius is: " Tc
50    PRINT "Corresponding temperature
      in Kelvin is: " Tk
60    PRINT
70    PRINT "Enter another value?"; A$
80    IF LEN (A$) = 0 THEN GOTO 70
90    A$ = LEFT$ (A$, 1)
100   IF A$ = "Y" OR A$ = "y" THEN GOTO 10
110   END
```

The algorithm here is to accept a numerical value from the user as the input variable, convert it to a new value, and then convert that to another new value. It then displays the calculated values, and prompts the user for a new input value. The program scans for the new input and if none is received it repeats sending the prompt and scanning for the new input until a keystroke is registered. If the keystroke is either an upper or lower case ''y' as the LEFT character of the words 'YES' or 'yes', the program cycles back to the beginning and runs again. Any other keystroke fails this test and the program ENDs. The $ attached to any variable indicates a text 'string', and is typically called 'string'' instead of 'dollar sign'.

WEAKNESSES OF BASIC

BASIC continues to be a popular and versatile language that is relatively easy to learn and to use. It does have some syntactical limitations that result in common errors. One of the most common errors made by novice users of BASIC is failing to include the END statement at the end of the program, causing the computer to idle as it awaits a next command to appear. The other major flaw is the ease with which an error in syntax can lock the computer into an infinite loop in which it performs an operation continuously without advancing the program operation. With attention to detail, however, the BASIC language is a very versatile programming language that is highly applicable in many fields.

—*Richard M. Renneboog MSc*

BIBLIOGRAPHY

Kemeny, John G. and Kurtz, Thomas E. (1985) *Back To BASIC: The History, Corruption, and Future of the Language.* Boston, MA: Addison-Wesley. Print.

Kumari, Ramesh (2005) *Computers and Their Applications to Chemistry.* 2nd ed. Oxford, UK: Alpha Science International. Print.

O'Regan, Gerard (2012) *A Brief History of Computing.* 2nd ed., New York, NY: Springer-Verlag. Print.

Scott,Michael L. (2016) *Programming Language Pragmatics.* 4th ed., Waltham, MA: Morgan Kaufmann. Print.

Vick, Paul (2004) *The Visual BASIC .NET Programming Language.* Boston, MA: Addison-Wesley. Print.

BINARY/HEXADECIMAL REPRESENTATIONS

FIELDS OF STUDY

Computer Science; Computer Engineering; Software Engineering

ABSTRACT

The binary number system is a base-2 number system. It is used by digital devices to store data and perform mathematical operations. The hexadecimal number system is a base-16 number system. It enables humans to work efficiently with large numbers stored as binary data.

PRINICIPAL TERMS

- **base-16:** a number system using sixteen symbols, 0 through 9 and A through F.
- **base-2 system:** a number system using the digits 0 and 1.
- **bit:** a single binary digit that can have a value of either 0 or 1.
- **byte:** a group of eight bits.
- **nibble:** a group of four bits.

UNDERSTANDING THE BINARY NUMBER SYSTEM

A mathematical number system is a way of representing numbers using a defined set of symbols. Number systems take their names from the number of symbols the system uses to represent numbers. For example, the most common mathematical number system is the decimal system, or base-10 system. *Deci-* means "ten." It uses the ten digits 0 through 9 as symbols for numbers. Number systems can be based on any number of unique symbols, however. For example, the number system based on the use of two digit symbols (0 and 1) is called the binary or base-2 system.

Both the decimal and binary number systems use the relative position of digits in a similar way when representing numbers. The value in the rightmost, or first, position is multiplied by the number of digits used in the system to the zero power. For the decimal system, this value is 10^0. For the binary system, this value is 2^0. Both 10^0 and 2^0 are equal to 1. Any number x raised to the zero power is equal to 1. The power used increases by one for the second position and so on.

Position	8	7	6	5	4	3	2	1
	Seventh Power	Sixth Power	Fifth Power	Fourth Power	Third Power	Second Power	First Power	Zero Power
Decimal	10,000,000 or 10^7	1,000,000 or 10^6	100,000 or 10^5	10,000 or 10^4	1,000 or 10^3	100 or 10^2	10 or 10^1	1 or 10^0
Binary	128 or 2^7	64 or 2^6	32 or 2^5	16 or 2^4	8 or 2^3	4 or 2^2	2 or 2^1	1 or 2^0

Using the decimal number system, the integer 234 is represented by placing the symbols 2, 3, and 4 in positions 3, 2, and 1, respectively.

Position	3	2	1
Decimal	100 or 10^2	10 or 10^1	1 or 10^0
Digits	2	3	4

In the decimal system, $234 = (2 \times 100) + (3 \times 10) + (4 \times 1)$, or $(2 \times 10^2) + (3 \times 10^1) + (4 \times 10^0)$. The binary system uses the relative position of the symbols 0 and 1 to express the integer 234 in a different manner.

Position	8	7	6	5	4	3	2	1
Binary	128 or 2^7	64 or 2^6	32 or 2^5	16 or 2^4	8 or 2^3	4 or 2^2	2 or 2^1	1 or 2^0
Bit	1	1	1	0	1	0	1	0

DECIMAL	HEXADECIMAL	BINARY
0	00	00000000
1	01	00000001
2	02	00000010
3	03	00000011
4	04	00000100
5	05	00000101
6	06	00000110
7	07	00000111
8	08	00001000
9	09	00001001
10	0A	00001010
11	0B	00001011
12	0C	00001100
13	0D	00001101
14	0E	00001110
15	0F	00001111
16	10	00010000
17	11	00010001
18	12	00010010
19	13	00010011
20	14	00010100
21	15	00010101
22	16	00010110
23	17	00010111
24	18	00011000
25	19	00011001
26	1A	00011010
27	1B	00011011
28	1C	00011100
29	1D	00011101
30	1E	00011110
31	1F	00011111
32	20	00100000
33	21	00100001
34	22	00100010
35	23	00100011
36	24	00100100
37	25	00100101
38	26	00100110
39	27	00100111
40	28	00101000
41	29	00101001
42	2A	00101010
43	2B	00101011
44	2C	00101100
45	2D	00101101
46	2E	00101110
47	2F	00101111
48	30	00110000
49	31	00110001
50	32	00110010

The American Standard Code for Information Interchange (ASCII) was an early system used to translate basic characters into a numerical code readable by computers. The common characters on a keyboard are provided with decimal and hexadecimal codes. EBSCO illustration.

SAMPLE PROBLEM

To work with binary numbers in digital applications, it is important to be able to translate numbers from their binary values to their decimal values. Translate the following binary byte to its decimal value: 10111001

Answer:

The decimal value of the binary byte 10111001 is 185. The decimal value can be determined using a chart and then calculating.

128 or 2^7	64 or 2^6	32 or 2^5	16 or 2^4	8 or 2^3	4 or 2^2	2 or 2^1	1 or 2^0
1	0	1	1	1	0	0	1

$$= (1 \times 2^7) + (0 \times 2^6) + (1 \times 2^5) + (1 \times 2^4) + (1 \times 2^3) + (0 \times 2^2) + (0 \times 2^1) + (1 \times 2^0)$$
$$= (1 \times 128) + (0 \times 64) + (1 \times 32) + (1 \times 16) + (1 \times 8) + (0 \times 4) + (0 \times 2) + (1 \times 1)$$
$$= 185$$

In the binary system, $234 = (1 \times 128) + (1 \times 64) + (1 \times 32) + (0 \times 16) + (1 \times 8) + (0 \times 4) + (1 \times 2) + (0 \times 1)$, or $234 = (1 \times 2^7) + (1 \times 2^6) + (1 \times 2^5) + (0 \times 2^4) + (1 \times 2^3) + (0 \times 2^2) + (1 \times 2^1) + (0 \times 2^0)$.

THE IMPORTANCE OF THE BINARY NUMBER SYSTEM

The binary number system is used to store numbers and perform mathematical operations in computers systems. Such devices store data using transistors, electronic parts that can each be switched between two states. One state represents the binary digit 0 and the other, the binary digit 1. These binary digits are bits, the smallest units of data that can be stored and manipulated. A single bit can be used to store the value 0 or 1. To store values larger than 1, groups of bits are used. A group of four bits is a nibble. A group of eight bits is a byte.

USING HEXADECIMAL TO SIMPLIFY BINARY NUMBERS

The hexadecimal number system is a base-16 system. It uses the digits 0 through 9 and the letters A through F to represent numbers. The hexadecimal digit, or hex digit, A has a decimal value of 10. Hex digit B equals 11, C equals 12, D equals 13, E equals 14, and F equals 15. In hexadecimal, the value 10 is equal to 16 in the decimal system. Using hexadecimal, a binary nibble can be represented by a single symbol. For example, the hex digit F can be used instead of the binary nibble 1111 for the decimal value 15. Sixteen different combinations of bits are possible in a binary nibble. The hexadecimal system, with sixteen different symbols, is therefore ideal for working with nibbles.

Position	8	7	6	5	4	3	2	1
	Seventh Power	Sixth Power	Fifth Power	Fourth Power	Third Power	Second Power	First Power	Zero Power
Decimal	10,000,000 or 10^7	1,000,000 or 10^6	100,000 or 10^5	10,000 or 10^4	1,000 or 10^3	100 or 10^2	10 or 10^1	1 or 10^0
Hexa-decimal	268,435,456 or 16^7	16,777,216 or 16^6	1,048,576 or 16^5	65,536 or 16^4	4,096 or 16^3	256 or 16^2	16 or 16^1	1 or 16^0

One disadvantage of using binary is that large numbers of digits are needed to represent large integers. For example, 1,000,000 is shown in binary digits as 11110100001001000000. The same number is shown in hex digits as F4240, which is equal to $(15 \times 65{,}536) + (4 \times 4{,}096) + (2 \times 256) + (4 \times 16) + (0 \times 1)$.

Position	5	4	3	2	1
Hexadecimal	65,536 or 16^4	4,096 or 16^3	256 or 16^2	16 or 16^1	1 or 16^0
Hex digits	F or 15	4	2	4	0

Computers can quickly and easily work with large numbers in binary. Humans have a harder time using binary to work with large numbers. Binary uses many more digits than hexadecimal does to represent large numbers. Hex digits are therefore easier for humans to use to write, read, and process than binary.

—*Maura Valentino, MSLIS*

BIBLIOGRAPHY

Australian National University. *Binary Representation and Computer Arithmetic.* Australian National U, n.d. Digital file.

Cheever, Erik. "Representation of Numbers." *Swarthmore College.* Swarthmore College, n.d. Web. 20 Feb. 2016.

Govindjee, S. *Internal Representation of Numbers.* Dept. of Civil and Environmental Engineering, U of California Berkeley, Spring 2013. Digital File.

Glaser, Anton. *History of Binary and Other Nondecimal Numeration.* Rev. ed. Los Angeles: Tomash, 1981. Print.

Lande, Daniel R. "Development of the Binary Number System and the Foundations of Computer Science." *Mathematics Enthusiast* 1 Dec. 2014: 513–40. Print.

"A Tutorial on Data Representation: Integers, Floating-Point Numbers, and Characters." *NTU. edu.* Nanyang Technological U, Jan. 2014. Web. 20 Feb. 2016.

BIOCHEMICAL ENGINEERING

FIELDS OF STUDY

Biotechnology

ABSTRACT

Biochemical engineering is a subfield of engineering focused on the creation of substances to be used in the production of food or industrial materials, using biological as well as manufactured chemical ingredients. Some biochemical engineers, for example, have discovered how to get bacteria to help break down oil spills in order to minimize the environmental impact of such disasters.

PRINICIPAL TERMS

- **affinity chromatography:** a technique for separating a particular biochemical substance from a mixture based on its specific interaction with another substance.
- **blotting:** a method of transferring RNA, DNA, and other proteins onto a substrate for analysis.
- **interferometry:** a technique for studying biochemical substances by superimposing light waves, typically one reflected from the substance and one reflected from a reference point, and analyzing the interference.
- **nuclear magnetic resonance (NMR) spectroscopy:** a technique for studying the properties of atoms or molecules by applying an external magnetic field to atomic nuclei and analyzing the resulting difference in energy levels.
- **polymerase chain reaction (PCR) machine:** a machine that uses polymerase chain reaction to amplify segments of DNA for analysis; also called a thermal cycler.

ANALYZING AND CREATING MATERIALS

Much of a biochemical engineer's time is spent studying different substances and biological processes to better understand how they work. With this knowledge, these substances and processes can be repurposed or redirected to meet society's needs for advanced materials.

The main purpose of DNA and similar proteins in an organism is to produce different chemical substances to keep the organism functioning. Biochemical engineers can cause DNA strands to act like miniature factories, producing the chemicals needed for a particular application. If a certain type of enzyme is needed to treat a disease, for example, it may be possible to locate a DNA segment in an organism that either naturally creates this material or can be "reprogrammed" to create it. In order to do so, biochemical engineers must be able to study the structure and composition of biomolecules (biological molecules).

METHODS OF ANALYSIS

Scientists use various techniques to study and analyze substances of interest. First, a sample of the substance must be prepared. This usually begins with blotting, which isolates DNA and other proteins for further study. When studying DNA, the sample may then be amplified in a polymerase chain reaction (PCR) machine. Polymerase chain reaction is a technique for producing many copies of a DNA strand. This allows close analysis or multiple tests to be performed on even a very small sample.

Once the sample is ready, there are several different methods that may be used to study it. Which

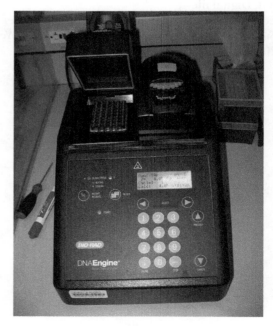

A PCR machine is responsible for reading DNA code and replicating the code to make many copies through a series of very specific chemical reactions. Public domain, via Wikimedia Commons.

method to use depends in part on the type of information needed. Nuclear magnetic resonance (NMR) spectroscopy involves exposing a substance to a magnetic field in order to alter the spin of its atomic nuclei. This change in spin results in a change in the energy level of the nuclei. The difference in energy levels provides information about the physical and chemical properties of the atoms. NMR spectroscopy is most often used to study the structure of organic compounds.

Interferometry refers to several related techniques involving wave superposition. In most cases, identical electromagnetic waves are reflected from both the sample material and a reference, then superimposed. The reference may be another substance, or it may be a mirror that reflects the wave unaltered. The resulting interference pattern will reveal any structural differences between the two samples. Interferometry can be used to identify materials, study their molecular structures, or provide detailed tissue imaging.

Affinity chromatography is a method for separating specific substances from mixtures. "Affinity" refers to the fact that certain biomolecules have a strong tendency to bind to other specific types of molecules. For example, an antigen is a harmful biomolecule that elicits an immune response in an organism. The immune system responds by producing antibodies, which are proteins that are "designed" to bind to that specific antigen. Thus, if a researcher wanted to separate antibodies from a biochemical mixture, they would introduce antigens specific to those antibodies into the mixture. The antibody molecules would separate from the mixture and bind to the antigens instead. Affinity chromatography can be used to purify or concentrate a substance or to identify substances in a mixture.

Duties of a Biochemical Engineer
The task of a biochemical engineer is to look at real-world problems and at the chemical and biological properties of organisms and materials as if they were pieces of a jigsaw puzzle, and find matches where two pieces fit together. They match up problems with solutions.

Biochemical engineers spend much of their time on design and analysis, but they are also involved in other tasks. Much of their time is spent working on product development. The overarching goal of the design

work they do is to further the creation of products that will benefit society and generate profits for the company that markets them. Product development requires a biochemical engineer to develop a thorough understanding of the needs and goals of a project in order to research potential solutions that fit with those needs. In the case of products with medical applications, there are also elaborate safety precautions that must be taken, as well as lengthy approval processes requiring cooperation with regulatory agencies.

Documentation and scholarly publishing are also important duties for biochemical engineers. Each stage of their research must be recorded and analyzed. Biochemical engineers are encouraged to share their findings with the global research community whenever possible. While this sometimes raises concerns about the confidentiality of proprietary information, in the long run the sharing of research helps advance the field and encourage innovation. Many discoveries in the field of biochemical engineering are carefully guarded because of their profit potential, as patents on biochemically engineered substances can be worth large amounts of money. This is why the patent system tries to balance monetary rewards for inventors against society's need for access to information.

Applications of Biochemical Engineering
One of the greatest applications of biochemical engineering is in the field of drug manufacturing. Traditional pharmaceutical manufacturing processes involve multiple types of chemical reactions, each a laborious step in the process of creating a batch of medicinal compound. Biochemical engineers reinvent this process by putting nature to work for them. They identify biological processes that can be tweaked to produce the same drugs without having to mix chemicals.

One newer area of research in biochemical engineering is the artificial production of human organs through processes similar to 3-D printing. By studying the ways that different types of tissue cells reproduce, biochemical engineers have been able to grow rudimentary structures such as replacement ears. Scientists anticipate that they will eventually be able to produce fully functioning hearts, livers, and other organs, avoiding the need for organ donation.

—*Scott Zimmer, JD*

Bibliography

Katoh, Shigeo, Jun-ichi Horiuchi, and Fumitake Yoshida. *Biochemical Engineering: A Textbook for Engineers, Chemists and Biologists.* 2nd rev. and enl. ed. Weinheim: Wiley, 2015. Print.

Kirkwood, Patricia Elaine, and Necia T. Parker-Gibson. *Informing Chemical Engineering Decisions with Data, Research, and Government Resources.* San Rafael: Morgan, 2013. Digital file.

Pourhashemi, Ali, ed. *Chemical and Biochemical Engineering: New Materials and Developed Components.* Rev. Gennady E. Zaikov and A. K. Haghi. Oakville: Apple Acad., 2015. Print.

Zeng, An-Ping, ed. *Fundamentals and Application of New Bioproduction Systems.* Berlin: Springer, 2013. Print.

Zhong, Jian-Jiang, ed. *Future Trends in Biotechnology.* Berlin: Springer, 2013. Print.

Zhou, Weichang, and Anne Kantardjieff, eds. *Mammalian Cell Cultures for Biologics Manufacturing.* Berlin: Springer, 2014. Print

BIOMEDICAL ENGINEERING

FIELDS OF STUDY

Biotechnology; Computer Engineering

ABSTRACT

Biomedical engineering (BME) applies engineering design principles to biology in order to improve human health. It encompasses low-tech tools, such as crutches and simple prostheses, as well as highly sophisticated devices, such as pacemakers and x-ray machines. Computer science is crucial to the design and production of many BME devices.

PRINICIPAL TERMS

- **bioinstrumentation:** devices that combine biology and electronics in order to interface with a patient's body and record or monitor various health parameters.
- **biomechanics:** the mechanics (structure and function) of living things in response to various forces.
- **bioMEMS:** short for "biomedical microelectromechanical system"; a microscale or nanoscale self-contained device used for various applications in health care.
- **genetic modification:** direct manipulation of an organism's genome, often for the purpose of engineering useful microbes or correcting for genetic disease.
- **telemedicine:** health care provided from a distance using communications technology, such as video chats, networked medical equipment, smartphones, and so on.

What Is Biomedical Engineering?

Biomedical engineering (BME) applies engineering design principles to the study of biology in order to improve health care. It developed from the overlap between engineering, biology, and medical research into its own branch of engineering. BME and related fields, such as biomedical research, remain distinct, as the case of the antibiotic penicillin shows. The accidental discovery that the *Penicillium* mold has antibacterial properties was a by-product of basic medical research, not BME. However, efforts to design and implement system to mass-produce and distribute the drug were a textbook example of BME. Products of BME include prosthetics, imaging equipment, regenerative medicine, medical implants, advances in drug production, and even telecommunications.

Although many innovations in BME are high tech, not all are. Consider crutches, which assist people with mobility after a foot or leg injury. Modern crutches are designed to be strong, lightweight, and comfortable during extended use. They are informed by knowledge of human biology, especially biomechanics, to minimize strain on the patient's bones and muscles.

Developing the Tools of Medicine

Many BME projects involve instrumentation, the various devices used in medical care. Such devices include imaging tools such as magnetic resonance

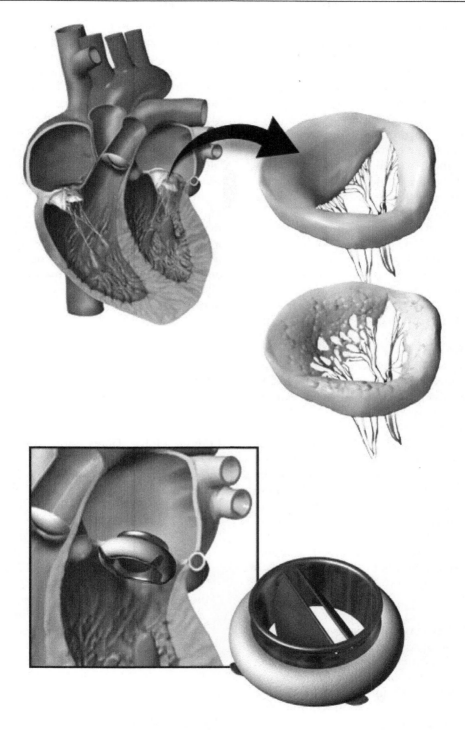

To build an artificial organ, live cells are placed on a scaffolding (an exact replica mold of the organ) and subjected to specific conditions to induce growth to develop the desired organ. By BruceBlaus, CC BY 3.0 (http://creativecommons.org/licenses/by/3.0), via Wikimedia Commons.

imaging (MRI) machines, positron emission tomography (PET) scanners, and x-ray machines. It also includes bioinstrumentation, devices specifically designed to connect to the human body. These either monitor health parameters or provide a therapeutic benefit. Implanted pacemakers, for example, use low-voltage electrical impulses to keep the heart pumping in proper rhythm. Different bioMEMS can act as sensors, analyze blood or genes, or deliver drugs at a much smaller scale than traditional devices. Some bioMEMS are known as "labs-on-a-chip" because they incorporate the functions of a full-sized biology lab on a computer chip.

The development of smaller, highly portable, easy-to-use instrumentation is a keystone of telemedicine. With smartphone apps, bioMEMS, and cell-phone networks, it is increasingly possible to diagnose and treat patients far from hospitals and clinics. Telemedicine is especially helpful developing countries that lack a strong infrastructure.

ENGINEERING BIOLOGY
Another major area of BME involves the engineering of life itself. This includes regenerative medicine, in which organs and tissues are grown in a lab, often from a combination of synthetic and organic biomaterials, to replace damaged or diseased ones. It also includes genetic engineering, in which an organism or cell's genes are manipulated toward a desired end. Genetic engineering of bacteria can produce useful drugs. Mice and other animals have been engineered to manifest traits such as cancer susceptibility to make them better study subjects. Genetic engineering could one day correct genetic conditions such as Huntington's disease in humans. Traditional breeding for desired traits among study organisms is also used in BME, though it is a slower, less powerful technique.

COMPUTERS IN BIOMEDICAL ENGINEERING
Computer use is widespread in BME design and device creation. Consider the design of a custom prosthetic. Computers scan the amputation site to ensure a secure fit, model the limb and a prosthetic to the correct proportions, and even print components of the prosthesis using a 3-D printer. Computer science techniques are also used to process the growing amount of patient data, searching out patterns of disease incidence.

Instrumentation is often built around computers. Medical imaging devices depend on embedded microcomputers and specialized software to process patient data and display it in a way that is useful to doctors. Smaller devices such as digital thermometers depend on microprocessors. An electrocardiograph (EKG) monitors a patient's heart rate and alerts the medical team if the patient goes into cardiac arrest. This is made possible through BME that combines biological knowledge (normal human heart rates) with sensors (electrodes that transmit electrical impulses from a patient's skin to a display) and computer science (processing the information, programming the device response). Advances in computer technology are driving advances in BME instrumentation. For instance, short-range wireless signals enable double amputees to walk by activating motorized joints and allowing the prostheses to communicate with one another.

—*Kenrick Vezina, MS*

BIBLIOGRAPHY
Badilescu, Simona, and Muthukumaran Packirisamy. *BioMEMS: Science and Engineering Perspectives.* Boca Raton: CRC, 2011. Print.

Enderle, John Denis, and Joseph D. Brozino. *Introduction to Biomedical Engineering.* 3rd ed. Burlington: Elsevier, 2012. Print.

"Examples and Explanations of BME." *Biomedical Engineering Society.* Biomedical Engineering Soc., 2012–14. Web. 23 Jan. 2016.

"Biomedical Engineers." *Occupational Outlook Handbook, 2016–2017 Edition.* Bureau of Labor Statistics, US Dept. of Labor, 17 Dec. 2015. Web. 23 Jan. 2016.

"Milestones of Innovation." *American Institute for Medical and Biological Engineering.* Amer. Inst. for Medical and Biological Engineering, 2016. Web. 25 Jan. 2016.

Pavel, M., et al. "The Role of Technology and Engineering Models in Transforming Healthcare." *IEEE Reviews in Biomedical Engineering.* IEEE, 2013. Web. 25 Jan. 2016.

Saltzman, W. Mark. "Lecture 1—What Is Biomedical Engineering?" *BENG 100: Frontiers of Biomedical Engineering.* Yale U, Spring 2008. Web. 23 Jan. 2016.

BIOMETRICS

FIELDS OF STUDY

Biotechnology; Security; Computer Science

ABSTRACT

Biometrics is the study of biology and metrics, or measurement, especially for the purpose of identifying individuals based on unique characteristics. Digital scanning, measurement, and matching tools are used to create automated systems for verifying identity through the comparison of biometric samples.

PRINICIPAL TERMS

- **bioinformatics:** the scientific field focused on developing computer systems and software to analyze and examine biological data.
- **biosignal processing:** the measurement and monitoring of a biological signal, such as heart rate or blood oxidation, rather than an image.
- **false match rate:** the probability that a biometric system incorrectly matches an input to a template contained within a database.
- **Hamming distance:** a measurement of the difference between two strings of information.
- **identifiers:** measurable characteristics used to identify individuals.

THE SCIENCE OF IDENTIFYING INDIVIDUALS

Biometrics is a scientific field that focuses on using human characteristics as identifiers to distinguish between different individuals. These include facial geometry, dermal patterns, biorhythms, and vocal character. Fingerprint identification was one of the first biometric techniques. It was first used in antiquity before becoming a global standard in forensic identification in the 1800s. Advances in computer science have allowed for the development of automated systems capable of identifying and matching other biometric identifiers. These include retinal blood vessel patterns, iris morphology, voices, and DNA.

BIOMETRICS BASICS

The ability to identify individuals is an important facet of human culture and society. Most people are instinctively able to identify each other by external features such as facial geometry and voice recognition. However, in some settings these methods are problematic and can lead to high false match rates. Individuals with similar voices or appearances may be mistaken for one another. Biometric science attempts to reduce the probability of false matches by determining the most accurate characteristics to use for identification and by creating systems for increasing the reliability of biometric security systems.

The skin of the human finger contains a pattern of ridges that are unique to each individual. The unique nature of fingerprints was discovered in antiquity, and in some cultures, including ancient Egypt and Babylonia, fingerprints were used in place of written signatures on documents or artworks. In the late 1800s, scientists began experimenting with the use of fingerprints to identify individuals for the purpose of forensic analysis. Fingerprinting quickly became the most widespread form of biometric classification. However, fingerprints can be accidentally or purposefully altered and can be imitated in a variety of ways, limiting their reliability for security and identification. In the twentieth and twenty-first centuries, biometrics specialists identified a number of other unique characteristics that can be used for identification. These include patterns of blood vessels in the retina, the structure of the iris, the geometry of the human face, and the unique genetic code in each individual's DNA. Biometrics became closely tied to advances in computer science that allowed such identifiers to be measured and recorded.

SAMPLING AND MATCHING

Biometric identification involves two primary processes: sampling and comparing biometric characteristics. Bioinformatics is a multidisciplinary field that develops and uses technology to analyze or model biological data. Bioinformatics software enables a scanned sample, such as a fingerprint, to be converted into a digital signal. A fingerprint scan creates a geometric map representing length, curvature, and other measurements, for instance. Various digital maps taken from samples can then be compared to determine if two samples came from the same individual. In the case of iris identification, the pattern of the iris can be mapped geometrically and then compressed into a set of binary codes. Two binary

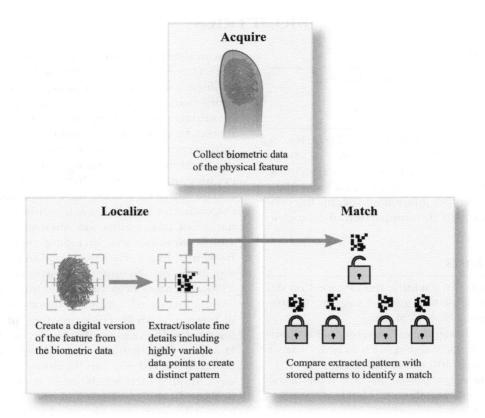

An attribute is scanned to make a digital representation, isolated details of that attribute are used to create an ID template for a particular person, and the template is matched to the "key" ID to retrieve the correct "key" for an individual. The number and location of isolated details must have a large enough hamming distance to ensure there is no duplication of templates. EBSCO illustration.

codes can be compared, and a measurement called Hamming distance can be used to determine how different they are. Similar measurement methods are used in DNA matching. Patterns of genetic codes can be compared, and algorithms automatically calculate the potential that one or more samples represent the same individual.

When an individual's identifiers are added to a database, the process is called "enrollment." Future biometric samples can be compared against all samples previously enrolled in the database. In biometric security technology, authorized individuals must first create a template of biometric samples matched with their identity. Any user who attempts to access the system is then asked to submit a biometric sample for validation. If the submitted sample matches an approved template according to a matching algorithm, access is granted. If it does not match, access can be denied.

APPLIED BIOMETRICS
The twenty-first century has witnessed the commercialization of biometric identification for personal security. Most individuals use passwords, physical keys, or other types of codes to protect property and data. However, these security methods have significant disadvantages. They can be lost, forgotten, stolen, or duplicated. Biometric characteristics, meanwhile, are unique, more difficult to forge, and cannot be lost or forgotten.

Improvements in computer technology have put biometrics in the reach of consumers. By the early

2000s, a number of companies began offering fingerprint scanning for security on personal computers and home security systems. Companies have also produced consumer versions of facial recognition, voice recognition, and iris/retinal scanning technology for home or commercial use. The most common use of biometric systems has been in personal computer security. For instance, technology company Apple's Touch ID system uses fingerprint scanning. Certain smart devices using the Android operating system can use a facial recognition program to allow user access. Some companies have introduced fingerprint or facial recognition as a key for electronic payment and debit transactions. Supporters claim that such systems offer greater security than traditional debit or credit cards, which can be stolen and used by other individuals.

Each available biometric system has its own advantages and disadvantages. For example, DNA analysis is the most accurate biometric characteristic. However, it depends on the ability to collect and analyze DNA samples, which is a time consuming and technologically advanced process. Opponents argue that if fingerprints, for instance, are lifted and stolen, the identity theft victim has no recourse like a password reset. Thus, biometrics' use for security may in fact present a great risk. In the 2010s, biometrics systems shifted toward a multimodal approach using multiple identifiers simultaneously to identify individuals. For instance, a number of researchers have suggested that biosignal processing could be used in conjunction with other biometric measures to produce more accurate biometric security and identification systems.

—*Micah L. Issitt*

BIBLIOGRAPHY

Biometric Center of Excellence. Federal Bureau of Investigation, 2016. Web. 21 Jan. 2016.

"Book—Understanding Biometrics." *Griaule Biometrics.* Griaule Biometrics, 2008. Web. 21 Jan. 2016.

"Introduction to Biometrics." *Biometrics.gov.* Biometrics.gov, 2006. Web. 21 Jan. 2016.

Jain, Anil K., Arun A. Ross, and Karthik Nandakumar. *Introduction to Biometrics.* New York: Springer, 2011. Print.

Modi, Shimon K. *Biometrics in Identity Management.* Boston: Artech House, 2011. Print.

Shahani, Aarthi. "Biometrics May Ditch the Password, But Not the Hackers." *All Things Considered.* NPR, 26 Apr. 2015. Web. 21 Jan. 2016.

Van den Broek, Egon L. "Beyond Biometrics." *Procedia Computer Science* 1.1 (2010): 2511–19. Print.

BIOTECHNOLOGY

FIELDS OF STUDY

Biotechnology; Robotics

ABSTRACT

Biotechnology uses living organisms or their derivatives to make products. Farming and animal husbandry are early examples. Modern examples include biomedicines, gene therapies, prostheses, and skin grafts.

PRINICIPAL TERMS

- **bioinformatics:** the use of computer-based tools and techniques to obtain and evaluate biological data.
- **bioinstrumentation:** the use of instruments to record data about an organism's physiology and functions.
- **biomaterials:** natural or synthetic materials that can be used to replace, repair, or modify organic tissues or systems.
- **biomechanics:** the various mechanical processes such as the structure, function, or activity of organisms.
- **bionics:** the use of biologically based concepts and techniques to solve mechanical and technological problems.

BIOLOGICAL TECHNOLOGY

"Technology" can be broadly defined as the use of scientific knowledge to solve practical problems. All devices, from the earliest stone tools to modern computers, are examples of technology. Biotechnology is the use of organisms or biological products in this problem-solving process.

Farming and animal husbandry are among the earliest forms of biotechnology. They represent the application of knowledge to increase food production. In the twenty-first century, biotechnology spans a broad range of industrial and scientific processes, from gene therapies to skin grafts.

STUDYING BIOLOGICAL STRUCTURES AND FUNCTIONS

Using animal skins for clothing, or plant material for roofing and food, may seem intuitive by modern standards. However, it took considerable experimentation and careful study to learn how to use natural products. Secondary natural products such as wine and cheese require deeper knowledge of organic interactions. A wine or cheese producer must know how chemical processes and microorganisms modify other biological products (grape juice, milk).

Basic science research into organisms' physiology and behavior drives biotechnology and its subfields. It was first recognized as a unique field of study in the 1970s and soon became one of the fastest-growing industrial fields. Among the more controversial early advancements was the genetic engineering of crops, which became common in the US food market in the 1990s.

Bioinformatics combines basic science, engineering, math, and computer science to advance the study of organisms. Thanks to twenty-first-century computer technology, bioinformatics is a leading field in biological research. Bioinformatics data is often directly used to develop biotechnology.

Bioinstrumentation develops specialized tools to record data about living organisms. Bioinstruments that record physiological processes such as blood pressure and brain activity have been crucial to advances in medicine. They have also led to products such as wearable heart-rate monitors, which became a popular fitness tool in the 2010s.

IMITATING NATURE

Another facet of biotechnology is the development of synthetic technology that mimics biological systems. The subfield of biomechanics has resulted in a variety of medical and industrial products. It also informs the industrial science of bionics, which develops synthetic tools that mirror the form or function of organic structures. For example, Velcro was designed to mimic the way the burrs of some plants cling to soft material such as clothing. Newer adhesive technology uses hydrogen bonding, the same phenomenon that allows geckos to cling even to smooth surfaces.

Bionics is actually a very old field. Many machines and devices, including the earliest flying machines and even building structures, have been inspired by similar biological structures. Bionics and biomechanics are also key to robotics and prosthetics.

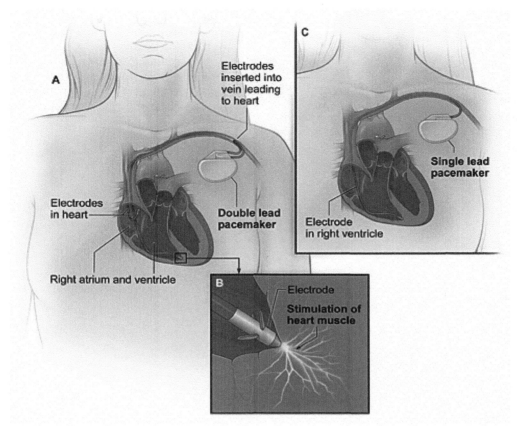

A pacemaker is an example of technology working with a biological system. Biotechnology can enhance the efficiency of a biological system or replace nonfunctional components of the system. By National Heart Lung and Blood Institute (NIH), public domain, via Wikimedia Commons.

MODIFYING NATURE

In some cases, biotechnology can involve creating tools, techniques, and processes to modify living organisms. For instance, biomaterials science creates synthetic or biological tissues to modify organisms. Skin grafts, artificial tooth replacement, and biosynthetic heart valves are examples of biomaterials used in medicine.

Biotechnology may also involve creating new hybrid or modified organisms. Though genetic modification of plants and animals is still in its infancy, it could have dramatic impact on agriculture in coming decades. The ability to modify specific genes and genetic systems also has implications for the future of medicine. For instance, it might lead to the eventual elimination of many genetically transmitted conditions. In addition, the emerging field of pharmacogenomics studies patient DNA to learn how different patients react differently to medications. It may one day be possible to tailor medicine to a patient's particular genetic complement.

BIOTECHNOLOGY IN THE FUTURE

The development of genetics, genomics, and other gene-related fields has opened a wealth of new possibilities for biotechnology research. From the earliest wooden tools and animal-skin clothing to robotic systems that mimic biomechanical movement, biotechnology has long played an important role in human life. Today, with scientists studying biological forms and functions at the level of individual genes and cells, biotechnology has become a highly profitable field of industrial and medical research.

—*Micah L. Issitt*

BIBLIOGRAPHY

"Biotechnology." *ACS.* Amer. Chemical Soc., n.d. Web. 27 Jan. 2016.

Godbey, W. T. *An Introduction to Biotechnology: The Science, Technology and Medical Applications.* Waltham: Academic, 2014. Print.

Intro to Biotechnology: Techniques and Applications. Cambridge: NPG Educ., 2010. *Scitable.* Web. 20 Jan. 2016.

"Introduction to Biotechnology." *Center for Bioenergy and Photosynthesis.* Arizona State U, 13 Feb. 2006. Web. 20 Jan 2016.

Pele, Maria, and Carmen Cimpeanu. *Biotechnology: An Introduction.* Billerica: WIT, 2012. Print.

Thieman, William J., and Michael A. Palladino. *Introduction to Biotechnology.* 3rd ed. San Francisco: Benjamin, 2012. Print.

C

C++

FIELDS OF STUDY

Programming languages; Software; Software engineering

ABSTRACT

C++ is an object-oriented programming language developed by Bjarne Stroustrup as an extension of the C programming language that preceded it. Both languages are very versatile and widely used. C++ programs use the same header-body format as C programs, but all input-output operations have been streamlined into a single header called "iostream," with input and output controlled by the "cin >>" and "cout <<" operators, respectively. C and C++ are the principal languages that form the basis of the Java universal programming language.

PRINICIPAL TERMS

- **class:** a collection of independent objects that share similar properties and behaviors.
- **function:** instructions read by a computer's processor to execute specific events or operations.
- **object:** an element with a unique identity and a defined set of attributes and behaviors.
- **object-oriented programming:** a type of programming in which the source code is organized into objects, which are elements with a unique identity that have a defined set of attributes and behaviors.
- **pseudocode:** a combination of a programming language and a spoken language, such as English, that is used to outline a program's code.
- **software:** the sets of instructions that a computer follows in order to carry out tasks. Software may be stored on physical media, but the media is not the software.
- **source code:** the set of instructions written in a programming language to create a program.
- **syntax:** rules that describe how to correctly structure the symbols that comprise a language.
- **variable:** a symbol representing a quantity with no fixed value.

HISTORICAL ASPECTS AND CHARACTERISTICS OF C++

The C++ programming language was developed by Bjarne Stroustrup in the early 1980s, at Bell Laboratories as an advanced form of the C programming language. The name C++ comes from the use of the "++" modifier within the C++ language to incrementally increase the value of the variable that it modifies. The C language was developed in its turn from the BCPL language (1967) that was used for writing operating system software and compilers, and from the B language (1970) that was used to create early versions of the UNIX operating system. It is also capable of incorporating program segments written in other languages such as Assembly. C and C++ form the basis of the Java language, which is used on almost all electronic devices. C++ is an object-oriented programming language, and is "backwards compatible" with C; any program or part of a program written in C will compile and run under C++. Object-oriented programs are easier to understand, correct and modify than programming techniques characteristic of languages like BASIC and FORTRAN. A C++ program consists of two parts: a "head" and a "body." The head of the program is a collection of class definitions and function statements, while the body contains the functions and variables that will be used to manipulate data.

CLASSES, FUNCTIONS AND OBJECTS

The variables and functions of a C++ program are assigned to defined classes. As an analogy, the assignment of classes in a C++ program has the same relationship as the different materials used in building a house. Each function exists independently of the program that uses it, just as the bricks used to build a wall of a house exist independently of the wall they are used to build. Different shapes and sizes of bricks and blocks may be used in the construction, but they can all be assigned to the same class of "stone and stone-like building materials" based on their common features. Different shapes and sizes of wood pieces may be assigned to another class of building materials, and so on. In the same way, variables and functions that have certain common characteristics belong to the same class. Just as a brick is an object used to construct a wall, a function is an object used to construct a program. In general, the structure of a C++ program has a "head" section consisting of standard class statements that provide standard functionalities such as control of input and output, mathematical functions and operations, and so on, that are specified by "include" statements. This is followed by a main program called "main," which is essentially just a list of functions that the "main" program will use, as well as some variable designations and relations. Following the "main" function are the functions that "main" will utilize. The general format of a C++ program is thus

```
#include statement
#include statement
main()
{variables
function1
function2
function3, etc
decision statements, relations, etc
}
function1 ()
{function1 statements, etc
}
function2 ()
{function2 statements, etc
}
function3 ()
{function3 statements, etc
}
```

This description shows how the program is assembled from various independent objects rather than being entirely self-contained as would be the case

with a program written in a language such as BASIC. It should also be apparent that any of the individual functions could be utilized as is or with minor alterations in any other program.

DESIGNING A C++ PROGRAM

Much as a building is designed and constructed from various standard and customized components to achieve the desired end product, a C++ program is designed and constructed from a variety of standard and customized components. The foundation parts of the program are the libraries of standard class definitions and functions, and the syntax of the programming language. The program is typically first designed using pseudocode, which is just an easily modified "blueprint" for the construction of the actual program. An example of pseudocode might be

```
header components (iostream)
main ()
{  define graphics interface
   cout<< graphics interface
   {data input ()
      {  prompt for data input
         cin>> data
      }
   }
   calculate output values()
   {  calculate power loss ()
      calculate output frequency ()
      calculate integral of frequency equation ()
      calculate first derivative of frequency equa-
      tion ()
      calculate second derivative of frequency
      equation ()
      return values
   }
   cout<< calculated values to file
   cout<< calculated values to graphic display
   new data input?
   If yes then data input () else end
end
}
calculate power loss()
calculate output frequency ()
calculate integral of frequency equation ()
calculate first derivative of frequency equation ()
calculate second derivative of frequency equation ()
```

Pseudocode will not compile and run. Each of the functions that is specified in the above example after the end of the main () function may in fact be pieces of independent source code that are much longer than the actual program source code. Some features of note are the designations "iostream," "cin >>" and "cout<<." "iostream" is the standard C++ header and includes all of the standard definitions of file and variable types, input and output functions used in C++ programs. Including this header eliminates that need to repeat the corresponding information when writing out a program, and inclusion of the iostream header is actually a standardization requirement. As their names suggest "cin>>" is the standard statement for input of all data types and formats in C++, and "cout<<" is the corresponding output statement. They take the place of several different input and output functions that were used in the C language before C++ was developed.

SAMPLE PROBLEM

Write a simple C++ program that asks for the users name and outputs a simple greeting.

Answer:

```
#include <iostream>
int main ()
{
    std::cout << "What is your name?\n";
    std::cin >> name;
    cout << "Hello, " name;
}
```

The iostream header is included at the beginning of every program to define the methods and formats of input and output used in C++ programs. The main() function defines the order of operations to be carried out in the program. The standard output (std::cout <<) displays the text within quotation marks on the screen exactly as they are stated. The standard input (std::cin >>) accepts all subsequent keystrokes in order until the "enter" key is pressed, and assigns them to a variable called "name." The output function then prints the text phrase "Hello," to the screen followed

by the value of the variable "name" (which in this case is supposed to be the name of the person using the program, but in reality could be any string of characters that was entered after the input prompt).

IMPORTANCE AND DEVELOPMENT OF C++
The C and C++ languages are essentially universal programming languages. It is possible to write C++ programs that will compile and run on any other computer. A number of variants of the language have been developed, including Visual C++ and C#. The language is well suited to game design and programming due to the extensive libraries of calculation and graphics functions that have been developed, and its ability to incorporate modules written in compatible languages. Perhaps the most important development of C++ is that it is the foundation of the Java programming language which is used by almost every electronic device in the world as a universal standard. A program written in Java will run on any Java-enabled device of any kind, such as smartphones, calculators, programmable logic controllers, gaming consoles, tablets, guidance control systems, and many others, including computers.

—*Richard M. Renneboog M.Sc.*

BIBLIOGRAPHY
Davis, Stephen R. (2015) *Beginning Programming with C++ for Dummies* 2nd ed. Joboken, NJ: John Wiley & Sons. Print.
Deitel, H.M. And Deitel, P.J. (2009) *C++ for Programmers* Upper Saddle River, NJ: Pearson Education Incorporated. Print.
Graham W. Seed (2012) *An Introduction to Object-Oriented Programming in C++ with Applications in Computer Graphics* New York, NY: Springer Science+Business Media. Print.
Kernighan, Brian W. and Ritchie, Dennis M. (1978) *The C Programming Language* Englewood Cliffs, NJ: Prentice-Hall. Print
Lippman, Stanley B, Lajoie, Josée and Moo, Barbara E. (2013) *C++ Primer* 5th ed. Upper Saddle River, NJ: Addison-Wesley. Print.
McGrath, Mike (2015) *C++ Programming in Easy Steps* 4th ed. Leamington Spa, UK: Easy Steps Limited. Print.
Stroustrup, Bjarne (2013) *The C++ Programming Language* 4th ed. Upper Saddle River, NJ: Addison-Wesley. Print.

CAD/CAM

FIELDS OF STUDY

Applications; Graphic Design

ABSTRACT

Computer-aided design (CAD) and computer-aided manufacturing (CAM) are software that enable users to design products and, through the use of computer-guided machinery, manufacture them according to the necessary specifications. CAD/CAM programs are used in a wide range of industries and play a key role in rapid prototyping, a process that allows companies to manufacture and test iterations of a product.

PRINICIPAL TERMS

- **four-dimensional building information modeling (4-D BIM):** the process of creating a 3-D model that incorporates time-related information to guide the manufacturing process.
- **rapid prototyping:** the process of creating physical prototype models that are then tested and evaluated.
- **raster:** a means of storing, displaying, and editing image data based on the use of individual pixels.
- **solid modeling:** the process of creating a 3-D representation of a solid object.
- **vector:** a means of storing, displaying, and editing image data based on the use of defined points and lines.

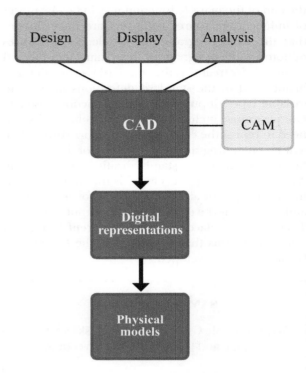

Computer-aided design (CAD) and computer-aided manufacturing (CAM) are used in many industries to fulfill the same basic goals. 3-D items are scanned and analyzed, new items are designed, and those designs can then be translated into manufactured items through CAM, which can develop the program necessary for machines to properly create the new item. EBSCO illustration.

APPLICATIONS OF CAD/CAM

The term "CAD/CAM" is an acronym for "computer-aided design" and "computer-aided manufacturing." CAD/CAM refers collectively to a wide range of computer software products. Although CAD software and CAM software are considered two different types of programs, they are frequently used in concert and thus associated strongly with each other. Used primarily in manufacturing, CAD/CAM software enables users to design, model, and produce various objects—from prototypes to usable parts—with the assistance of computers.

CAD/CAM software originated in the 1960s, when researchers developed computer programs to assist professionals with design and modeling. Prior to that point, designing objects and creating 3-D models was a time-consuming process. Computer programs designed to aid with such tasks represented a significant time savings. By the late 1960s, CAD programs began to be used alongside early CAM software. CAM enabled users to instruct computer-compatible machinery to manufacture various objects according to digital designs. The use of CAD/CAM software became widespread over the following decades.

CAD/CAM is now a key part of the manufacturing process for numerous companies, from large corporations to small start-ups. Industries in which CAD/CAM proved particularly useful include the automotive and computer technology industries. However, CAM software has also been widely used in less obvious fields, including dentistry and textile manufacturing.

In addition to its use alongside CAM software, CAD software functions alone in a number of fields. CAD software allows users to create 3-D models of objects or structures that do not need to be manufactured by machine. For instance, specialized CAD software are used in architecture to design floor plans and 3-D models of buildings.

COMPUTER-AIDED DESIGN

Using CAD/CAM software is a two-part process that begins with design. In some cases, the user begins designing an object by using CAD software to create 2-D line drawings of the object. This process is known as "drafting." He or she may then use tools within the CAD software to transform those 2-D plans into a 3-D model. As CAD/CAM is used to create physical objects, the modeling stage is the most essential stage in the design process. In that stage, the user creates a 3-D representation of the item. This item may be a part for a machine, a semiconductor component, or a prototype of a new product, among other possibilities.

In some cases, the user may create what is known as a "wire-frame model," a 3-D model that resembles the outline of an object. However, such models do not include the solid surfaces or interior details of the object. Thus, they are not well suited for CAM, the goal of which is to manufacture a solid object. As such, those using CAD software in a CAD/CAM context often focus more on solid modeling. Solid modeling is the process of creating a 3-D model of an object that includes the object's edges as well as its internal structure. CAD software typically allows the user to rotate or otherwise manipulate the created model. With CAD, designers can ensure that all the separate parts of a product will fit together as intended. CAD also enables users to modify the digital model. This is less time-consuming and produces less waste than modifying a physical model.

When designing models with the intention of manufacturing them through CAM technology, users must be particularly mindful of their key measurements. Precision and accurate scaling are crucial. As such, users must be sure to use vector images when designing their models. Unlike raster images, which are based on the use of individual pixels, vector images are based on lines and points that have defined relationships to one another. No matter how much a user shrinks or enlarges a vector image, the image will retain the correct proportions in terms of the relative placement of points and lines.

COMPUTER-AIDED MANUFACTURING

After designing an object using CAD software, a user may use a CAM program to manufacture it. CAM programs typically operate through computer numerical control (CNC). In CNC, instructions are transmitted to the manufacturing machine as a series of numbers. Those instructions tell the machine how to move and what actions to perform in order to construct the object. The types of machines used in that process vary and may include milling machines, drills, and lathes.

In the early twenty-first century, 3-D printers, devices that manufacture objects out of thin layers of plastic or other materials, began to be used in CAM. Unlike traditional CNC machinery, 3-D printers are typically used by individuals or small companies for whom larger-scale manufacturing technology is excessive.

SPECIALIZED APPLICATIONS

As CAD/CAM technology has evolved, it has come to be used for a number of specialized applications. Some CAD software, for instance, is used to perform four-dimensional building information modeling (4-D BIM). This process enables a user to incorporate information related to time. For instance, the schedule for a particular project can be accounted for in the modeling process with 4-D BIM.

Another common CAD/CAM application is rapid prototyping. In that process, a company or individual can design and manufacture physical prototypes of an object. This allows the designers to make changes in response to testing and evaluation and to test different iterations of the product. The resulting prototypes are often manufactured using 3-D printers. Rapid prototyping results in improved quality control and a reduced time to bring a product to market.

—*Joy Crelin*

BIBLIOGRAPHY

Bryden, Douglas. *CAD and Rapid Prototyping for Product Design.* London: King, 2014. Print.

Chua, Chee Kai, Kah Fai Leong, and Chu Sing Lim. *Rapid Prototyping: Principles and Applications.* Hackensack: World Scientific, 2010. Print.

"Computer-Aided Design (CAD) and Computer-Aided Manufacturing (CAM)." *Inc.* Mansueto Ventures, n.d. Web. 31 Jan. 2016.

"Design and Technology: Manufacturing Processes." *GCSE Bitesize.* BBC, 2014. Web. 31 Jan. 2016.

Herrman, John. "How to Get Started: 3D Modeling and Printing." *Popular Mechanics.* Hearst Digital Media, 15 Mar. 2012. Web. 31 Jan. 2016.

Krar, Steve, Arthur Gill, and Peter Smid. *Computer Numerical Control Simplified.* New York: Industrial, 2001. Print.

Sarkar, Jayanta. *Computer Aided Design: A Conceptual Approach.* Boca Raton: CRC, 2015. Print.

CLOUD COMPUTING

FIELDS OF STUDY

Information Technology; Computer Science; Software

ABSTRACT

Cloud computing is a networking model in which computer storage, processing, and program access are handled through a virtual network. Cloud computing is among the most profitable IT trends. A host of cloud-oriented consumer products are available through subscription.

PRINICIPAL TERMS

- **hybrid cloud:** a cloud computing model that combines public cloud services with a private cloud platform linked through an encrypted connection.
- **infrastructure as a service:** a cloud computing platform that provides additional computing resources by linking hardware systems through the Internet; also called "hardware as a service."
- **multitenancy:** a software program that allows multiple users to access and use the software from different locations.
- **platform as a service:** a category of cloud computing that provides a virtual machine for users to develop, run, and manage web applications.
- **software as a service:** a software service system in which software is stored at a provider's data center and accessed by subscribers.
- **third-party data center:** a data center service provided by a separate company that is responsible for maintaining its infrastructure.

CLOUD NETWORK DESIGN

Cloud computing is a networking model that allows users to remotely store or process data. Several major Internet service and content providers offer cloud-based storage for user data. Others provide virtual access to software programs or enhanced processing capabilities. Cloud computing is among the fastest-growing areas of the Internet services industry. It has also been adopted by government and research organizations.

TYPES OF CLOUD NETWORKS

Private clouds are virtual networks provided to a limited number of known users. These are often used in corporations and research organizations. Operating a private cloud requires infrastructure (software, servers, etc.), either on-site or through a third party. Public clouds are available to the public or to paying subscribers. The public-cloud service provider owns and manages the infrastructure. Unlike private clouds, public clouds provide access to an unknown pool of users, making them less secure. Public clouds tend to be based on open-source code, which is free and can be modified by any user.

The hybrid cloud lies somewhere between the two. It offers access to private cloud storage or software services, such as database servers, while keeping some services or components in a public cloud. Setup costs may be lower with hybrid cloud services. A group using a hybrid cloud outsources some aspects of infrastructure investment and maintenance but still enjoys greater security than with a public cloud. Hybrid clouds have become widespread in the health care, law, and investment fields, where sensitive data must be protected on-site.

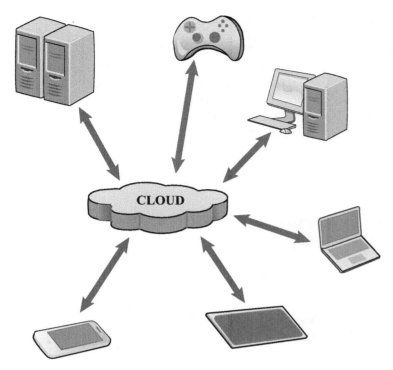

Cloud computing refers to the use of processors, memory, and other peripheral devices offsite, connected by a network to one's workstation. Use of the cloud protects data by storing it and duplicating it offsite and reduces infrastructure and personnel needs. EBSCO illustration.

CLOUD COMPUTING AS A SERVICE

The infrastructure as a service (IaaS) model offers access to virtual storage and processing capability through a linked network of servers. Cloud-based storage has become popular, with services such as Apple iCloud and Dropbox offering storage alternatives beyond the memory on users' physical computers. IaaS can also give users greater computing power by allowing certain processes to run on virtual networks, rather than on the hardware of a single system. Using IaaS enables companies to create a corporate data center through third-party data centers. These third-party centers provide expert IT assistance and server resources, generally for subscription fees.

The platform as a service (PaaS) model mainly offers access to a specific platform that multiple users can use to develop software applications, or apps. Many apps require access to specific development programs. The Google App Engine and IBM's developer Works Open provide an environment that stores, supports, and runs web apps. PaaS allows software developers to create apps without investing in infrastructure and data center support. Providers may also offer virtual storage, access to virtual networks, and other services.

The software as a service (SaaS) model offers users subscription-based or shared access to software programs through a virtual network. Adobe Systems' Creative Cloud provides access to programs such as Photoshop, Illustrator, and Lightroom for a monthly fee. Users pay a smaller amount over time rather than paying a higher cost up front to purchase the program. SaaS supports multitenancy, in which a single copy of a program is available to multiple clients. This allows software providers to earn revenue from multiple clients through a single instance of a software program.

ADVANTAGES AND DISADVANTAGES OF THE CLOUD

Cloud networking allows small companies and individuals access to development tools, digital storage, and software that once were prohibitively expensive or required significant management and administration. By paying subscription fees, users can gain monthly, yearly, or as-used access to software or other computing tools with outsourced administration. For service providers, cloud computing is cost effective because it eliminates the cost of packaging and selling individual programs and other products.

Data security is the chief concern among those considering cloud computing. The private and hybrid cloud models provide a secure way for companies to reap the benefits of cloud computing. Firewalls and encryption are common means of securing data in these systems. Providers are working to increase the security of public clouds, thus reducing the need for private or hybrid systems.

—*Micah L. Issitt*

BIBLIOGRAPHY

Beattie, Andrew. "Cloud Computing: Why the Buzz?" *Techopedia*. Techopedia, 30 Nov. 2011. Web. 21 Jan. 2016.

Huth, Alexa, and James Cebula. *The Basics of Cloud Computing*. N.p.: Carnegie Mellon U and US Computer Emergency Readiness Team, 2011. PDF file.

Kale, Vivek. *Guide to Cloud Computing for Business and Technology Managers*. Boca Raton: CRC, 2015. Print.

Kruk, Robert. "Public, Private and Hybrid Clouds: What's the Difference?" *Techopedia*. Techopedia, 18 May 2012. Web. 21 Jan. 2016.

Rountree, Derrick, and Ileana Castrillo. *The Basics of Cloud Computing*. Waltham: Elsevier, 2014. Print.

Ryan, Janel. "Five Basic Things You Should Know about Cloud Computing." *Forbes*. Forbes.com, 30 Oct. 2013. Web. 30 Oct. 2013.

Sanders, James. "Hybrid Cloud: What It Is, Why It Matters." *ZDNet*. CBS Interactive, 1 July 2014. Web. 10 Jan. 2016.

COMBINATORICS

FIELDS OF STUDY

Information Technology; Algorithms; System Analysis

ABSTRACT

Combinatorics is a branch of mathematics that is concerned with sets of objects that meet certain conditions. In computer science, combinatorics is used to study algorithms, which are sets of steps, or rules, devised to address a certain problem.

PRINICIPAL TERMS

- **analytic combinatorics:** a method for creating precise quantitative predictions about large sets of objects.
- **coding theory:** the study of codes and their use in certain situations for various applications.
- **combinatorial design:** the study of the creation and properties of finite sets in certain types of designs.
- **enumerative combinatorics:** a branch of combinatorics that studies the number of ways that certain patterns can be formed using a set of objects.
- **graph theory:** the study of graphs, which are diagrams used to model relationships between objects.

BASICS OF COMBINATORICS

Combinatorics is a branch of mathematics that studies counting methods and combinations, permutations, and arrangements of sets of objects. For instance, given a set of fifteen different objects, combinatorics studies equations that determine how many different sets of five can be created from the original set of fifteen. The study of combinatorics is crucial to the study of algorithms. Algorithms are sets of rules, steps, or processes that are linked together to address a certain problem.

THE SCIENCE OF COUNTING AND COMBINATIONS

Combinatorics is often called the "science of counting." It focuses on the properties of finite sets of objects, which do not have infinite numbers of objects and so are theoretically countable. The process of describing or counting all of the items in a specific set is called "enumeration." Combinatorics also includes the study of combinations, a process of selecting items from a set when the order of selection does not matter. Finally, combinatorics also studies permutations. Permutations involve selecting or arranging items in a list, when the order of arrangement is important. Combinatorics also studies the relationships between objects organized into sets in various ways.

There are numerous subfields of combinatorics used to study sets of objects in different ways. Enumerative combinatorics is the most basic branch of the field. It can be described as the study of counting methods used to derive the number of objects in a given set. By contrast, analytic combinatorics is a subfield of enumerative combinatorics. It deals with predicting the properties of large sets of objects, using quantitative analysis. All combinatorics analysis requires detailed knowledge of calculus. Many subfields make extensive use of probability theory and predictive analysis.

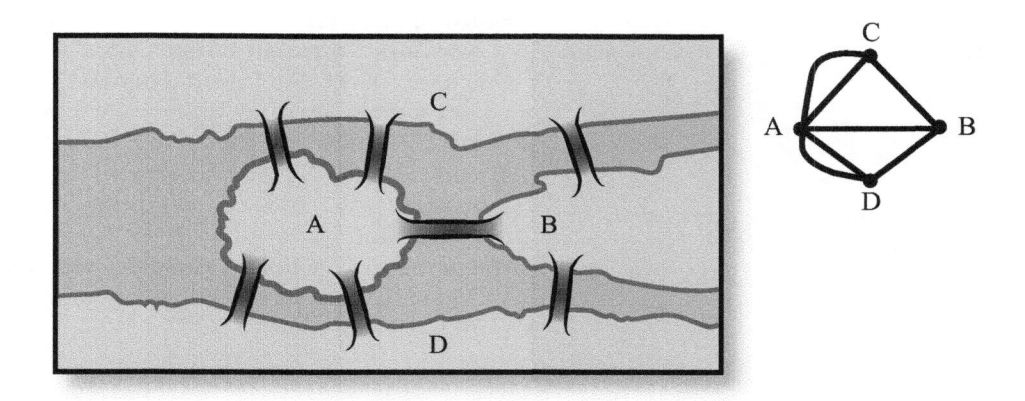

The Konigsberg bridge problem is a common example of combinatorial structures used to identify and quantify possible combinations of values. Each landmass becomes a vertex or node, and each bridge becomes an arc or edge. In computer science, these graphs are used to represent networks or the flow of computation. EBSCO illustration.

COMBINATORICS APPLICATIONS

There are many different applications for combinatorics in analytic mathematics, engineering, physics, and computer science. Among the most familiar basic examples of combinatorics is the popular game sudoku. The game challenges players to fill in the blanks in a "magic square" diagram with specific column and row values. Sudoku puzzles are an example of combinatorial design, which is a branch of combinatorics that studies arrangements of objects that have symmetry or mathematical/geometric balance between the elements.

Combinatorics is also crucial to graph theory. Graph theory is a field of mathematics that deals with graphs, or representations of objects in space. Graphs are used in geometry, computer science, and other fields to model relationships between objects. In computer science, for instance, graphs are typically used to model computer networks, computational flow, and the structure of links within websites. Combinatorics is used to study the enumeration of graphs. This can be seen as counting the number of different possible graphs that can be used for a certain application or model.

COMBINATORICS IN ALGORITHM DESIGN

In computer science, combinatorics is used in the creation and analysis of algorithms. Algorithms are sets of instructions, computing steps, or other processes that are linked together to address a certain computational problem. As algorithms are essentially sets, the steps within algorithms can be studied using combinatorial analysis, such as enumeration or permutation. As combinatorics can help researchers to find more efficient arrangements of objects, or sets, combinatorial analysis is often used to test and assess the efficiency of computer algorithms. Combinatorics is also key to the design of sorting algorithms. Sorting algorithms allow computers to sort objects (often pieces of data, web pages, or other informational elements), which is a necessary step in creating effective algorithms for web searching.

COMBINATORICS IN CODING

Combinatorics is also used in coding theory, the study of codes and their associated properties and characteristics. Codes are used for applications, including cryptography, compressing or translating data, and correcting errors in mathematical, electrical, and information systems. Coding theory emerged from combinatorial analysis. These two branches of mathematics are distinct but share theories and techniques.

Combinatorics is an advanced field of study. The arrangement, organization, and study of relationships between objects provides analytical information applicable to many academic and practical fields. Combinatorics influences many aspects of computer design and programming including the development of codes and the precise study of information and arrangement of data.

—*Micah L. Issit*

BIBLIOGRAPHY

Beeler, Robert A., *How to Count: An Introduction to Combinatorics.* New York: Springer, 2015. Print.

"Combinatorics." *Mathigon.* Mathigon, 2015. Web. 10 Feb. 2016.

Faticoni, Theodore G., *Combinatorics: An Introduction.* New York: Wiley, 2014. Digital file.

Guichard, David. "An Introduction to Combinatorics and Graph Theory." *Whitman.* Whitman Coll., 4 Jan 2016. Web. 10 Feb. 2016.

Roberts, Fred S., and Barry Tesman. *Applied Combinatorics.* 2nd ed. Boca Raton: Chapman, 2012. Print.

Wagner, Carl. "Choice, Chance, and Inference." *Math.UTK.edu.* U of Tennessee, Knoxville, 2015. Web. 10 Feb. 2016.

COMMUNICATION TECHNOLOGY

FIELDS OF STUDY

Digital Media; Information Technology

ABSTRACT

Communication technology is a broad field that includes any type of tool or process that makes it possible for individuals and groups to communicate over large areas. Many types of communication technology have been used throughout recorded history—smoke signals, carved signs and tokens, letters, radio waves, telegraph, television, and microwave signals, to name only a few.

PRINICIPAL TERMS

- **attenuation:** the loss of intensity from a signal being transmitted through a medium.
- **broadcast:** an audio or video transmission sent via a communications medium to anyone with the appropriate receiver.
- **multiplexing:** combining multiple data signals into one in order to transmit all signals simultaneously through the same medium.
- **receiver:** a device that reads a particular type of transmission and translates it into audio or video.
- **transmission medium:** the material through which a signal can travel.
- **transmitter:** a device that sends a signal through a medium to be picked up by a receiver.

SENDING AND RECEIVING SIGNALS

Communication technology at its most basic involves sending a signal from a transmitter to a receiver. The signal carries information in audio, video, or some other data format. It travels in the form of energy waves through a transmission medium. This medium might be a device designed to carry the signal, such as a fiber-optic cable, or it might simply be the air through which a sound wave travels.

The purpose of most communication technology is to translate human language into a signal that can be transmitted through the appropriate medium, then revert the signal to the original format once it is received. For example, audio is sent through a digital telephone line in the form of binary data. When one person speaks, the sound waves they produce are converted into a stream of ones and zeros that represents a digital model of the analog waveform. This stream is transmitted in the form of electronic pulses through an electrical network. (If the network is optical rather than electrical, it is transmitted as light pulses instead.) When the data stream reaches its destination, a receiver reads the binary data and converts it back into analog sound waves that the person on the other end of the phone line can understand. In short, communication technology is all about encoding and decoding messages so they can be sent from one place to another.

Advances in technology have provided a wide variety of modes of communication. Phones, tablets, computers, and other devices allow people to write, talk, and send images and videos all around the globe. Public domain, via Wikimedia Commons.

CONSIDERATIONS OF COMMUNICATION TECHNOLOGY

Communication technology is not just about the transfer of information from one individual to another. Mass communication involves the transmission of information on what is called a "one-to-many" basis. A broadcast is the transmission of a single signal to multiple receivers in a given area. This signal can be received by anyone within range who possesses the necessary receiver, such as a radio, a television, or some other device. When a broadcast is transmitted over a very large area, such as a radio broadcast in a major city, the signal may experience some attenuation, or weakening of intensity. This attenuation can be overcome by installing repeaters along the transmission route to help boost the signal. These repeaters work by receiving the original transmission and then rebroadcasting it from a new location. A chain of repeaters can carry a signal significantly farther than it would otherwise travel.

One fact of modern communications is that at any given moment, there are far more messages in need of transmission than there is transmission capacity. This problem is handled in a number of different ways. First, different types of communication technology are used, so that not all communications need to share the same medium. Second, some types of communication technology, such as radio, make it possible to transmit at different frequencies at the same time, so that multiple transmissions can occur simultaneously. Combining transmissions in this way is known as multiplexing. Multiplexing can be accomplished either by dividing up frequencies or by dividing transmission times into very small intervals. These approaches are called frequency-division multiplexing and time-division multiplexing, respectively. In the United States, there is stiff competition for broadcast frequency space among communications technology companies seeking to control as much of the broadcast spectrum as possible.

INTERNET COMMUNICATIONS

The fact that the Internet can carry so many different types of communications—video, audio, images, text, and so on—sometimes causes confusion about what type of communication technology it is. Fundamentally, the Internet is a tool for digital communication. Regardless of the type of data contained in an Internet transmission, it is all transmitted in ones and zeros. Only once it arrives at its destination is the binary code translated back into a video, an image, or some other format.

IT AND COMMUNICATION OVERLAP

The end of the twentieth century and the beginning of the twenty-first have seen a growing convergence of the fields of information technology (IT) and communication technology. IT has largely "taken over" communication technology, since most modern communication technology being used and developed involves computers and similar types of technology. The digital revolution has transformed the way society thinks about information and its transmission. It is difficult to imagine a new type of communication technology that would not rely either wholly or in part on computers.

—*Scott Zimmer, JD*

BIBLIOGRAPHY

Adams, Ty, and Stephen A. Smith. *Communication Shock: The Rhetoric of New Technology.* Newcastle upon Tyne: Cambridge Scholars, 2015. Print.

Dor, Daniel. *The Instruction of Imagination: Language as a Social Communication Technology.* New York: Oxford UP, 2015. Print.

Englander, Irv. *The Architecture of Computer Hardware, Systems Software, & Networking: An Information Technology Approach.* 5th ed. Hoboken: Wiley, 2014. Print.

Fuchs, Christian, and Marisol Sandoval, eds. *Critique, Social Media and the Information Society.* New York: Routledge, 2014. Print.

Hart, Archibald D., and Sylvia Hart Frejd. *The Digital Invasion: How Technology Is Shaping You and Your Relationships.* Grand Rapids: Baker, 2013. Print.

Tosoni, Simone, Matteo Tarantino, and Chiara Giaccardi, eds. *Media and the City: Urbanism, Technology and Communication.* Newcastle upon Tyne: Cambridge Scholars, 2013. Print.

COMPTIA A+ CERTIFICATION

FIELDS OF STUDY

Computer Science; Information Technology

ABSTRACT

The CompTIA A+ certification is one of many certifications in the field of information technology (IT) granted by the Computing Technology Industry Association (CompTIA). An IT professional may gain the certification after passing a pair of exams that test their knowledge of computer hardware, operating systems, troubleshooting, and other essential topics.

PRINICIPAL TERMS

- **American National Standards Institute (ANSI):** a nonprofit organization that oversees the creation and use of standards and certifications such as those offered by CompTIA.
- **computer technician:** a professional tasked with the installation, repair, and maintenance of computers and related technology.
- **information technology:** the use of computers and related equipment for the purpose of processing and storing data.
- **Initiative for Software Choice (ISC):** a consortium of technology companies founded by CompTIA, with the goal of encouraging governments to allow competition among software manufacturers.

CERTIFYING IT

The CompTIA A+ certification is a professional certification in information technology (IT). IT is a broad field that deals with the use of computers and related devices. IT professionals known as computer technicians must know a variety of hardware, software, and other technological information. College degrees in IT or computer science and hands-on IT experience are typically important to employers. Some companies seek job candidates who have earned professional certifications. In the United States, the Computer Technology Industry Association (CompTIA), an IT trade association, is one of the chief providers of such certifications.

CompTIA was founded in 1982 as the Association of Better Computer Dealers. The organization began offering certifications in 1992, two years after changing its name to CompTIA. CompTIA's certifications are vendor neutral. This means that they test knowledge of computer hardware, operating systems, and peripherals from multiple manufacturers, not just one. The organization also offers certifications in computer networking, servers, and subfields such as healthcare IT. CompTIA offers educational and professional development opportunities for IT professionals as well. In the early twenty-first century, the organization became increasingly involved in public policy advocacy. It cofounded the Initiative for Software Choice (ISC) in 2002. The ISC is concerned with government policies regarding software use.

Understanding the CompTIA A+ Certification

The CompTIA A+ certification is a popular credential among IT professionals. As of early 2016, it had been awarded to more than one million people globally, according to CompTIA. The certification shows that the holder is skilled in the key areas of IT. These areas include computer troubleshooting and repair, setup and installation, and maintenance. To earn the certification, a professional must prove their knowledge of a number of specific topics, including the use of operating systems, computer networking procedures, and computer security. In addition to desktop and laptop computers, the certification signals the holder's skill with mobile devices, such as smartphones, and peripherals, such as printers. The A+ certification and two other CompTIA certifications have been accredited by the American National Standards Institute (ANSI). ANSI oversees the granting of certifications for many professions.

Obtaining the CompTIA A+ Certification

To earn the CompTIA A+ certification, one must pass two exams. These exams are periodically updated to keep up with changes in technology. An updated set of exams, CompTIA A+ 220-901 and 220-902, were introduced on December 15, 2015. Those exams replaced the 220-801 and 220-802 exams, which were set to be retired on June 30, 2016, for English-speaking test takers, and December 31 of that year for all others.

Each exam is ninety minutes long and has up to ninety questions. Some questions are multiple-choice, while others are performance based. Performance-based questions simulate real-world problems that the test taker must solve. Both tests are scored out of a possible 900 points. The test taker must score at least 675 points to pass the first exam and 700 to pass the second. Exams must be completed in person at a testing center approved by CompTIA.

Together, the two exams cover a wide range of topics generally considered essential basic knowledge for a computer technician. The first exam deals mostly with hardware, including computers, mobile devices, and peripherals. The exam also covers troubleshooting. The second exam focuses on operating systems, including Widows and Apple systems as well as open-source operating systems such as Linux. CompTIA does not require test takers to complete formal schooling before taking the exams, but it recommends that those seeking A+ certification have at least six months of practical experience with the covered topics. To aid IT professionals in preparing for the exams, CompTIA offers optional preparatory courses and self-guided training materials.

CompTIA A+ in the Field

While the computer skills evaluated by the CompTIA A+ exams are not possessed solely by those professionals who are certified, the certification provides clear evidence of an IT professional's understanding of the field's core concepts and procedures. As such, some employers prefer to hire employees who have already attained that certification. However, other employers may prefer to hire computer technicians with formal education in computer science or IT or with significant hands-on experience. Likewise, some employers may not require job seekers to have obtained A+ certification or may pay for their computer technicians to take the exams once employed. Because of these varying requirements, individuals seeking work in IT should consider their options carefully before pursuing the CompTIA A+ or any other certification.

In 2010, CompTIA announced that all A+ certifications, which had previously been issued for life, would in future need to be renewed every three years. The organization later ruled that previously issued certifications would be exempt from this requirement. However, all A+ certifications issued on or after January 1, 2011, expire after three years unless renewed. To remain certified, a professional holding the certification must pay a fee and complete continuing-education programs to remain current in the field.

—Joy Crelin

BIBLIOGRAPHY

"About ANSI." *ANSI.* American Natl. Standards Inst., 2016. Web. 31 Jan. 2016.

Anderson, Nate. "CompTIA Backs Down; Past Certs Remain Valid for Life." *Ars Technica.* Condé Nast, 26 Jan. 2010. Web. 31 Jan. 2016.

"ANSI Accredits Four Personnel Certification Programs." *ANSI.* American Natl. Standards Inst., 8 Apr. 2008. Web. 31 Jan. 2016.

"CompTIA A+." *CompTIA.* Computing Technology Industry Assn., 2015. Web. 31 Jan. 2016.

"Our Story." *CompTIA.* Computing Technology Industry Assn., n. d. Web. 31 Jan. 2016.

"What Is the CompTIA A+ Certification?" *Knowledge Base.* Indiana U, 15 Jan. 2015. Web. 31 Jan. 2016.

COMPUTER ANIMATION

FIELDS OF STUDY

Digital Media; Graphic Design

ABSTRACT

Computer animation is the creation of animated projects for film, television, or other media using specialized computer programs. As animation projects may range from short, simple clips to detailed and vibrant feature-length films, a wide variety of animation software is available, each addressing the particular needs of animators. The computer animation process includes several key steps, including modeling, keyframing, and rendering. These stages are typically carried out by a team of animators.

PRINICIPAL TERMS

- **animation variables (avars):** defined variables that control the movement of an animated character or object.
- **keyframing:** the process of defining the first and last—or key—frames in an animated transition.
- **render farms:** large computer systems dedicated to rendering animated content.
- **3-D rendering:** the process of creating a 2-D animation using 3-D models.

- **virtual reality:** a form of technology that enables the user to view and interact with a simulated environment.

HISTORY OF COMPUTER ANIMATION

Since the early twentieth century, the field of animation has been marked by frequent, rapid change. Innovation in the field has been far reaching, filtering into film, television, advertising, video games, and other media. It was initially an experimental method and took decades to develop. Computer animation revitalized the film and television industries during the late twentieth and early twenty-first centuries, in many ways echoing the cultural influence that animation had decades before.

Prior to the advent of computer animation, most animated projects were created using a process that later became known as "traditional," or "cel," animation. In cel animation, the movement of characters, objects, and backgrounds was created frame by frame. Each frame was drawn by hand. This time-consuming and difficult process necessitated the creation of dozens of individual frames for each second of film.

As computer technology developed, computer researchers and animators began to experiment with creating short animations using computers. Throughout the 1960s, computers were used to

From designing the original animation model to creating algorithms that control the movement of fluids, hair, and other complex systems, computer software has drastically changed the art of animation. Through software that can manipulate polygons, a face can be rendered and further manipulated to create a number of images much more efficiently than with hand-drawn illustrations. Thus, the detail of the imaging is increased, while the time needed to develop a full animation is reduced. By Diego Emanuel Viegas, CC BY-SA 3.0 (http://creativecommons.org/licenses/by-sa/3.0), via Wikimedia Commons.

create 2-D images. Ed Catmull, who later founded the studio Pixar in 1986, created a 3-D animation of his hand using a computer in 1972. This was the first 3-D computer graphic to be used in a feature film when it appeared in *Futureworld* (1976). Early attempts at computer animation were found in live-action films. The 1986 film *Labyrinth*, for instance, notably features a computer-animated owl flying through its opening credits. As technology improved, computer animation became a major component of special effects in live-action media. While cel animation continued to be used in animated feature films, filmmakers began to include some computer-generated elements in such works. The 1991 Walt Disney Studios film *Beauty and the Beast*, for instance, featured a ballroom in one scene that was largely created using a computer.

In 1995, the release of the first feature-length computer-animated film marked a turning point in the field of animation. That film, *Toy Story*, was created by Pixar, a pioneer in computer animation. Over the following decades, Pixar and other studios, including Disney (which acquired Pixar in 2006) and DreamWorks, produced numerous computer-animated films. Computer animation became a common process for creating animated television shows as well as video games, advertisements, music videos, and other media.

In the early twenty-first century, computer animation also began to be used to create simulated environments accessed through virtual reality equipment such as the head-mounted display Oculus Rift. Much of the computer-animated content created during this time featured characters and surroundings that appeared 3-D. However, some animators opted to create 2-D animations that more closely resemble traditionally animated works in style.

THREE-DIMENSIONAL COMPUTER ANIMATION

Creating a feature-length computer-animated production is a complex and time-intensive process that is carried out by a large team of animators, working with other film-industry professionals. When creating a 3-D computer-animated project, the animation team typically begins by drawing storyboards. Storyboards are small sketches that serve as a rough draft of the proposed scenes.

Next, animators transform 2-D character designs into 3-D models using animation software. They use animation variables (avars) to control the ways in which the 3-D characters move, assigning possible directions of movement to various points on the characters' bodies. The number of avars used and the areas they control can vary widely. The 2006 Pixar film *Cars* reportedly used several hundred avars to control the characters' mouths alone. Using such variables gives animated characters a greater range of motion and often more realistic expressions and gestures. After the characters and objects are modeled and animated, they are combined with backgrounds as well as lighting and special effects. All of the elements are then combined to transform the 3-D models into a 2-D image or film. This process is known as 3-D rendering.

TWO-DIMENSIONAL COMPUTER ANIMATION

Animating a 2-D computer-animated work is somewhat different from its 3-D counterpart, in that it does not rely on 3-D modeling. Instead, it typically features the use of multiple layers, each of which contains different individual elements. This method of animating typically features keyframing. In this procedure, animators define the first and last frames in an animated sequence and allow the computer to fill in the movement in between. This process, which in traditionally animated films was a laborious task done by hand, is often known as "inbetweening," or "tweening."

TOOLS

Various animation programs are available to animators, each with its own strengths and weaknesses. Some animation software, such as Maya and Cinema 4D, are geared toward 3-D animation. Others, such as Adobe Flash, are better suited to 2-D animation. Adobe Flash in particular has commonly been used to produce 2-D cartoons for television, as it is considered a quick and low-cost means of creating such content. Animation studios such as Pixar typically use proprietary animation software, thus ensuring that their specific needs are met.

In addition to animation software, the process of computer animation relies heavily on hardware, as many steps in the process can be taxing for the systems in use. Rendering, for example, often demands a sizable amount of processing power. As such, many studios make use of render farms, large, powerful computer systems devoted to that task.

—*Joy Crelin*

BIBLIOGRAPHY

Carlson, Wayne. "A Critical History of Computer Graphics and Animation." *Ohio State University.* Ohio State U, 2003. Web. 31 Jan. 2016.
Highfield, Roger. "Fast Forward to Cartoon Reality." *Telegraph.* Telegraph Media Group, 13 June 2006. Web. 31 Jan. 2016.
Parent, Rick. *Computer Animation: Algorithms and Techniques.* Waltham: Elsevier, 2012. Print.
"Our Story." *Pixar.* Pixar, 2016. Web. 31 Jan. 2016.
Sito, Tom. *Moving Innovation: A History of Computer Animation.* Cambridge: MIT P, 2013. Print.
Winder, Catherine, and Zahra Dowlatabadi. *Producing Animation.* Waltham: Focal, 2011. Print.

COMPUTER CIRCUITRY: FLIP FLOPS

FIELDS OF STUDY

Computer Engineering; Computer Science

ABSTRACT

Flip-flops are bistable multivibrator devices because they have two stable states. The device holds a one-bit signal, either high or low, that is sent as output from the device when it is triggered by a clock signal. The device is used to construct counters, accumulators and registers in central processing chips, and is often used as the triggering device for hardware interruption processes. When triggered, a flip-flop outputs its saved state and sets a new state according to its input signals.

PRINICIPAL TERMS

- **clock speed:** the speed at which a microprocessor can execute instructions; also called "clock rate."
- **counter:** a digital sequential logic gate that records how many times a certain event occurs in a given amount of time.
- **hardware:** the physical parts that make up a computer. These include the motherboard and processor, as well as input and output devices such as monitors, keyboards, and mice.
- **hardware interruption:** a device attached to a computer sending a message to the operating system to inform it that the device needs attention, thereby "interrupting" the other tasks that the operating system was performing.
- **inverter:** a logic gate whose output is the inverse of the input; also called a NOT gate.
- **negative-AND (NAND) gate:** a logic gate that produces a false output only when both inputs are true
- **transistor:** a computing component generally made of silicon that can amplify electronic signals or work as a switch to direct electronic signals within a computer system.
- **vibrator:** an electronic component that (oscillates) or switches between electronic states

TRANSISTORS AND GATES

A flip-flop is a digital electronic device that holds a particular output state until it is triggered by a "clock signal" to "flip-flop" the output state. A typical flip-flop is constructed from a set of transistor "gates" in such a way that the output signal from each gate is used as an input signal to the other gate. Each gate is formed by a specific interconnection of transistor structures. There are three basic types of transistor-based gates, designated as AND, OR and NOT. A NOT gate is also known as an "inverter," because it inverts the value of the digital signal passing through it. The AND gate produces a "high" output signal only when all of its input signals are "high," but a "low" signal if this condition is not met. The OR gate produces a "high" output signal only if some, but not all, of its input signals are "high," and a "low" output signal otherwise. The combination of a NOT gate and an AND or OR gate produces a negative-AND (NAND) gate and a negative-OR (NOR) gate, respectively. Accordingly, digital logic circuits can be designed using AND, OR, NOT, NAND and NOR gates as needed. However, transistor-transistor logic is more compact and efficient when circuits are constructed in a negative sense using NAND and NOR gates instead of in the positive sense using AND, OR and NOT gates. Flip-flops are constructed exclusively from NAND gates and NOR gates.

TYPES OF FLIP-FLOPS

There are two basic types of flip-flops used in the construction of digital computer hardware devices, termed J-K and R-S flip-flops. In both types, the output signal from each gate is used as an input signal to the other gate. Another input signal required for the operation of a flip-flop is the "clock" signal. The "clock" signal may be either a regular square wave signal that determines the clock speed of the particular device, or an irregular square wave signal from another segment of the digital circuitry. Square wave signals are nominally instantaneous changes between high and low signals. However, the tiny amount of time required for the change to take place as the semiconductor responds to the new condition produces a "ramping" effect on both the leading and trailing edge of the signal change. The state that exists on the ramped portion of the signal change is not "allowed," since it is neither "high" nor "low." Flip-flops are designed to trigger their output signal when one of the disallowed states is detected. The difference in the manner of changing state determines whether the structure functions as a flip-flop or as a latch. Triggering either one causes the device to output the state that it has been holding and allows the device to read the states of its input signals. This sets a new state that the device will hold until it is again triggered.

SAMPLE PROBLEM

A flip-flop constructed from two NAND gates has four inputs. The states of one pair of inputs is both high, while the other pair of inputs is one high and one low. What is the output from the flip-flop?

Answer:

A NAND gate outputs a low signal only when both inputs are high, and a low signal otherwise. In this example, the NAND gate with the two high inputs would output a low signal. This is the low input to the other NAND gate, which must therefore output a low signal. Since the output signal of this NAND gate is the second input to the other NAND gate in the device, and that NAND gate has two high input signals, this situation is not allowed.

APPLICATIONS OF FLIP-FLOPS

Flip-flops and latches are termed bistable multivibrator devices: bistable because they have only two stable states, and multivibrators because they can switch back and forth between those two states. A series of latch structures in combination can be used to hold a corresponding series of bits in a parallel data stream and pass that set of data on at each clock cycle. This feature is the basic structure of an accumulator, counter or a register in central processing chips. Flip-flops are also typically used as the triggering signal generator for hardware interruption. Every central processing chip must have a signal input to indicate that another hardware device, such as a modem or a printer, requires attention. When that signal is received at the appropriate terminal on the chip, the process being carried out by the central processing unit is interrupted, allowing the function of the device to take precedence temporarily. Flip-flops can also be interconnected in a "master-slave" combination, in which input is registered on the leading edge of the signal and output is registered on the trailing edge.

—*Richard M. Renneboog M.Sc.*

BIBLIOGRAPHY

Bishop, Owen (2011) *Electronics. Circuits and Systems* 4th ed., New York, NY: E;sevier. Print.

Brindley, Keith (2011) *Starting Electronics* 4th ed., New York, NY: Elsevier. Print.

Clements, Alan (2006) *Principles of Computer Hardware* 4th ed., New York, NY: Oxford University Press. Print.

Gibson, J.R. (2011) *Electronic Logic Circuits* 3rd ed., New York, NY: Routledge. Print.

Hsu, John Y. (2002) *Computer Logic Design Principles and Applications* New York, NY: Springer. Print.

COMPUTER CIRCUITRY: SEMICONDUCTORS

FIELDS OF STUDY

Computer Engineering

ABSTRACT

Semiconductor materials are the principal component of computer technology, and are primarily based on silicon and germanium. Pure silicon does not have enough 'free' valence electrons to conduct electrical current adequately, and small percentages of other elements are blended into the silicon to enhance the conductivity of the material. Current flows in semiconductors as electrons moving through the material driven by a voltage difference, and as 'holes' equivalent to positive charge. Production of silicon-based semiconductor material and the subsequent manufacture of integrated circuit chips is a complex, multistep process requiring stringent quality control measures.

PRINICIPAL TERMS

- **band gap:** the energy difference between the ground state of electrons in a material and the conduction band.
- **central processing unit (CPU):** electronic circuitry that provides instructions for how a computer handles processes and manages data from applications and programs.
- **conduction band:** the region of atomic and molecular orbitals in a material through which free electrons move in an electrical current.
- **hole:** a vacant location in the normally-filled valence shell of an atom, created when an electron is excited to a higher energy level in the same atom; a hole behaves like a positive charge by attracting an available electron of the corresponding energy.
- **semiconductor intellectual property (SIP) block:** a quantity of microchip layout design that is owned by a person or group; also known as an "IP core."
- **solid-state storage:** computer memory that stores information in the form of electronic circuits, without the use of disks or other read/write equipment.
- **transistor:** a computing component generally made of silicon that can amplify electronic

signals or work as a switch to direct electronic signals within a computer system.

CONDUCTIVITY AND ELECTRICAL CURRENT

The ability of any material to conduct electricity is a function of the distribution of electrons and energy levels in the atoms and molecules of that material. Materials in which electrons can easily enter the conduction band of the material across a small band gap between the normal ground state of the electrons and the conduction band generally are good conductors of electrical current. Materials that have a large band gap between the normal ground state of the electrons and the conduction band are generally poor conductors of electrical current. Materials that have a band gap too large to be good conductors, and too small to be non-conductors, are called semiconductors. Good conductors have little energy difference between the occupied valence shell of electrons and the vacant orbitals in the next highest electron shell, and the valence electrons are able to move between them and from atom to atom through them relatively easily. In poor conductors and non-conductors, this energy difference is much greater, making it more difficult for electrons to move through the material. The essential semiconductor materials used in computer technology are silicon and germanium. The electron distribution in the atoms of these materials is very stable and not easily disrupted.

CHARGE CARRIERS

All electrical current results from the displacement of charge. Current flow in semiconductors is described in terms of charge carriers. Electrons moving though the conduction band of the material is the conventional definition of an electrical current. However, in semiconductor applications, electrical current is also produced by the movement of a hole. Heat energy or an applied voltage can excite electrons from their ground state in the valence shell of atoms to higher energy levels within those same atoms, where they become free electrons that can move through the conduction band of the material. The orbital position in the valence shell that the electrons leave behind, called 'holes', can then act as though they are positive charges by attracting any nearby electron of the corresponding energy from adjacent atoms. This creates another hole in the adjacent atom. The progression of holes through the material is equivalent to the movement of positive charge as electrons shift from one atom to another without flowing through the conduction band.

PREPARATION OF SEMICONDUCTOR MATERIALS

The principal material for the production of semiconductor-based devices is silicon. Pure silicon, however, does not function well as a semiconducting material for the production of transistor structures used for all kinds of integrated circuit (IC) chips, and especially for central processing unit (CPU) chips and solid-state storage devices. To improve the desired characteristics of the material, a small percentage of another element can be added as a "dopant" to either increase or decrease the number of available electrons in the resulting alloy. This silicon blend is prepared by melting a quantity of pure silicon together with the proper amount of dopant. A single, large cylindrical crystal is then drawn slowly from the molten mass and then cut into thin wafers for use in the manufacture of chips.

TRANSISTOR STRUCTURES

The thin wafers obtained from slicing up the large silicon crystal are polished to produce a highly uniform surface. Millions of transistor structures are then etched onto the surface of each wafer in a multistep process. The pattern and order of etching can constitute a proprietary or patented segment of the final product, known as a semiconductor intellectual property (SIP) block within the overall design of the integrated circuit. The overall process requires several individual steps that are carried out with stringent quality control at all stages of production. The

number of transistor structures that can be etched onto the surface of a silicon chip, and hence the computing capabilities of computers, has followed the empirical observation known as Moore's Law quite well for several decades. Moore's Law states that the number of possible transistor structures and the corresponding\ computing power doubles approximately every eighteen months. Logically, this means that there is a finite limit to both, determined by the physical minimum size of the structures themselves, and when that limit is reached no further advance can be made. However,current research with novel materials such as graphene and nanotubes, and toward the successful development of quantum computers may eventually end dependency on traditional semiconductor materials entirely.

—*Richard M. Renneboog M.Sc.*

BIBLIOGRAPHY

Haug, Hartmut, and Stephan W. Koch. *Quantum Theory of the Optical and Electronic Properties of Semiconductors.* New Jersey [u.a.]: World Scientific, 2009. Print.

Köhler, Anna, and Heinz Bässler. *Electronic Processes in Organic Semiconductors: An Introduction.* Wiley-VHC Verlag, 2015. Print.

Könenkamp, Rolf. *Photoelectric Properties and Applications of Low-Mobility Semiconductors.* Berlin: Springer, 2000. Print.

Mishra, Umesh. *Semiconductor Device Physics and Design.* Place of publication not identified: Springer, 2014. Print.

Rockett, Angus. *The Materials Science of Semiconductors.* New York, NY: Springer, 2010. Print.

Yu, P.Y, and M Cardona. *Fundamentals of Semiconductors: Physics and Materials Properties.* Berlin: Springer, 2010. Print.

COMPUTER MEMORY

FIELDS OF STUDY

Computer Science; Computer Engineering; Information Technology

ABSTRACT

Computer memory is the part of a computer used for storing information that the computer is currently working on. It is different from computer storage space on a hard drive, disk drive, or storage medium such as CD-ROM or DVD. Computer memory is one of the determining factors in how fast a computer can operate and how many tasks it can undertake at a time.

PRINICIPAL TERMS

- **flash memory:** nonvolatile computer memory that can be erased or overwritten solely through electronic signals, i.e. without physical manipulation of the device.
- **nonvolatile memory:** memory that stores information regardless whether power is flowing to the computer.
- **random access memory (RAM):** memory that the computer can access very quickly, without regard to where in the storage media the relevant information is located.
- **virtual memory:** memory used when a computer configures part of its physical storage (on a hard drive, for example) to be available for use as additional RAM. Information is copied from RAM and moved into virtual memory whenever memory resources are running low.
- **volatile memory:** memory that stores information in a computer only while the computer has power; when the computer shuts down or power is cut, the information is lost.

OVERVIEW OF COMPUTER MEMORY

Computer memory is an extremely important part of configuring and using computers. Many users find themselves confused by the concepts of computer memory and computer storage. After all, both store information. Storage serves a very different purpose from memory, however. Storage is slower, because it uses a series of spinning platters of magnetic disks and a moving read/write head. These create a tiny magnetic field that can modify the polarity of tiny sections of the platters in order to record or read information. The benefit of storage is that it is nonvolatile memory, meaning that its contents remain even when the power is turned off. Computer memory, by contrast, is volatile memory because it only stores information while power is flowing through the computer.

This can be a drawback, especially when a user has vital information, such as a newly created but unsaved document, that is lost when a power surge shuts down the system. However, memory is also

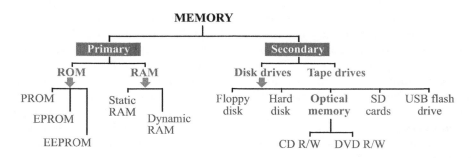

Computer memory comes in a number of formats. Some are the primary CPU memory, such as RAM and ROM; others are secondary memory, typically in an external form. External memory formats have changed over the years and have included hard drives, floppy drives, optical memory, flash memory, and secure digital (SD) memory. Adapted from the Its All About Embedded blog.

incredibly useful because it allows the computer to store information that it is currently working on. For example, if a user wished to update a résumé file saved on their hard drive, they would first tell the computer to access its storage to find a copy of the file. The computer would locate the file and then copy the file's contents into its volatile memory and open the document. As the user makes changes to the file, these changes are reflected only in the memory, until the user saves them to storage. Some have compared storage and memory to a librarian. Storage is like the thousands of books the librarian has in the library, which they can consult if given enough time. Memory is like the knowledge the librarian carries around in their head. It is accessible very quickly but holds a smaller amount of information than the books.

RANDOM AND VIRTUAL MEMORY
One feature of memory that makes it much faster than other types of information storage is that it is a type of random access memory (RAM). Any address in the memory block can be accessed directly, without having to sort through all of the other entries in the memory space. This contrasts with other types of memory, such as magnetic tape, which are sequential access devices. In order to get to a certain part of the tape, one must move forward through all the other parts of the tape that come first. This adds to the time it takes to access that type of memory.

Memory, more than almost any other factor, determines how fast a computer responds to requests and completes tasks. This means that more memory is constantly in demand, as consumers want systems that are faster and can on more tasks at the same time. When a computer is running low on memory because too many operations are going on at once, it may use virtual memory to try to compensate. Virtual memory is a technique in which the computer supplements its memory space by using some of its storage space. If the computer is almost out of memory and another task comes in, the computer copies some contents of its memory onto the hard drive. Then it can remove this information from memory and make space for

the new task. Once the new task is managed, the computer pulls the information that was copied to the hard drive and loads it back into memory.

FLASH AND SOLID STATE MEMORY
Some newer types of memory straddle the line between memory and storage. Flash memory, used in many mobile devices, can retain its contents even when power to the system is cut off. However, it can be accessed or even erased purely through electrical signals. Wiping all the contents of flash memory and replacing them with a different version is sometimes called "flashing" a device.

Many newer computers have incorporated solid state disks (SSDs), which are similar in many respects to flash memory. Like flash memory, SSDs have no moving parts and can replace hard drives because they retain their contents after system power has shut down. Many advanced users of computers have adopted SSDs because they are much faster than traditional computer configurations. A system can be powered on and ready to use in a matter of seconds rather than minutes.

—*Scott Zimmer, JD*

BIBLIOGRAPHY
Biere, Armin, Amir Nahir, and Tanja Vos, eds. *Hardware and Software: Verification and Testing.* New York: Springer, 2013. Print.

Englander, Irv. *The Architecture of Computer Hardware and System Software: An Information Technology Approach.* 5th ed. Hoboken: Wiley, 2014. Print.

Kulisch, Ulrich. *Computer Arithmetic and Validity: Theory, Implementation, and Applications.* 2nd ed. Boston: De Gruyter, 2013. Print.

Pandolfi, Luciano. *Distributed Systems with Persistent Memory: Control and Moment Problems.* New York: Springer, 2014. Print.

Patterson, David A., and John L. Hennessy. *Computer Organization and Design: The Hardware/Software Interface.* 5th ed. Waltham: Morgan, 2013. Print.

Soto, María, André Rossi, Marc Sevaux, and Johann Laurent. *Memory Allocation Problems in Embedded Systems: Optimization Methods.* Hoboken: Wiley, 2013. Print.

COMPUTER MODELING

FIELDS OF STUDY

Computer Science; Computer Engineering; Software Engineering

ABSTRACT

Computer modeling is the process of designing a representation of a particular system of interacting or interdependent parts in order to study its behavior. Models that have been implemented and executed as computer programs are called computer simulations.

PRINICIPAL TERMS

- **algorithm:** a set of step-by-step instructions for performing computations.
- **data source:** the origin of the information used in a computer model or simulation, such as a database or spreadsheet.
- **parameter:** a measurable element of a system that affects the relationships between variables in the system.
- **simulation:** a computer model executed by a computer system.
- **system:** a set of interacting or interdependent component parts that form a complex whole.
- **variable:** a symbol representing a quantity with no fixed value.

UNDERSTANDING COMPUTER MODELS

A computer model is a programmed representation of a system that is meant to mimic the behavior of the system. A wide range of disciplines, including meteorology, physics, astronomy, biology, and economics, use computer models to analyze different types of systems. When the program representing the system is executed by a computer, it is called a simulation.

One of the first large-scale computer models was developed during the Manhattan Project by scientists designing and building the first atomic bomb. Early computer models produced output in the form of tables or matrices that were difficult to analyze. It was later discovered that humans can see data trends more easily if the data is presented visually. For example, humans find it easier to analyze the output of a storm-system simulation if it is presented as graphic symbols on a map rather than as a table of meteorological data. Thus, simulations that produced graphic outputs were developed.

Computer models are used when a system is too complex or hard to study using a physical model. For example, it would be difficult if not impossible to create a physical model representing the gravitational effects of planets and moons on each other and on other objects in space.

There are several different types of models. Static models simulate a system at rest, such as a building design. Dynamic models simulate a system that changes over time. A dynamic model could be used to simulate the effects of changing ocean temperatures on the speed of ocean currents throughout the year. A continuous model simulates a system that changes constantly, while a discrete model simulates a system that changes only at specific times. Some models contain both discrete and continuous elements. A farming model might simulate the effects of both weather patterns, which constantly change, and pesticide spraying, which occurs at specified times.

HOW COMPUTER MODELS WORK

To create a computer model, one must first determine the boundaries of the system being modeled and what aspect of the system is being studied. For example, if the model is of the solar system, it might be used to study the potential effect on the orbits of the existing planets if another planet were to enter the solar system.

To create such a model, a computer programmer would develop a series of algorithms that contain the equations and other instructions needed to replicate the operation of the system. Variables are used to represent the input data needed. Examples of variables that might be used for the solar system model include the mass, diameter, and trajectory of the theoretical new planet. The values that define the system, and thus how the variables affect each other, are the parameters of the system. The parameters control the outputs of the simulation when it is run. Different values can be used to test different scenarios related to the system and problem being studied. Example parameters for the solar system

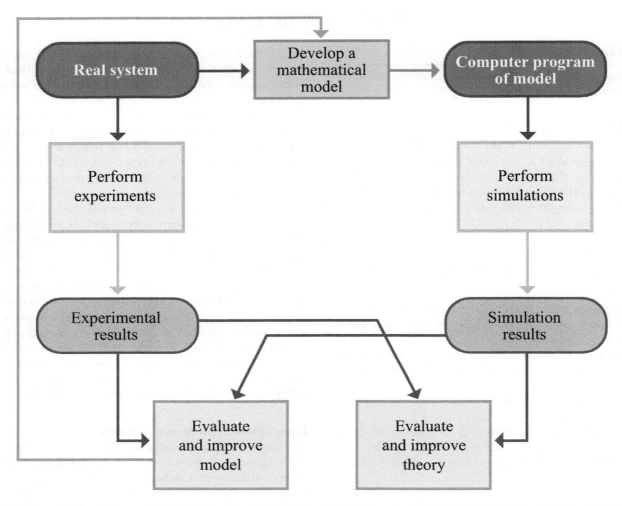

Mathematical models are used to identify and understand the details influencing a real system. Computer programs allow one to evaluate many variations in a mathematical model quickly and accurately, thus efficiently evaluating and improving models. Adapted from Allen and Tildesly, "Computer Simulation of Liquids."

model might include the orbits of the known planets, their distance from the sun, and the equations that relate an object's mass to its gravity. Certain parameters can be changed to test different scenarios each time a simulation is run. Because parameters are not always constant, they can be difficult to distinguish from variables at times.

The model must also have a data source from which it will draw the input data. This data may be directly entered into the program or imported from an external source, such as a file, database, or spreadsheet.

WHY COMPUTER MODELS ARE IMPORTANT
Computer models have provided great benefits to society. They help scientists explore the universe, understand the earth, cure diseases, and discover and test new theories. They help engineers design buildings, transportation systems, power systems, and other items that affect everyday life. With the development of more powerful computer systems, computer models will remain an important mechanism for understanding the world and improving the human condition.

—*Maura Valentino, MSLIS*

BIBLIOGRAPHY

Agrawal, Manindra, S. Barry Cooper, and Angsheng Li, eds. *Theory and Applications of Models of Computation: 9th Annual Conference, TAMC 2012, Beijing, China, May 16–21, 2012.* Berlin: Springer, 2012. Print.

Edwards, Paul N. *A Vast Machine: Computer Models, Climate Data, and the Politics of Global Warming.* Cambridge: MIT P, 2010. Print.

Kojić, Miloš, et al. *Computer Modeling in Bioengineering: Theoretical Background, Examples and Software.* Hoboken: Wiley, 2008. Print.

Law, Averill M. *Simulation Modeling and Analysis.* 5th ed. New York: McGraw, 2015. Print.

Morrison, Foster. *The Art of Modeling Dynamic Systems: Forecasting for Chaos, Randomness, and Determinism.* 1991. Mineola: Dover, 2008. Print.

Seidl, Martina, et al. *UML@Classroom: An Introduction to Object-Oriented Modeling.* Cham: Springer, 2015. Print.

COMPUTER PROGRAMMING: IMAGE EDITING

FIELDS OF STUDY

Digital Media; Graphic Design; Software Engineering

ABSTRACT

Image editing software uses computing technology to change digital images. Image editing can involve altering the appearance of an image, such as showing a hot air balloon underwater, or improving the quality of a low-resolution image. Images may also be compressed so that they require less computer storage space.

PRINICIPAL TERMS

- **interpolation:** a process of estimating intermediate values when nearby values are known; used in image editing to "fill in" gaps by referring to numerical data associated with nearby points.
- **lossless compression:** a method of reducing image file size that maintains image quality without degradation.
- **lossy compression:** a method of decreasing image file size by discarding some data, resulting in some image quality being irreversibly sacrificed.
- **nondestructive editing:** a mode of image editing in which the original content of the image is not destroyed because the edits are made only in the editing software.
- **rendering:** the production of a computer image from a 2-D or 3-D computer model.

OVERVIEW OF IMAGE EDITING

There are as many ways of digitally altering images as there are uses for digital art. The first step in image editing is to obtain an image in digital format. The easiest method is to use a digital camera to take a photograph and then transfer the photograph to a computer for editing. Another approach that is occasionally necessary is to scan a print photograph or film negative. This converts the photo to a digital image ready for editing. It is even possible to create an image by hand in native digital format, by using

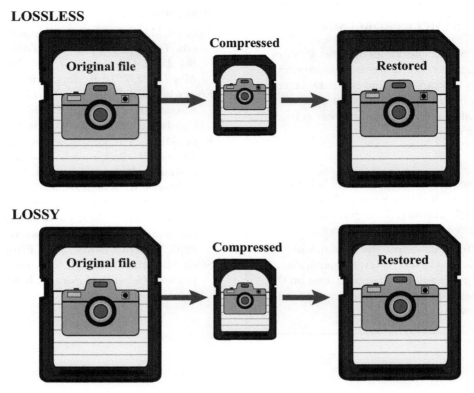

Image formats store varying levels and arrangements of data. Choosing the appropriate file format can make a difference when images are manipulated, resized, or compressed and restored EBSCO illustration.

a tablet and stylus to draw and paint. Finally, rendering makes it possible for a computer to produce a digital image from a 2-D or 3-D model.

Once the digital image is available, the next step is to determine what will be done to it. Most often, the image will be enhanced (improving the image quality through interpolation or other techniques), compressed (decreasing the file size by sacrificing some image quality or clarity), or altered (made to depict something that was not originally there). These changes may be through destructive or nondestructive editing. In destructive editing, the changes are applied to the original file. By contrast, in nondestructive editing, they are saved in a separate version file.

In the early days of the Internet, image compression was an especially important type of image editing. Bandwidth was limited then, and it could take several minutes to transmit even a medium sized image file. Image compression algorithms were invented to help reduce the size of these files, with some loss of quality. Lossless compression can avoid degradation of the image, but in most cases, it does not reduce file size as much as lossy compression does.

HOW COMPRESSION WORKS

Computers store image data as sets of numeric values. Each pixel onscreen is lit in a particular way when an image is displayed, and the colors of each pixel are stored as numbers. For example, if the color black were represented by the number eight, then anywhere in a picture that has three black pixels in a row would be stored as 8, 8, 8. Because an image is composed of thousands of pixels, all of the numbers needed to describe the colors of those pixels, when combined, take up a lot of storage space. One way to store the same information in less space is to create substitutions for recurring groups of numbers. The symbol $q1$ could be used to represent three black pixels in a row, for instance. Thus, instead of having to store three copies of the value "8" to represent each of the three black pixels, the computer could simply store the two-letter symbol *q1*, thus saving one-third of the storage space that otherwise would be required. This is the basis for how digital images are compressed.

Most images are compressed using lossy compression algorithms, such as JPEG. Compression thus usually requires that the sacrifice of some image quality. For most purposes, the reduction in quality is not noticeable and is made up for by the convenience of more easily storing and transmitting the smaller file. It is not uncommon for compression algorithms to reduce the file size of an image by 75 to 90 percent, without noticeably affecting the image's quality.

A NUMBERS GAME

Image enhancement typically relies on the mathematical adjustment of the numeric values that represent pixel hues. For instance, if an image editor were to desaturate a photograph, the software would first recognize all of the pixel values and compare them to a grayscale value. It would then interpolate new values for the pixels using a linear operation. Similarly, a filtering algorithm would find and apply a weighted average of the pixel values around a given pixel value in order to identify the new color codes for each pixel being adjusted. The median or the mode (most common) value could also be used. The type of filter being applied determines which mathematical operation is performed. Filters are often used to correct for noise, or unwanted signal or interference.

IMAGE EDITING GOES MOBILE

Image editing is now even possible on mobile platforms. Certain programs work only on PCs, others strictly on mobile devices, and still others on both. Besides the well-known Adobe Photoshop, other programs, including iPhoto, Apple Photos, Google's Picasa, and Gimp, also provide image editing for desktop computers. Fotor and Pixlr Editor work across platforms, giving users flexibility between their desktop, mobile device, and the Web. Similarly, photo collaging software abounds. Among these programs are Photoshop CC and CollageIt on desktops, Ribbet and Fotojet online, and Pic Stitch and BeFunky on mobile devices.

Nokia and Apple have developed the capability to create "live photos" with their smartphones. These are a hybrid of video and still images in which a few seconds of video are recorded prior to the still photo being taken. This feature represents yet another direction for image capture, alteration, and presentation.

—Scott Zimmer, JD

BIBLIOGRAPHY

Busch, David D. *Mastering Digital SLR Photography.* 2nd ed. Boston: Thompson Learning, 2008. Print.

Freeman, Michael. *Digital Image Editing & Special Effects: Quickly Master the Key Techniques of Photoshop & Lightroom.* New York: Focal, 2013. Print.

Galer, Mark, and Philip Andrews. *Photoshop CC Essential Skills: A Guide to Creative Image Editing.* New York: Focal, 2014. Print.

Goelker, Klaus. *Gimp 2.8 for Photographers: Image Editing with Open Source Software.* Santa Barbara: Rocky Nook, 2013. Print.

Holleley, Douglas. *Photo-Editing and Presentation: A Guide to Image Editing and Presentation for Photographers and Visual Artists.* Rochester: Clarellen, 2009. Print.

Xue, Su. *Data-Driven Image Editing for Perceptual Effectiveness.* New Haven: Yale U, 2013. Print.

COMPUTER PROGRAMMING: MUSIC EDITING

FIELDS OF STUDY

Digital Media; Software Engineering

ABSTRACT

Music editing involves the use of computer technology to alter files of recorded sound. Music files can be compressed, enhanced, combined, or separated through editing software. Both the recording and film industries rely on music editing software to create their products. Compression has enabled the music industry to move away from physical media and has driven the growth of online music transfer. The sharing of audio files has also led to a more collaborative music scene.

PRINICIPAL TERMS

- **audio codec:** a program that acts as a "coder-decoder" to allow an audio stream to be encoded for storage or transmission and later decoded for playback.
- **lossy compression:** a method of reducing the size of an audio file while sacrificing some of the quality of the original file.
- **mastering:** the creation of a master recording that can be used to make other copies for distribution.
- **mixing:** the process of combining different sounds into a single audio recording.
- **nondestructive:** a form of editing that alters a digital audio file without destroying the file's original form.
- **scrubbing:** navigating through an audio recording repeatedly in order to locate a specific cue or word.

Digital audio formats that store varying amounts and arrangements of data. Choosing the appropriate file format can make a difference when the audio speed amplitude is manipulated, or compressed and restored. EBSCO illustration.

Mixing and Mastering

Music editing software is used by sound engineers working with musicians to record and produce music for commercial purposes. Performers record their music in studios, often in separate sessions. Once all of the recording has been completed, sound engineers begin mixing the tracks together. Mixing combines different sounds into a single audio recording. One track might include the vocals, another the backup singers, and a third the instruments. Recording these separately can help preserve the full depth of sound produced by each source (or input). Sound engineers mix these recordings together such that they support and enhance one another instead of competing. Mixers allow them to adjust the volume of each input (channel), equalize frequencies, reduce noise, add effects, and control channel subgroups (buses). Music editing software usually includes a virtual mixer and a timeline with a scroll bar, allowing the editor to move back and forth along an audio track in a process called scrubbing. Scrubbing is particularly helpful for aligning vocal tracks with other audio tracks or with film visuals.

When the mixing stage has been completed for an album, the next step is mastering. This is the creation of a master recording that can be used to produce copies for sale thereafter. Mastering was much more complex when music was mainly sold on physical media, such as record albums or cassette tapes. The recording master had to be protected from theft, damage, and the ravages of time. Today, most music editing is done entirely digitally. In digital mastering, the mixed audio recording is checked for errors, reformatted for distribution, and then subtly adjusted for the best sound quality possible.

Compression and Transmission

Music editing can be destructive, meaning that changing a recording destroys the original version of the recording. Or it can be nondestructive, in which editing preserves the original recording. The technology that made it possible for music to move away from physical media was compression, specifically the MP3 compression algorithm and file type. MP3 was developed in the early 1990s as a type of lossy compression for audio files. It can greatly reduce the size of an ordinary audio file while losing only a negligible amount of sound quality. MP3 is an audio codec that makes music portable by compressing it into files small enough for transfer over the Internet and for digital storage. A codec can reduce the size of audio files by using shorthand to describe repeated data within the file more concisely. When a compressed recording is played back, it is decoded by the device playing it, provided that the device is equipped with the proper codec. The recording's fidelity (similarity to the original) depends on how much information is kept and how much of the original frequency and dynamic range are captured. For instance, 16 bits of data are encoded for a high fidelity CD as compared to 24 bits in a studio recording.

Motion Picture Music Editing

Motion pictures rely on the skills of sound engineers to mix the film's dialogue, musical score and soundtrack, and sound effects into a single audio accompaniment to the visuals. The process is like that used in a recording studio. However, there is the added complication that not only must the audio tracks all be in harmony, they must also follow the timing of the movie's scenes. Usually the dialogue takes priority with the music and sound effects being mixed in later.

Culture of Remixing

Unfortunately, the development and adoption of the MP3 audio codec made it easier and more common than ever for music to be shared illegally. People found it preferable to make illegal copies of music than to buy it, since those copies essentially retain the same quality as the original. Despite this, some praise the sharing of music online as giving rise to a new culture of remixing. Artists release their creative output into the world for audiences to enjoy it and other artists to build off it and even transform it. In remixing, other artists add their own perspectives to produce a collaborative work that is greater than the sum of its parts. Consumer audio editing software, such as Audacity, Adobe Audition, or Lexis Audio Editor, enables such artists to record and mix music using just their PC instead of a studio.

—*Scott Zimmer, JD*

BIBLIOGRAPHY

Collins, Mike. *Pro Tools 11: Music Production, Recording, Editing, and Mixing.* Burlington: Focal, 2014. Print.

Cross, Mark. *Audio Post Production for Film and Television.* Boston: Berklee, 2013. Print.

Kefauver, Alan P., and David Patschke. *Fundamentals of Digital Audio.* Middleton: A-R Editions, 2007. Print.

Pinch, T. J., and Karin Bijsterveld. *The Oxford Handbook of Sound Studies.* New York: Oxford UP, 2013. Print.

Saltzman, Steven. *Music Editing for Film and Television: The Art and the Process.* Burlington: Focal, 2015. Print.

Shen, Jialie, John Shepherd, Bin Cui, and Ling Liu. *Intelligent Music Information Systems: Tools and Methodologies.* Hershey: IGI Global, 2008. Print.

COMPUTER PROGRAMMING: VIDEO EDITING

FIELDS OF STUDY

Digital Media; Graphic Design

ABSTRACT

Video editing is the process of altering, combining, or otherwise manipulating video frames or sequences of video frames after recording. Early video editing was done with time-consuming and visually limited linear editing techniques. Revolutionary advances in digital technology ushered in the era of nonlinear video editing.

PRINICIPAL TERMS

- **bit rate:** the amount of data encoded for each second of video; often measured in kilobits per second (kbps) or kilobytes per second (Kbps).
- **edit decision list (EDL):** a list that catalogs the reel or time code data of video frames so that the frames can be accessed during video editing.
- **frames per second (FPS):** a measurement of the rate at which individual video frames are displayed.
- **nonlinear editing:** a method of editing video in which each frame of video can be accessed, altered, moved, copied, or deleted regardless of the order in which the frames were originally recorded.
- **video scratching:** a technique in which a video sequence is manipulated to match the rhythm of a piece of music.
- **vision mixing:** the process of selecting and combining multiple video sources into a single video.

WHAT IS VIDEO EDITING?

Video editing is the process of altering, combining, or otherwise manipulating video frames or sequences of video frames after recording. Cameras create video by recording a series of still images, or "frames," which when rapidly displayed in sequence, create the appearance of a moving picture. The speed at which video frames are recorded and displayed is the frame rate. This is measured in frames per second (FPS). Early recording technologies used frame rates as low as 16 FPS. By contrast, digital video cameras such as the Go Pro Hero 4 can record at speeds as high as 240 FPS. To achieve the semblance of smooth motion, frame rates of 24 to 30 FPS are used. Higher frame rates are said to provide clear, more lifelike images, while lower ones produce jerkier motion or blurrier video.

The frames recorded digitally are composed of bits. A bit is the smallest unit of data that can be stored and manipulated. The amount of data encoded, in bits, for each second of video is the bit rate. The greater the bit rate is, the higher the resolution and, thus, the quality. However, higher bit rates also result in larger video files.

VIDEO EDITING IN PRACTICE

In the past, video was edited using linear video editing, a technique for reordering sequences of video frames captured on analog video tape. Linear video editing was initially achieved by physically cutting videotape into sections and then splicing the sections together in the desired sequence. By the 1960s, electronic tape editing technology allowed different sections of videotape to be combined mechanically onto a single videotape. Even with this advance, however, linear video editing remained difficult and time consuming.

The advent of digital technology revolutionized video editing. Computer hardware and video-editing software enabled the use of nonlinear editing. Nonlinear editing allows individual video frames to be accessed, altered, moved, copied, or deleted regardless of the order in which they were recorded. The frames can then be combined into a single video sequence using software.

Nonlinear editing enabled video editors to employ a range of new techniques that would have been impossible to achieve previously. One such technique is video scratching, where a sequence of video frames is manipulated to match the rhythm of a piece of music. Another is digital compositing. In this process, video frames from different sources can be combined with each other and with still images and computer graphics to create the illusion that they were all recorded at once. This is related to vision mixing, in which various video sources are combined into a single recording. An example is the combining of footage from multiple cameras at a live event, such as a concert, for a seamless broadcast viewing experience.

An important tool used in video editing is the edit decision list (EDL). An EDL is an ordered list of video

Digital video formats store varying quantities and qualities of data. Choosing the appropriate file format can make a difference when the audio speed amplitude is manipulated, or compressed and restored. With so many video formats available, it is important to choose editing software that is compatible with one's file format of choice. EBSCO illustration.

SAMPLE PROBLEM

When editing video footage in a digital environment, the size and quality of the final file must be considered. Video file size and video quality are directly related. As the quality of a digital video increases, so does its file size (measured in bits). A video's file size can be calculated using the following formulas:

video size = bit rate × video length

bit rate = pixel density × color depth × frame rate

Calculate the maximum frame rate in FPS that can be used to ensure that a video's file size will be 7,200,000 bits if the pixel density is 1,000 pixels per frame, the color depth is 24 bits per pixel, and the video is 15 seconds long.

Note that this formula only takes the visual portion of a video into account and that to obtain the actual final file size, one would need to calculate the size of the related audio data and add the two together.

Answer:

First, substitute the formula for bit rate into the formula for video size:

video size = bit rate × video length

video size = (pixel density × color depth × frame rate) × video length

Then rewrite the equation in terms of frame rate:

frame rate = video size / (pixel density × color depth × video length)

Next substitute in the known values for video size, pixel density, color depth, and video length, and solve:

frame rate = 7,200,000 bits / [(1,000 pixels/frame)(24 bits/pixel)(15 seconds)]

= (7,200,000 bits / 24,000 bits/frame)(1/15 seconds)

= (300 frames)(1/15 seconds)

= 20 frames/second (FPS)

---

frames or frame sequences ("clips") that contains information such as time data. That information allows the video editor to find individual clips quickly and efficiently during editing.

VIDEO EDITING IN A DIGITAL WORLD

The development of affordable PCs, digital video cameras, and smartphones has brought video editing from the studio to the realm of ordinary people. Video-editing software ranges from applications for PCs, such as Adobe's Premiere, to those designed for smartphones, such as Corel's Pinnacle Studio Pro. Others, such as Apple's iMovie, work on both PC and mobile devices. When used along with social media websites, such as YouTube and Facebook, video-editing software allows users to easily share videos with anyone who has an Internet connection. Thus, anyone with a smartphone can be a video writer, director, editor, and movie studio. This has led to a powerful new form of personal and artistic expression that will continue to grow and change as new video technologies are developed and adopted.

—*Maura Valentino, MSLIS*

BIBLIOGRAPHY

Anderson, Gary H. *Video Editing and Post Production: A Professional Guide.* Woburn: Focal, 1999. Print.

Ascher, Steven. *The Filmmaker's Handbook: A Comprehensive Guide for the Digital Age.* New York: Plume, 2012. Print.

Dancyger, Ken. *The Technique of Film and Video Editing: History, Theory, and Practice.* Burlington: Focal, 2013. Digital file.

Goodman, Robert, and Patrick McGrath. *Editing Digital Video: The Complete Creative and Technical Guide.* New York: McGraw, 2003. Print.

Ohanian, Thomas. *Digital Nonlinear Editing: Editing Film and Video on the Desktop.* Woburn: Focal, 1998. Print.

Rubin, Michael. *Nonlinear—A Field Guide to Digital Video and Film Editing.* Gainesville: Triad, 2000, Print.

Tarantola, Andrew. "Why Frame Rate Matters." *Gizmodo.* Gizmodo, 14 Jan. 2015. Web. 11 Mar. 2016.

COMPUTER SECURITY

FIELDS OF STUDY

Information Technology; Security

ABSTRACT

The goal of computer security is to prevent computer and network systems from being accessed by those without proper authorization. It encompasses different aspects of information technology, from hardware design and deployment to software engineering and testing. It even includes user training and workflow analysis. Computer security experts update software with the latest security patches, ensure that hardware is designed appropriately and stored safely, and train users to help protect sensitive information from unauthorized access.

PRINICIPAL TERMS

- **backdoor:** a hidden method of accessing a computer system that is placed there without the knowledge of the system's regular user in order to make it easier to access the system secretly.
- **device fingerprinting:** information that uniquely identifies a particular computer, component, or

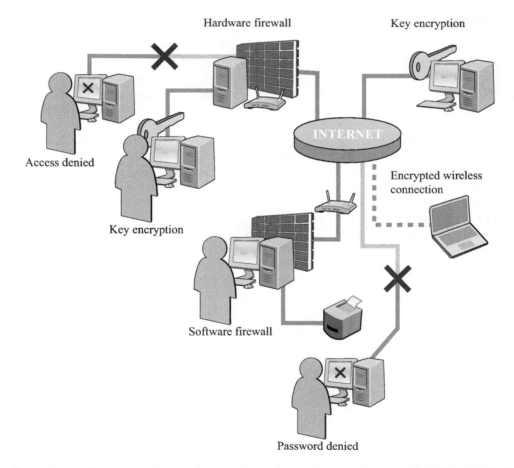

Computer security can come in many forms to ensure data and programs are protected. Passwords limit who can access a computer, key encryption limits who can read transmitted data, and firewalls can limit intrusion from the Internet. EBSCO illustration.

piece of software installed on the computer. This can be used to find out precisely which device accessed a particular online resource.

- **intrusion detection system:** a system that uses hardware, software, or both to monitor a computer or network in order to determine when someone attempts to access the system without authorization.
- **phishing:** the use of online communications in order to trick a person into sharing sensitive personal information, such as credit card numbers or social security numbers.
- **principle of least privilege:** a philosophy of computer security that mandates users of a computer or network be given, by default, the lowest level of privileges that will allow them to perform their jobs. This way, if a user's account is compromised, only a limited amount of data will be vulnerable.
- **trusted platform module (TPM):** a standard used for designing cryptoprocessors, which are special chips that enable devices to translate plain text into cipher text and vice versa.

HARDWARE SECURITY

The first line of defense in the field of computer security concerns the computer hardware itself. At a basic level, computer hardware must be stored in secure locations where it can only be accessed by authorized personnel. Thus, in many organizations, access to areas containing employee workstations is restricted. It may require a badge or other identification to gain access. Sensitive equipment such as an enterprise server is even less accessible, locked away in a climate-controlled vault. It is also possible to add hardware security measures to existing computer systems to make them more secure. One example of this is using biometric devices, such as fingerprint scanners, as part of the user login. The computer will only allow logins from people whose fingerprints are authorized. Similar restrictions can be linked to voice authentication, retina scans, and other types of biometrics. Computer security can come in many forms to ensure data and programs are protected. Passwords limit who can access a computer, key encryption limits who can read transmitted data, and firewalls can limit intrusion from the Internet.

Inside a computer, a special type of processor based on a trusted platform module (TPM) can manage encrypted connections between devices. This ensures that even if one device is compromised,

the whole system may still be protected. Another type of security, device fingerprinting, can make it possible to identify which device or application was used to access a system. For example, if a coffee shop's wireless access point was attacked, the access point's logs could be examined to find the machine address of the device used to launch the attack. One highly sophisticated piece of security hardware is an intrusion detection system. These systems can take different forms but generally consist of a device through which all traffic into and out of a network or host is filtered and analyzed. The intrusion detection system examines the flow of data in order to pinpoint any attempt at hacking into the network. The system can then block the attack before it can cause any damage.

NETWORK SECURITY

Network security is another important aspect of computer security. In theory, any computer connected to the Internet is vulnerable to attack. Attackers can try to break into systems by exploiting weak points in the software's design or by tricking users into giving away their usernames and passwords. The latter method is called phishing, because it involves "fishing" for information. Both of these methods can be time consuming, however. So once a hacker gains access, they may install a backdoor. Backdoors allow easy, undetected access to a system in future.

One way of preventing attackers from tricking authorized users into granting access is to follow the principle of least privilege. According to this principle, user accounts are given the minimum amount of access rights required for each user to perform their duties. For instance, a receptionist's account would be limited to e-mail, scheduling, and switchboard functions. This way, a hacker who acquired the receptionist's username and password could not do things such as set their own salary or transfer company assets to their own bank account. Keeping privileges contained thus allows an organization to minimize the damage an intruder may try to inflict.

SOFTWARE SECURITY

Software represents another vulnerable point of computer systems. This is because software running on a computer must be granted certain access privileges in order to function. If the software is not written in a secure fashion, then hackers may be able to enhance the software's privileges. Hackers can then use these enhanced privileges to perform unintended functions or even take over the computer running the software. In the vernacular of hackers, this is known as "owning" a system.

A Moving Target

Computer security professionals have an unenviable task. They must interfere with the way users wish to use their computers, in order to make sure that hardware and software vulnerabilities are avoided as much as possible. Often, the same users whom they are trying to protect attempt to circumvent those protective measures, finding them inconvenient or downright burdensome. Computer security in these cases can become a balancing act between safety and functionality.

—Scott Zimmer, JD

Bibliography

Boyle, Randall, and Raymond R. Panko. *Corporate Computer Security.* 4th ed. Boston: Pearson, 2015. Print.4 Science Reference Center™ Computer Security

Brooks, R. R. *Introduction to Computer and Network Security: Navigating Shades of Gray.* Boca Raton: CRC, 2014. Digital file.

Jacobson, Douglas, and Joseph Idziorek. *Computer Security Literacy: Staying Safe in a Digital World.* Boca Raton: CRC, 2013. Print.

Schou, Corey, and Steven Hernandez. *Information Assurance Handbook: Effective Computer Security and Risk Management Strategies.* New York: McGraw, 2015. Print.

Vacca, John R. *Computer and Information Security Handbook.* Amsterdam: Kaufmann, 2013. Print.

Williams, Richard N. *Internet Security Made Easy: Take Control of Your Computer.* London: Flame Tree, 2015. Print.

COMPUTER-ASSISTED INSTRUCTION

FIELDS OF STUDY

Computer Science; Information Systems

ABSTRACT

Computer-assisted instruction is the use of computer technology as a means of instruction or an aid to classroom teaching. The instructional content may or may not pertain to technology. Computer-assisted instruction often bridges distances between instructor and student and allows for the instruction of large numbers of students by a few educators.

PRINICIPAL TERMS

- **learner-controlled program:** software that allows a student to set the pace of instruction, choose which content areas to focus on, decide which areas to explore when, or determine the medium or difficulty level of instruction; also known as a "student-controlled program."
- **learning strategy:** a specific method for acquiring and retaining a particular type of knowledge, such as memorizing a list of concepts by setting the list to music.
- **learning style:** an individual's preferred approach to acquiring knowledge, such as by viewing visual stimuli, reading, listening, or using one's hands to practice what is being taught.
- **pedagogy:** a philosophy of teaching that addresses the purpose of instruction and the methods by which it can be achieved.
- **word prediction:** a software feature that recognizes words that the user has typed previously and offers to automatically complete them each time the user begins typing.

Computer-Assisted vs. Traditional Instruction

In a traditional classroom, a teacher presents information to students using basic tools such as pencils, paper, chalk, and a chalkboard. Most lessons consist of a lecture, group and individual work by students, and the occasional hands-on activity, such as a field trip or a lab experiment. For the most part, students must adapt their learning preferences to

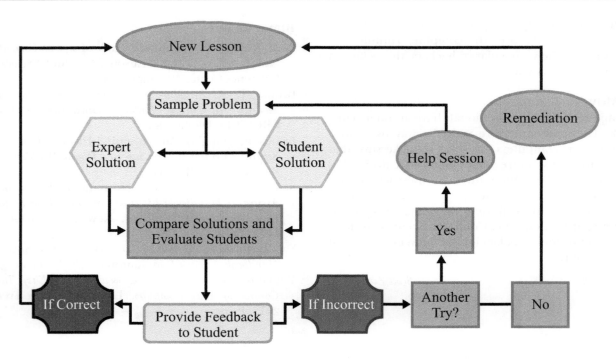

Computer-assisted instruction uses programming to determine whether a student understands the lesson (correctly answers sample problems) or needs more help or remediation (incorrectly answers sample problems). This flowchart indicates a general path of computer-assisted instruction. Adapted from S. Egarievwe, A. O. Ajiboye, and G. Biswas, "Internet Application of LabVIEW in Computer Based Learning," 2000.

the teacher's own pedagogy, because it would be impractical for one teacher to try to teach to multiple learning styles at once.

Computer-assisted instruction (CAI) supplements this model with technology, namely a computing device that students work on for some or all of a lesson. Some CAI programs offer limited options for how the material is presented and what learning strategies are supported. Others, known as learner-controlled programs, are more flexible.

A typical CAI lesson focuses on one specific concept, such as long division or the history of Asia. The program may present information through audio, video, text, images, or a combination of these. It then quizzes the student to make sure that they have paid attention and understood the material. Such instruction has several benefits. First, students often receive information through several mediums, so they do not have to adapt to the teacher's preferred style. Second, teachers can better support students as they

move through the lessons without having to focus on presenting information to a group.

ADVANTAGES AND DISADVANTAGES
CAI has both benefits and drawbacks. Certain software features make it easier to navigate the learning environment, such as word prediction to make typing easier and spell-checking to help avoid spelling mistakes. Copy-and-paste features save users time that they would otherwise spend reentering the same information over and over. Speech recognition can assist students who are blind, have physical disabilities, or have a learning disability that affects writing. Other helpful features are present despite not having been intended to benefit students. For example, CAI video lessons include the option to pause playback, skip ahead or back, or restart. These functions can be vital to a student who is struggling to grasp an especially difficult lesson or is practicing note-taking. They can stop the video

at any time and restart it after catching up with the content that has been presented. By contrast, in a regular classroom, the lecturer often continues at the same pace regardless how many students may be struggling to understand and keep up.

Learning is rarely a "one size fits all" affair. Different topics pose greater or lesser challenges to different students, depending on their natural abilities and study habits. In regular classrooms, teachers can often sense when some students are not benefiting from a lesson and adapt it accordingly. With some CAI, this is not an option because the lesson is only presented in one way.

ADAPTIVE INSTRUCTION

Fortunately, some forms of CAI address different learning rates by using adaptive methods to present material. These programs test students' knowledge and then adapt to those parts of the lesson with which they have more difficulty. For instance, if a math program notices that a student often makes mistakes when multiplying fractions, it might give the student extra practice in that topic. Adaptive programs give teachers the means to better assess students' individual needs and track their progress. As the technology improves, more detailed and specific results may bolster teachers' efforts to tailor instruction further.

DISTANCE EDUCATION

CAI is especially important to the growing field of online education. Online instructors often use elements of CAI to supplement their curricula. For example, an online course might require students to watch a streaming video about doing library research so that they will know how to complete their own research paper for the course. Online education also enables just a few instructors to teach large numbers of students across vast distances. Tens of thousands of students may enroll in a single massive open online course (MOOC).

—*Scott Zimmer, JD*

BIBLIOGRAPHY

Abramovich, Sergei, ed. *Computers in Education.* 2 vols. New York: Nova, 2012. Print.

Erben, Tony, Ruth Ban, and Martha E. Castañeda. *Teaching English Language Learners through Technology.* New York: Routledge, 2009. Print.

Miller, Michelle D. *Minds Online: Teaching Effectively with Technology.* Cambridge: Harvard UP, 2014. Print.

Roblyer, M. D., and Aaron H. Doering. *Integrating Educational Technology into Teaching.* 6th ed. Boston: Pearson, 2013. Print.

Tatnall, Arthur, and Bill Davey, eds. *Reflections on the History of Computers in Education: Early Use of Computers and Teaching about Computing in Schools.* Heidelberg: Springer, 2014. Print.

Tomei, Lawrence A., ed. *Encyclopedia of Information Technology Curriculum Integration.* 2 vols. Hershey: Information Science Reference, 2008. Print.

CONNECTION MACHINE

FIELDS OF STUDY

Computer Science; Computer Engineering; Information Technology

ABSTRACT

The Connection Machines grew out of Daniel Hillis's Ph.D. Thesis research in Electrical Engineering and Computer Science at the Massachusetts Institute of Technology. The basic idea was that of having a massively parallel computer network, with each processor connected to all the others. The majority of digital computers have the so-called Von Neumann Architecture, in which a Central Processing Unit does the actual computation, taking intermediate results from specified locations and storing intermediate results for later use. But the human brain and other brains found in nature do not work in that way. Instead there is no central processor, and each neuron is connected to a great many others. The connection machine series was one of the first commercial attempts to capture the brain's way of thinking.

PRINCIPAL TERMS

- **central processing unit:** the main operating function of a conventional computer which can be programmed to read characters from a storage medium, combine them as instructed and store the output for later use.
- **Hebbian learning:** machine learning as a result of modifying the weights in a neural network.
- **Hopfield neural network:** a one layer neural net where each neuron knows the state of all other neurons at each time step.
- **parallel processing:** several computational steps done simultaneously
- **Von Neumann architecture:** The simplest computer organization, in which the central processing unit modifies one location in memory at a time

HISTORY & DEVELOPMENT OF CONNECTION MACHINES

The connection machines were among the first parallel processing computers to be commercially available. They were an outgrowth of Daniel Hillis's Ph. D. Thesis at MIT, built partially on the neural network ideas of John Hopfield and influenced in its early stages by Nobel Laureate Richard P. Feynman, a colorful figure who began his career as head of a computational group at Los Alamos and went on to be one of the founders of quantum electrodynamics.

In the early days of computation, McCullough and Pitts studied the behavior of collections of neurons whose output was a step function of a weighed sum of their inputs. If the weights were allowed to change based on the number of times a neuron fired, learning was found to take place. This was an example of Hebbian learning, so named after D. O. Hebb who promoted the concept. In the 1970's John Hopfield studied the behavior of a single network of such neurons. The Hopfield arrangement of neurons was the basis for the first versions of the connection machine. Hopfield's method proved ideally suited to the connection machine, with the weights represented as decimal numbers. This allowed all the processors to be used concurrently resulting in computing times orders of magnitude faster then in conventional machines.

Thinking Machines Corporation was founded by Daniel Hillis and Sheryl Handler in Waltham, Massachusetts in 1983 and later moved to Cambridge MA. In 1999 it was acquired by Sun Microsystems.

While it would in principle be desirable for a computer to simultaneously update all processors, it proved much too unwieldy for the first connection machines, which instead adopted a hypercube arrangement of processors. The CM-1 had as many as 65,536 processors each processing one bit at a time. The CM-2 launched in 1987 added floating point numeric coprocessors. In 1991 the CM 5 was announced, Thinking Machines went to a new architecture.

Much thinking had been devoted to the physical layout of the connection machines. The initial designs reflected the machines hypercube architecture. The CM-5 had large panels of red blinking light emitting diodes. Because of its design, a CM 5 was included in the control room in *Jurassic Park*.

The involvement of Richard Feynman, who shared in the Nobel Prize for Physics in 1965 is noteworthy for several reasons. Feynman, one of whose major contribution to quantum electrodynamics was in introducing a diagrammatic method for summing term in the solution of the basic equations of motion was a, graduate student in Physics at Princeton University when World War II began. While completing his thesis he was also head of the theoretical division at Los Alamos, which then consisted mainly of high school graduates working with adding machines or similar devices to solve problems which had grown out of the atomic bomb project. It was this great familiarity with the innards of machines that did arithmetic that may have led him to the ultimate techniques involving diagrams in the theory of electrons and photons. By the time the connection machine was contemplated, physicists had moved to the theory of particles within the nucleus, quantum chromodynamics, so named

after the quantum rules that explained which particles could be observed. And indeed one of the first problems to which the connection machine was applies involved the application of quantum chromodynamics.

—Donald Franceschetti, PhD

BIBLIOGRAPHY

Anderson, James A, Edward Rosenfeld, and Andras Pellionisz. *Neurocomputing.* Cambridge, Mass: MIT Press, 1990. Print.

Hillis, W. Daniel. "Richard Feynman and the Connection Machine." *Phys. Today Physics Today* 42.2 (1989): 78. Web.

Hopfield, J. J. "Neural Networks and Physical Systems with Emergent Collective Computational Abilities." *Proceedings of the National Academy of Sciences* 79.8 (1982): 2554-558. Web.

CONSTRAINT PROGRAMMING

FIELDS OF STUDY

Software Engineering; System-Level Programming

ABSTRACT

Constraint programming typically describes the process of embedding constraints into another language, referred to as the "host" language since it hosts the constraints. Not all programs are suitable for constraint programming. Some problems are better addressed by different approaches, such as logic programming. Constraint programming tends to be the preferred method when one can envision a state in which multiple constraints are simultaneously satisfied, and then search for values that fit that state.

PRINICIPAL TERMS

- **constraints:** limitations on values in computer programming that collectively identify the solutions to be produced by a programming problem.

- **domain:** the range of values that a variable may take on, such as any even number or all values less than –23.7.
- **functional programming:** a theoretical approach to programming in which the emphasis is on applying mathematics to evaluate functional relationships that are, for the most part, static.
- **imperative programming:** programming that produces code that consists largely of commands issued to the computer, instructing it to perform specific actions.

MODELS OF CONSTRAINT PROGRAMMING

Constraint programming tends to be used in situations where the "world" being programmed has multiple constraints, and the goal is to have as many of them as possible satisfied at once. The aim is to find values for each of the variables that collectively describe the world, such that the values fall within that variable's constraints.

To achieve this state, programmers use one of two main approaches. The first approach is the perturbation model. Under this model, some of

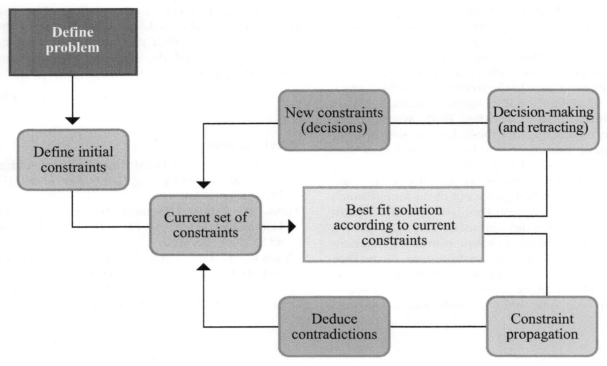

In constraint programming, the design of the program is dictated by specific constraints determined prior to development. After a problem is identified, the initial constraints are defined and condensed into the current program constraints. As solutions are developed and tested, decisions are made as to whether the constraints must be modified or retracted to remove duplications or whether new ones must be developed. The process cycles through current restraints and solutions until a solution is developed that meets all the constraints and solves the initial problem. Adapted from Philippe Baptiste, "Combining Operations Research and Constraint Programming to Solve Real-Life Scheduling Problems."

the variables are given initial values. Then, at different times, variables are changed ("perturbed") in some way. Each change then moves through the system of interrelated variables like ripples across the surface of a pond: other values change in ways that follow their constraints and are consistent with the relationships between values. The perturbation model is much like a spreadsheet. Changing one cell's value causes changes in other cells that contain formulas that refer to the value stored in the original cell. A single change to a cell can propagate throughout many other cells, changing their values as well.

A contrasting approach is the refinement model. Whereas the perturbation model assigns particular values to variables, under the refinement model,

each variable can assume any value within its domain. Then, as time passes, some values of one variable will inevitably be ruled out by the values assumed by other variables. Over time, each variable's possible values are refined down to fewer and fewer options. The refinement model is sometimes considered more flexible, because it does not confine a variable to one possible value. Some variables will occasionally have multiple solutions.

A DIFFERENT APPROACH TO PROGRAMMING

Constraint programming is a major departure from more traditional approaches to writing code. Many programmers are more familiar with imperative programming, where commands are issued to the computer to be executed, or functional programming, where the program receives certain values and then performs various mathematical functions using those values. Constraint programming, in contrast, can be less predictable and more flexible.

A classic example of a problem that is especially well suited to constraint programming is that of map coloring. In this problem, the user is given a map of a country composed of different states, each sharing one or more borders with other states. The user is also given a palette of colors. The user must find a way to assign colors to each state, such that no adjacent states (i.e., states sharing a border) are the same color. Map makers often try to accomplish this in real life so that their maps are easier to read.

Those experienced at constraint programming can immediately recognize some elements of this problem. The most obvious is the constraint, which is the restriction that adjacent states may not be the same color. Another element is the domain, which is the list of colors that may be assigned to states. The fewer the colors included in the domain, the more challenging the problem becomes. While this map-coloring problem may seem simplistic, it is an excellent introduction to the concept of constraint programming. It provides a useful situation for student programmers to try to translate into code.

FEASIBILITY VS. OPTIMIZATION

Constraint programming is an approach to problem solving and coding that looks only for a solution that works. It is not concerned with finding the optimal solution to a problem, a process known as "optimization." Instead, it seeks values for the variables that fit all of the existing constraints. This may seem like a limitation of constraint programming. However, its flexibility can mean that it solves a problem faster than expected.

Another example of a problem for which constraint programming is well suited is that of creating a work schedule. The department or team contains multiple variables (the employees), each with their own constraints. Mary can work any day except Friday, Thomas can work mornings on Monday through Thursday but only evenings on Friday, and so forth. The goal is to simply find a schedule that fits all of the constraints. It does not matter whether it is the best schedule, and in fact, there likely is no "best" schedule.

—*Scott Zimmer, JD*

BIBLIOGRAPHY

Baptiste, Philippe, Claude Le Pape, and Wim Nuijten. *Constraint-Based Scheduling: Applying Constraint Programming to Scheduling Problems*. New York: Springer, 2013. Print.

Ceberio, Martine, and Vladik Kreinovich. *Constraint Programming and Decision Making*. New York: Springer, 2014. Print.

Henz, Martin. *Objects for Concurrent Constraint Programming*. New York: Springer, 1998. Print.

Hofstedt, Petra. *Multiparadigm Constraint Programming Languages*. New York: Springer, 2013. Print.

Pelleau, Marie, and Narendra Jussien. *Abstract Domains in Constraint Programming*. London: ISTE, 2015. Print.

Solnon, Christine. *Ant Colony Optimization and Constraint Programming*. Hoboken: Wiley, 2010. Print.

CONTROL SYSTEMS

FIELDS OF STUDY

Embedded Systems; System Analysis

ABSTRACT

A control system is a device that exists to control multiple other systems or devices. For example, the control system in a factory would coordinate the operation of all of the factory's interconnected machines. Control systems are used because the coordination of these functions needs to be continuous and nearly instantaneous, which would be difficult and tedious for human beings to manage.

PRINICIPAL TERMS

- **actuator:** a motor designed to move or control another object in a particular way.
- **automatic sequential control system:** a mechanism that performs a multistep task by triggering a series of actuators in a particular sequence.
- **autonomous agent:** a system that acts on behalf of another entity without being directly controlled by that entity.
- **fault detection:** the monitoring of a system in order to identify when a fault occurs in its operation.
- **system agility:** the ability of a system to respond to changes in its environment or inputs without failing altogether.
- **system identification:** the study of a system's inputs and outputs in order to develop a working model of it.

TYPES OF CONTROL SYSTEMS

Different types of control systems are used in different situations. The nature of the task being performed usually determines the design of the system. At the most basic level, control systems are classified as either open-loop or closed-loop systems. In an open-loop system, the output is based solely on the input fed into the system. In a closed-loop system, the output generated is used as feedback to adjust the system as necessary. Closed-loop systems can make fault detection more difficult, as they are designed to minimize the deviations created by faults.

One of the first steps in designing a control system is system identification. This is the process of modeling the system to be controlled. It involves studying the inputs and outputs of the system and determining how they need to be manipulated in order to produce the desired outcome. Some control systems require a certain degree of human interaction or guidance during their operation. However, control-system designers generally prefer systems that can function as autonomous agents. The purpose of a control system is to reduce human involvement in the process as much as possible. This is partly so personnel can focus on other, more important tasks and partly because problems are more likely to arise from human error.

The downside of an autonomous agent is that it requires greater system agility, so that when problems are encountered, the control system can either continue to perform its role or "fail gracefully" rather than failing catastrophically or ceasing to function altogether. This could mean the difference between a jammed conveyor belt being automatically corrected and that same belt forcing an entire assembly line to shut down.

CONTROL SYSTEM PERFORMANCE

Control systems are more than just computers that monitor other processes. Often, they also control the physical movements of system components. A control system may cause an assembly-line robot to pivot an automobile door so that another machine can attach its handle, or it may cause a device to fold cardboard into boxes to be filled with candy. Such tasks are accomplished via actuators, which are motors that the control system manipulates to execute the necessary physical movements. Sometimes only a single movement is required from a device. At other times, a device may have to perform several different movements in a precisely timed sequence. If this is the case, an automatic sequential control system is used to tell the machine what to do and when.

The development of very small and inexpensive microprocessors and sensors has made it possible for them to be incorporated into control systems. Similar to the closed-loop approach, these tiny computers help a control system monitor its performance and

Feedback

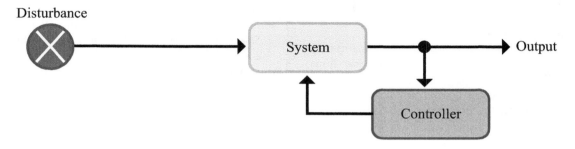

Feed Forward/Anticipation/Planning

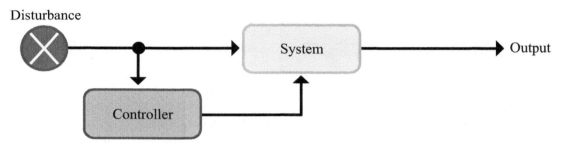

Control systems are used to alter a process pathway to reach a desired output. A controller senses some type of information along the path-way and causes a change in the system. Making changes to the system prior to system input is called a "feed-forward control"; making changes to the system based on the output is called "feedback control." Adapted from http://www.csci.csusb.edu/dick/cs372/a1.html. EBSCO illustration

provide detailed information about potential issues. For example, if a control system were using a new type of raw material that produced increased heat due to friction and caused the system to operate less efficiently, microsensors inside the machinery could detect this and alert those in charge of supervising the system. Such information could help avoid costly breakdowns of equipment and delays in production by allowing supervisors to address problems before they become severe.

LINEAR CONTROL SYSTEMS

Linear control systems are a type of closed-loop system. They receive linear negative feedback from their outputs and adjust their operating parameters in response. This allows the system to keep relevant variables within acceptable limits. If the

sensors in a linear control system detect excess heat, as in the example above, this might cause the system to initiate additional cooling. An example of this is when a computer's fan speeds up after an intensive application causes the system to begin generating heat.

BEHIND THE SCENES

Control systems are a vital part of everyday life in the industrialized world. Most of the products people use every day are produced or packaged using dozens of different control systems. Human beings have come to rely heavily on technology to assist them in their work, and in some cases to completely take over that work. Control systems are the mechanisms that help make this happen.

—*Scott Zimmer, JD*

BIBLIOGRAPHY

Ao, Sio-Iong, and Len Gelman, eds. *Electrical Engineering and Intelligent Systems*. New York: Springer, 2013. Print.

Chen, Yufeng, and Zhiwu Li. *Optimal Supervisory Control of Automated Manufacturing Systems*. Boca Raton: CRC, 2013. Print.

Janert, Philipp K. *Feedback Control for Computer Systems*. Sebastopol: O'Reilly, 2014. Print.

Li, Han-Xiong, and XinJiang Lu. *System Design and Control Integration for Advanced Manufacturing*. Hoboken: Wiley, 2015. Print.

Song, Dong-Ping. *Optimal Control and Optimization of Stochastic Supply Chain Systems*. London: Springer, 2013. Print.

Van Schuppen, Jan H., and Tiziano Villa, eds. *Coordination Control of Distributed Systems*. Cham: Springer, 2015. Print

CPU DESIGN

FIELDS OF STUDY

Computer Engineering; Information Technology

ABSTRACT

CPU design is an area of engineering that focuses on the design of a computer's central processing unit (CPU). The CPU acts as the "brain" of the machine, controlling the operations carried out by the computer. Its basic task is to execute the instructions contained in the programming code used to write software. Different CPU designs can be more or less efficient than one another. Some designs are better at addressing certain types of problems.

PRINICIPAL TERMS

- **control unit design:** describes the part of the CPU that tells the computer how to perform the instructions sent to it by a program.
- **datapath design:** describes how data flows through the CPU and at what points instructions will be decoded and executed.
- **logic implementation:** the way in which a CPU is designed to use the open or closed state of combinations of circuits to represent information.
- **microcontroller:** a tiny computer in which all of the essential parts of a computer are united on a single microchip—input and output channels, memory, and a processor.
- **peripherals:** devices connected to a computer but not part of the computer itself, such as scanners, external storage devices, and so forth.

- **protocol processor:** a processor that acts in a secondary capacity to the CPU, relieving it from some of the work of managing communication protocols that are used to encode messages on the network.

CPU DESIGN GOALS

The design of a CPU is a complex undertaking. The main goal of CPU design is to produce an architecture that can execute instructions in the fastest, most efficient way possible. Both speed and efficiency are relevant factors. There are times when having an instruction that is fast to execute is adequate, but there are also situations where it would not make sense to have to execute that simple instruction hundreds of times in order to accomplish a task.

Often the work begins by designers considering what the CPU will be expected to do and where it will be used. A microcontroller inside an airplane performs quite different tasks than one inside a kitchen appliance, for instance. The CPU's intended function tells what type of instruction-set architecture to use in the microchip that contains the CPU. Knowing what types of programs will be used most frequently also allows CPU designers to develop the most efficient logic implementation. Once this has been done, the control unit design can be defined. Defining the datapath design is usually the next step, as the CPU's handling of instructions is given physical form.

Often a CPU will be designed with additional supports to handle the processing load, so that the CPU itself does not become overloaded. Protocol processors, for example, may assist with communications

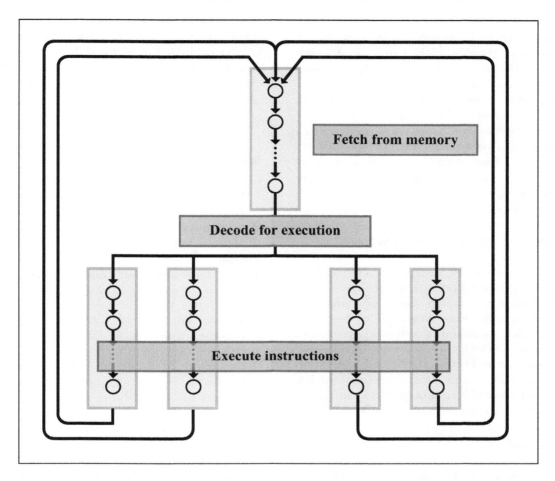

A generic state diagram shows the simple processing loop: fetch instructions from memory, decode instructions to determine the proper execute cycle, execute instructions, and then fetch next instructions from memory and continue the cycle. A state diagram is the initial design upon which data paths and logic controls can be designed. EBSCO illustration.

protocol translation involving the transmission of data between the computer and its peripherals or over the Internet. Internet communication protocols are quite complex. They involve seven different, nested layers of protocols, and each layer must be negotiated before the next can be addressed. Protocol processors take this workload off the CPU.

INSTRUCTION SETS
CPU design is heavily influenced by the type of instruction sets being used. In general, there are two approaches to instructions. The first is random logic. Sometimes this is referred to as "hardwired

instructions." Random logic uses logic devices, such as decoders and counters, to transport data and to perform calculations. Random logic can make it possible to design faster chips. The logic itself takes up space that might otherwise be used to store instructions, however. Therefore, it is not practical to use random logic with very large sets of instructions.

The second approach to instruction sets is microcode. Microcode is sometimes called "emulation" because it references an operations table and uses sets of microinstructions indexed by the table in order to execute each CPU instruction. Microcode can sometimes be slower to run than random logic, but it also has advantages that offset this weakness. Microcode breaks down complex instructions into sets of microinstructions. These microinstructions are used in several complex instructions. A CPU executing microcode would therefore be able to reuse microinstructions. Such reuse saves space on the microchip and allows more complex instructions to be added.

The most influential factor to consider when weighing random logic against microcode is memory speed. Random logic usually produces a speedier CPU when CPU speeds outpace memory speeds. When memory speeds are faster, microcode is faster than random logic.

REDUCED INSTRUCTION SET COMPUTER (RISC)

Early in the history of CPU design, it was felt that the best way to improve CPU performance was to continuously expand instruction sets to give programmers more options. Eventually, studies began to show that adding more complex instructions did not always improve performance, however. In response, CPU manufacturers produced reduced instruction set computer (RISC) chips. RISC chips could use less complex instructions, even though this meant that a larger number of instructions were required.

MOORE'S LAW

Moore's law is named after Gordon Moore, a cofounder of the computer manufacturer Intel. In 1975, Moore observed that the computing power of

an integrated circuit or microchip doubles, on average, every two years. This pace of improvement has been responsible for the rapid development in technological capability and the relatively short lifespan of consumer electronics, which tend to become obsolete soon after they are purchased.

—*Scott Zimmer, JD*

BIBLIOGRAPHY

Englander, Irv. *The Architecture of Computer Hardware, Systems Software, & Networking: An Information Technology Approach.* Hoboken: Wiley, 2014. Print.

Hyde, Randall. *Write Great Code: Understanding the Machine.* Vol. 1. San Francisco: No Starch, 2005. Print.

Jeannot, Emmanuel, and J. Žilinskas. *High Performance Computing on Complex Environments.* Hoboken: Wiley, 2014. Print.

Lipiansky, Ed. *Electrical, Electronics, and Digital Hardware Essentials for Scientists and Engineers.* Hoboken: Wiley, 2013. Print.

Rajasekaran, Sanguthevar. *Multicore Computing: Algorithms, Architectures, and Applications.* Boca Raton: CRC, 2013. Print.

Stokes, Jon. *Inside the Machine: An Illustrated Introduction to Microprocessors and Computer Architecture.* San Francisco: No Starch, 2015. Print.

Wolf, Marilyn. *High Performance Embedded Computing: Architectures, Applications, and Methodologies.* Amsterdam: Elsevier, 2014. Print.

CRYPTOGRAPHY

FIELDS OF STUDY

Computer Science; Computer Engineering; Algorithms

ABSTRACT

Cryptography is the process of encrypting messages and other data in order to transmit them in a form that can only be accessed by the intended recipients. It was initially applied to written messages. With the introduction of modern computers, cryptography became an important tool for securing many types of digital data.

PRINICIPAL TERMS

- **hash function:** an algorithm that converts a string of characters into a different, usually smaller, fixed-length string of characters that is ideally impossible either to invert or to replicate.
- **public-key cryptography:** a system of encryption that uses two keys, one public and one private, to encrypt and decrypt data.
- **substitution cipher:** a cipher that encodes a message by substituting one character for another.
- **symmetric-key cryptography:** a system of encryption that uses the same private key to encrypt and decrypt data.
- **transposition cipher:** a cipher that encodes a message by changing the order of the characters within the message.

WHAT IS CRYPTOGRAPHY?

The word "cryptography" comes from the Greek words *kryptos* ("hidden," "secret") and *graphein* ("writing"). Early cryptography focused on ensuring that written messages could be sent to their intended recipients without being intercepted and read by other parties. This was achieved through various encryption techniques. Encryption is based on a simple principle: the message is transformed in such a way that it becomes unreadable. The encrypted message is then transmitted to the recipient, who reads it by transforming (decrypting) it back into its original form.

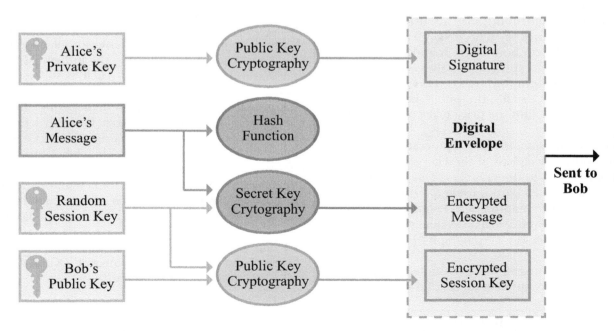

This is a diagram of cryptography techniques: public (asymmetric) key, private (symmetric) key, and hash functions. Each secures data in a different way and requires specific types of keys to encrypt and decrypt the data. Adapted from Gary C. Kessler, "An Overview of Cryptography."

Early forms of encryption were based on ciphers. A cipher encrypts a message by altering the characters that comprise the message. The original message is called the "plaintext," while the encrypted message is the "ciphertext." Anyone who knows the rules of the cipher can decrypt the ciphertext, but it remains unreadable to anyone else. Early ciphers were relatively easy to break, given enough time and a working knowledge of statistics. By the 1920s, electromechanical cipher machines called "rotor machines" were creating complex ciphers that posed a greater challenge. The best-known example of a rotor machine was the Enigma machine, used by the German military in World War II. Soon after, the development of modern computer systems in the 1950s would change the world of cryptography in major ways.

CRYPTOGRAPHY IN THE COMPUTER AGE
With the introduction of digital computers, the focus of cryptography shifted from just written language to any data that could be expressed in binary format. The encryption of binary data is accomplished through the use of keys. A key is a string of data that determines the output of a cryptographic algorithm. While there are many different types of cryptographic algorithms, they are usually divided into two categories. Symmetric-key cryptography uses a single key to both encrypt and decrypt the data. Public-key cryptography, also called "asymmetric-key cryptography," uses two keys, one public and one private. Usually, the public key is used to encrypt the data, and the private key is used to decrypt it.

When using symmetric-key cryptography, both the sender and the recipient of the encrypted message must have access to the same key. This key must be exchanged between parties using a secure channel, or else it may be compromised. Public-key cryptography does not require such an exchange. This is one reason that public-key cryptography is considered more secure.

Another cryptographic technique developed for use with computers is the digital signature. A digital signature is used to confirm the identity of the sender of a digital message and to ensure that no one has tampered with its contents. Digital signatures use public-key encryption. First, a hash function is used to compute a unique value based on the data contained in the message. This unique value is called a "message digest," or just "digest." The signer's private key is then used to encrypt the digest. The combination of the digest and the private key creates the signature. To verify the digital signature, the recipient uses the signer's public key to decrypt the digest. The same hash function is then applied to the data in the message. If the new digest matches the decrypted digest, the message is intact.

DECRYPTING A BASIC CIPHER
Among the earliest ciphers used were the transposition cipher and the substitution cipher. A transposition cipher encrypts messages by changing the order of the letters in the message using a well-defined scheme. One of the simplest transposition ciphers involves reversing the order of letters in each word. When encrypted in this fashion, the message "MEET ME IN THE PARK" becomes "TEEM EM NI EHT KRAP." More complicated transposition ciphers might involve writing out the message in a particular orientation (such as in stacked rows) and then reading the individual letters in a different orientation (such as successive columns).

A substitution cipher encodes messages by substituting certain letters in the message for other letters. One well-known early substitution cipher is the Caesar cipher, named after Julius Caesar. This cipher encodes messages by replacing each letter with a letter that is a specified number of positions to its right or left in the alphabet. For example, Caesar is reported to have used a left shift of three places when encrypting his messages.

Using this cipher, Julius Caesar's famous message "I came, I saw, I conquered" becomes "F ZXJB F PXT F ZLKNRBOBA" when encrypted.

Original	A	B	C	D	E	F	G	H	I	J	K	L	M	N	O	P	Q	R	S	T	U	V	W	X	Y	Z
Replacement	X	Y	Z	A	B	C	D	E	F	G	H	I	J	K	L	M	N	O	P	Q	R	S	T	U	V	W

SAMPLE PROBLEM

The following message has been encoded with a Caesar cipher using a left shift of five:

JSHWDUYNTS NX KZS

What was the original text of the message?

Answer:

The answer can be determined by replacing each letter in the encoded message with the letter five places to the left of its position in the alphabet, as shown in the following chart:

Original	A	B	C	D	E	F	G	H	I	J
Replacement	V	W	X	Y	Z	A	B	C	D	E

The original text read "ENCRYPTION IS FUN."

and individuals with the needs of law enforcement and government agencies. Businesses, governments, and consumers must deal with the challenges of securing digital communications for commerce and banking on a daily basis. The impact of cryptography on society is likely to increase as computers grow more powerful, cryptographic techniques improve, and digital technologies become ever more important.

—*Maura Valentino, MSLIS*

BIBLIOGRAPHY

Esslinger, Bernhard, et al. *The CrypTool Script: Cryptography, Mathematics, and More.* 11th ed. Frankfurt: CrypTool, 2013. *CrypTool Portal.* Web. 2 Mar. 2016.

Hoffstein, Jeffrey, Jill Pipher, and Joseph H. Silverman. *An Introduction to Mathematical Cryptography.* 2nd ed. New York: Springer, 2014. Print.

Katz, Jonathan, and Yehuda Lindell. *Introduction to Modern Cryptography.* 2nd ed. Boca Raton: CRC, 2015. Print.

Menezes, Alfred J., Paul C. van Oorschot, and Scott A. Vanstone. *Handbook of Applied Cryptography.* Boca Raton: CRC, 1996. Print.

Neiderreiter, Harald, and Chaoping Xing. *Algebraic Geometry in Coding Theory and Cryptography.* Princeton: Princeton UP, 2009. Print.

Paar, Christof, and Jan Pelzi. *Understanding Cryptography: A Textbook for Students and Practitioners.* Heidelberg: Springer, 2010. Print.

WHY IS CRYPTOGRAPHY IMPORTANT?

The ability to secure communications against interception and decryption has long been an important part of military and international affairs. In the modern age, the development of new computer-based methods of encryption has had a major impact on many areas of society, including law enforcement, international affairs, military strategy, and business. It has also led to widespread debate over how to balance the privacy rights of organizations

D

DEADLOCK

FIELDS OF STUDY

Information Systems; Operating Systems; System Analysis

ABSTRACT

In computing, a deadlock is when two or more processes are running simultaneously and each one needs access to a resource that the other is using. Neither process can terminate until the other one does, so they become stuck, or "locked." Deadlock can occur in several different areas of computer systems, including operating systems, database transaction processing, and software execution.

PRINICIPAL TERMS

- **circular wait:** a situation in which two or more processes are running and each one is waiting for a resource that is being used by another; one of the necessary conditions for deadlock.
- **livelock:** a situation in which two or more processes constantly change their state in response to one another in such a way that neither can complete.
- **mutual exclusion:** a rule present in some database systems that prevents a resource from being accessed by more than one operation at a time; one of the necessary conditions for deadlock.
- **resource holding:** a situation in which one process is holding at least one resource and is requesting further resources; one of the necessary conditions for deadlock.
- **transactional database:** a database management system that allows a transaction— a sequence of operations to achieve a single, self-contained task—to be undone, or "rolled back," if it fails to complete properly.

HOLDING RESOURCES

In order to understand how a deadlock can arise in a computer system, one must first understand how resources are used in a computer system. One way is to think of the system as a bank account. A database keeps records of debits and credits to the account, and each new transaction is recorded in the database. If money is transferred from one account to another—for example, from a customer's savings account to their checking account—the transfer must be recorded twice, as both a debit from the savings account and a credit to the checking account. Recording the debit and recording the credit are two separate actions. However, in a transactional database, they represent a single transaction, because together they form a logical unit: money cannot be added to one account without being subtracted from another. In this case, the balances of both accounts represent system resources. In order to prevent another process, such as a withdrawal from the checking account, from interrupting the transaction before both the debit and the credit are recorded, the system enforces a rule known as mutual exclusion. This rule means that no two processes can access the same resources at once. Instead, the transfer process places a hold on the account balances until the transaction is complete. This practice is called resource holding, or sometimes "hold and wait." Only after the transfer is recorded as both a debit and a credit will the transfer process release its hold on the account balance resources. At this point, the withdrawal process is free to access the checking account balance.

Resource holding is necessary in such cases because if resources were not locked, it would not be clear how a system would resolve simultaneous events. In the example above, if the system had tried to calculate a withdrawal from the checking account

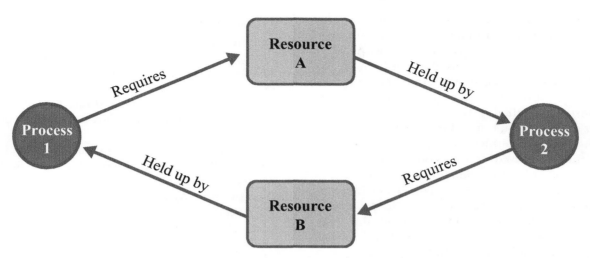

This is a diagram of a circular wait deadlock. Process 1 requires resource A to complete, resource A is held up by process 2, process 2 cannot complete without resource B, and resource B is held up by process 1. EBSCO illustration.

before the credit to the account was recorded, the customer might have overdrawn their account. Resource holding is used to make systems more consistent, since it defines rules that allow ambiguities to be resolved in a straightforward manner.

STANDOFF

Deadlock can be understood as an unintended consequence of resource holding. While resource holding serves a legitimate purpose, inevitably situations arise in which two processes each hold a resource that the other needs in order to finish its work. Process A needs a resource locked by process B in order to finish, while process B needs a resource locked by process A to finish, and neither can relinquish its resources until it is completed. This situation is called a circular wait.

In 1971, computer scientist Edward G. Coffman described the four conditions, now known as the Coffman conditions, necessary for deadlock to occur:

> Mutual exclusion,
> Resource holding,
> No preemption, and
> Circular wait

Mutual exclusion, resource holding, and circular wait are described above. Preemption is when one process takes control of a resource being used by another process. Some designated processes have

priority over other operations. As such, they can force other processes to stop and let them finish first. Preemption is typically reserved for processes that are fundamental to the operation of the system. If one of the conflicting processes can preempt another, deadlock will not occur. In fact, if any one of the Coffman conditions is not true, deadlock will not occur.

RESPONSES TO DEADLOCK

System designers have several options for dealing with deadlocks. One is to simply ignore the issue altogether, though this is not practical in most circumstances. A second approach is to focus on detecting deadlocks and then have a variety of options available for resolving them. A third approach is to try to avoid deadlocks by understanding which system operations tend to cause them. Finally, system designers may adopt a preventive approach in which they make changes to the system architecture specifically to avoid potential deadlocks.

Usually some type of outside intervention is required to resolve a deadlock situation. To accomplish this, many systems will include mechanisms that function to detect and resolve deadlock. Ironically, in trying to resolve deadlock, these systems may produce a similar situation known as livelock. Like a deadlock, a livelock results when two or more concurrent processes are unable to complete. Unlike a deadlock, this happens not because the processes are at a standstill but because they are continually responding to one another. The situation can be compared to two people traveling in opposite directions down a hallway. To avoid bumping into each other, one person moves to their left, while the other moves to the right. In response to the other's actions, the first person then moves to their right, but at the same time the other person moves to their left. In real life, this situation is eventually resolved. In a computer system, it can continue on an infinite loop until one or both processes are forced to end. If a deadlock detection algorithm can resolve a deadlock and the previously stuck processes begin to function again, they may produce another deadlock. This will trigger the detection algorithm again, and so on, creating a livelock. One way to avoid this would be to have the detection algorithm release one of the deadlocked processes and either abort the other or keep it on hold until it is completed.

—*Scott Zimmer, JD*

BIBLIOGRAPHY

Connolly, Thomas M., and Carolyn E. Begg. *Database Systems: A Practical Approach to Design, Implementation, and Management.* 6th ed. Boston: Pearson, 2015. Print.

Harth, Andreas, Katja Hose, and Ralf Schenkel, eds. *Linked Data Management.* Boca Raton: CRC, 2014. Print.

Hoffer, Jeffrey A., V. Ramesh, and Heikki Topi. *Modern Database Management.* 12th ed. Boston: Pearson, 2016. Print.

Kshemkalyani, Ajay D., and Mukesh Singhal. *Distributed Computing: Principles, Algorithms, and Systems.* New York: Cambridge UP, 2008. Print.

Rahimi, Saeed K., and Frank S. Haug. *Distributed Database Management Systems: A Practical Approach.* Hoboken: Wiley, 2010. Print.

Wills, Craig E. "Process Synchronization and Interprocess Communication." *Computing Handbook: Computer Science and Software Engineering.* Ed. Teofilo Gonzalez and Jorge Díaz-Herrera. 3rd ed. Boca Raton: CRC, 2014. 52-1–21. Print.

DEBUGGING

FIELDS OF STUDY

Computer Science; Software Engineering

ABSTRACT

Debugging is the process of identifying and addressing errors, known as "bugs," in computer systems. It is an essential step in the development of all kinds of programs, from consumer programs such as web browsers and video games to the complex systems used in transportation and infrastructure. Debugging can be carried out through a number of methods depending on the nature of the computer system in question.

PRINICIPAL TERMS

- **delta debugging:** an automated method of debugging intended to identify a bug's root cause while eliminating irrelevant information.
- **in-circuit emulator:** a device that enables the debugging of a computer system embedded within a larger system.
- **integration testing:** a process in which multiple units are tested individually and when working in concert.
- **memory dumps:** computer memory records from when a particular program crashed, used to pinpoint and address the bug that caused the crash.
- **software patches:** updates to software that correct bugs or make other improvements.

UNDERSTANDING DEBUGGING

Debugging is the process of testing software or other computer systems, noting any errors that occur, and finding the cause of those errors. Errors, or "bugs," in a computer program can seriously affect the program's operations or even prevent it from functioning altogether. The ultimate goal of debugging is to get rid of the bugs that have been identified. This should ensure the smooth and error-free operation of the computer program or system.

Computer programs consist of long strings of specialized code that tell the computer what to do and how to do it. Computer code must use specific vocabulary and structures in order to function properly. As such code is written by human programmers, there is always the possibility of human error, which is the cause of many common bugs. Perhaps the most common bugs are syntax errors. These are the result of small mistakes, such as typos, in a program's code. In some cases, a bug may occur because the programmer neglected to include a key element in the code or structured it incorrectly. For example, the code could include a command instructing the computer to begin a specific process but lack the corresponding command to end it.

Bugs fall into one of several categories, based on when and how they affect the program. Compilation errors prevent the program from running. Run-time errors, meanwhile, occur as the program is running. Logic errors, in which flaws in the program's logic produce unintended results, are a particularly common form of bug. Such errors come about when a program's code is syntactically correct but does not make logical sense. For instance, a string of code with flawed logic may cause the program to become caught in an unintended loop. This can cause it to become completely unresponsive, or freeze. In other cases, a logic error might result when a program's code instructs the computer to divide a numerical value by zero, a mathematically impossible task.

WHY DEBUG?

Bugs may interfere with a program's ability to perform its core functions or even to run. Not all bugs are related to a program's core functions, and some programs may be usable despite the errors they contain. However, ease of use is an important factor that many people consider when deciding which program to use. It is therefore in the best interest of software creators to ensure that their programs are as free of errors as possible. In addition to testing a program or other computer system in house prior to releasing them to the general public, many software companies collect reports of bugs from users following its release. This is often done through transfers of

collected data commonly referred to as memory dumps. They can then address such errors through updates known as software patches.

While bugs are an inconvenience in consumer computer programs, in more specialized computer systems, they can have far more serious consequences. In areas such as transportation, infrastructure, and finance, errors in syntax and logic can place lives and livelihoods at risk. Perhaps the most prominent example of such a bug was the so-called Y2K bug. This bug was projected to affect numerous computer systems beginning on January 1, 2000. The problem would have resulted from existing practices related to the way dates were written in computer programs. However, it was largely averted through the work of programmers who updated the affected programs to prevent that issue. As the example of the far-reaching Y2K bug shows, the world's growing reliance on computers in all areas of society has made thorough debugging even more important.

IDENTIFYING AND ADDRESSING BUGS

The means of debugging vary based on the nature of the computer program or system in question. However, in most cases bugs may be identified and addressed through the same general process. When a bug first appears, the programmer or tester must first attempt to reproduce the bug in order to identify the results of the error and the conditions under which they occur. Next, the programmer must attempt to determine which part of the program's code is causing the error to occur. As programs can be quite complex, the programmer must simplify this process by eliminating as much irrelevant data as possible. Once the faulty segment of code has been found, the programmer must identify and correct the specific problem that is causing the bug. If the cause is a typo or a syntax error, the programmer may simply need to make a small correction. If there is a logic error, the programmer may need to rewrite a portion of the code so that the program operates logically.

DEBUGGING IN PRACTICE

Programmers use a wide range of tools to debug programs. As programs are frequently complex and lengthy, automating portions of the debugging process is often essential. Automated debugging programs, or "debuggers," search through code line by line for syntax errors or faulty logic that could cause bugs. A technique known as delta debugging provides an automated means of filtering out irrelevant information when the programmer is looking for the root cause of a bug.

Different types of programs or systems often require different debugging tools. An in-circuit emulator is used when the computer system being tested is an embedded system (that is, one located within a larger system) and cannot otherwise be accessed. A form of debugging known as integration testing is often used when a program consists of numerous components. After each component is tested and debugged on its own, they are linked together and tested as a unit. This ensures that the different components function correctly when working together.

—*Joy Crelin*

BIBLIOGRAPHY

Foote, Steven. *Learning to Program*. Upper Saddle River: Pearson, 2015. Print.

McCauley, Renée, et al. "Debugging: A Review of the Literature from an Educational Perspective." *Computer Science Education* 18.2 (2008): 67–92. Print.

Myers, Glenford J., Tom Badgett, and Corey Sandler. *The Art of Software Testing*. Hoboken: Wiley, 2012. Print.

St. Germain, H. James de. "Debugging Programs." *University of Utah*. U of Utah, n.d. Web. 31 Jan. 2016.

"What Went Wrong? Finding and Fixing Errors through Debugging." *Microsoft Developer Network Library*. Microsoft, 2016. Web. 31 Jan. 2016.

Zeller, Andreas. *Why Programs Fail: A Guide to Systematic Debugging*. Burlington: Kaufmann, 2009. Print.

DEMON DIALING/WAR DIALING

FIELDS OF STUDY

Security

ABSTRACT

War dialing is the practice of autodialing a large range of phone numbers to find computer modems. It involves using a software program to call all of the phone numbers within an area code to see which ones are set up to accept incoming connections. Demon dialing is a synonym for war dialing, though it has also been used to describe making repeated calls to a single modem in a brute-force attempt to guess its password.

PRINICIPAL TERMS

- **cracker:** a criminal hacker; one who finds and exploits weak points in a computer's security system to gain unauthorized access for malicious purposes.
- **hacking:** the use of technical skill to gain unauthorized access to a computer system; also, any kind of advanced tinkering with computers to increase their utility.
- **port scanning:** the use of software to probe a computer server to see if any of its communication ports have been left open or vulnerable to an unauthorized connection that could be used to gain control of the computer.
- **voice over IP (VoIP):** short for "voice over Internet Protocol"; a telephone service designed for transmission over broadband Internet connections rather than telephone wires.
- **wardriving:** driving around with a device such as a laptop that can scan for wireless networks that may be vulnerable to hacking.

THE EARLY INTERNET

Prior to the mid-1990s, the vast majority of people wishing to go online did so using dial-up modems. A dial-up modem connected a computer to the telephone network via a telephone wall jack. It accessed the Internet by making a phone call to another computer or a server and transmitting data over the telephone line. Most early Internet connections were made to limited, self-contained services, such as local area networks (LANs) and bulletin-board systems (BBSs) hosted on computer systems running terminal programs. In order for a connection to be established, the computer user would have to have the phone number of the host computer, and the host computer would have to be available. If the host computer was not online or had too many connections already, it could not accept another connection. Eventually the first Internet service providers (ISPs) were established, allowing many computers to access the public Internet simultaneously.

As the number of computers with modems increased, computer hackers began to try to discover and connect to them. For some, hacking became a hobby of sorts. It presented a challenge that was exciting and intellectually stimulating. For crackers, or criminal hackers, it was a way to steal data or commit other malicious acts. Regardless of the hackers' motives, they all needed some way to find the phone numbers of computers with modems.

One such way was war dialing. Hackers wrote software that would use a computer's modem to dial every phone number in a given area code. The software ran through each number, dialing and then hanging up after two rings. Most modems were set up to pick up after one ring, so if a number rang twice it most likely did not have a modem. The war-dialing software recorded which numbers had modems so that hackers could try to connect to them later.

"Demon dialing" originally meant making repeated calls to a single modem in a brute-force attempt to guess its password. The practice was named for the Demon Dialer, a telephone dialer once sold by Zoom Telephonics. This device could automatically redial a busy phone number until the call went through. Over time, demon dialing became a synonym for war dialing.

MODERN TAKES ON WAR DIALING
There have been many changes in technology that make the old methods of war dialing obsolete. The number of computers still connecting to the Internet via dial-up modems decreased sharply after broadband Internet access and wireless networking became mainstream in the mid-2000s. However, war dialing itself is still practiced; it simply requires different techniques. For example, the open-source software WarVOX is a war-dialing tool that connects via voice over IP (VoIP) systems instead of landline telephones. It uses signal-processing techniques to probe and analyze telephone systems. Some information technology (IT) security personnel use VoIP-based

war dialers to find unauthorized modems and faxes on their organization's computer networks.

Another modern technique similar to war dialing is called port scanning. Computers connect to the Internet using different ports, which are like virtual connection points. Some ports are traditionally used for certain connections, such as printers or web browsing. Other ports are left open for whichever application needs to create a connection. When a computer has been secured, ports not in use are kept closed to prevent unauthorized connections. Port scanners bombard computers with connection attempts on many different ports at once, then report vulnerable port numbers back to the hacker. To protect against port scanning, many companies now use intrusion-detection systems that can identify when a port scan is underway. This security measure has in turn motivated hackers to develop port-scanning methods that can gather information without openly trying to connect to each port.

WI-FI WAR DIALING
Some hackers target wireless networks instead of wired ones. One technique for doing so is wardriving, in which hackers drive around a neighborhood with a laptop running Wi-Fi scanning software. The software identifies wireless networks as it passes through them and collects information about the type of security each wireless access point is using. Hackers can then sort through this information to find vulnerable networks. A hacker may exploit the network to gain free wireless Internet access or to disguise their online identity.

—*Scott Zimmer, JD*

BIBLIOGRAPHY
Coleman, E Gabriella. *Coding Freedom: The Ethics and Aesthetics of Hacking.* Princeton: Princeton UP, 2013. Print.

Haerens, Margaret, and Lynn M. Zott, eds. *Hacking and Hackers.* Detroit: Greenhaven, 2014. Print.

Kizza, Joseph Migga. *Guide to Computer Network Security.* 3rd ed. London: Springer, 2015. Print.

Morselli, Carlo, ed. *Crime and Networks.* New York: Routledge, 2014. Print.

Naraine, Ryan. "Metasploit's H. D. Moore Releases 'War Dialing' Tools." *ZDNet.* CBS Interactive, 6 Mar. 2009. Web. 15 Mar. 2016.

Netzley, Patricia D. *How Serious a Problem Is Computer Hacking?* San Diego: ReferencePoint, 2014. Print.

Shakarian, Paulo, Jana Shakarian, and Andrew Ruef. *Introduction to Cyber-Warfare: A Multidisciplinary Approach.* Waltham: Syngress, 2013. Print.

DEVICE DRIVERS

FIELDS OF STUDY

Computer Engineering; Software Engineering

ABSTRACT

Device drivers are software interfaces that allow a computer's central processing unit (CPU) to communicate with peripherals such as disk drives, printers, and scanners. Without device drivers, the computer's operating system (OS) would have to come preinstalled with all of the information about all of the devices it could ever need to communicate with. OSs contain some device drivers, but these can also be installed when new devices are added to a computer.

PRINICIPAL TERMS

- **device manager:** an application that allows users of a computer to manipulate the device drivers installed on the computer, as well as adding and removing drivers.
- **input/output instructions:** instructions used by the central processing unit (CPU) of a computer when information is transferred between the CPU and a device such as a hard disk.
- **interface:** the function performed by the device driver, which mediates between the hardware of the peripheral and the hardware of the computer.
- **peripheral:** a device that is connected to a computer and used by the computer but is not part of the computer, such as a printer.
- **virtual device driver:** a type of device driver used by the Windows operating system that handles communications between emulated hardware and other devices.

How Device Drivers Work

The main strength of device drivers is that they enable programmers to write software that will run on a computer regardless of the type of devices that are connected to that computer. Using device drivers allows the program to simply command the computer to save data to a file on the hard drive. It needs no specific information about what type of hard drive is installed in the computer or connections the hard drive has to other hardware in the computer. The device driver acts as an interface between computer components.

When a program needs to send commands to a peripheral connected to the computer, the program communicates with the device driver. The device driver receives the information about the action that the device is being asked to perform. It translates this information into a format that can be input into the device. The device then performs the task or tasks requested. When it finishes, it may generate output that is communicated to the driver, either as a message or as a simple indication that the task has been completed. The driver then translates this information into a form that the original program can understand. The device driver acts as a kind of translator between the computer and its peripherals, conveying input/output instructions between the two. Thus, the computer program does not need to include all of the low-level commands needed to make the device function. The program only needs to be able to tell the device driver what it wants the device to do. The device driver takes care of translating this into concrete steps.

How Device Drivers Are Made

Writing device drivers is a highly technical undertaking. It is made more challenging by the fact that device drivers can be unforgiving when a mistake is made in their creation. This is because higher-level applications do not often have unlimited access to all of the computer's functionality. Issuing the wrong command with unrestricted privileges can cause serious damage to the computer's operating system (OS) and, in some cases, to the hardware. This is a real possibility with device drivers, which usually need to have unrestricted access to the computer.

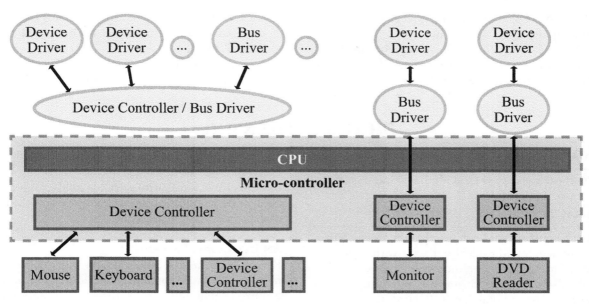

Each device connected to a CPU is controlled by a device driver, software that controls, manages, and monitors a specific device (e.g., keyboard, mouse, monitor, DVD reader). Device drivers may also drive other software that drives a device (e.g., system management bus, universal serial bus controller). EBSCO illustration.

Because writing a device driver requires a lot of specialized information, most device drivers are made by software engineers who specialize in driver development and work for hardware manufacturers. Usually the device manufacturer has the most information about the device and what it needs in order to function properly. The exception to this trend is the impressive amount of driver development accomplished by the open-source movement. Programmers all over the world have volunteered their own time and talent to write drivers for the Linux OS.

Often development is separated into logical and physical device driver development. Logical device driver development tends to be done by the creator of the OS that the computer will use. Physical device driver development, meanwhile, is handled by the device manufacturer. This division of labor makes sense, but it does require coordination and a willingness to share standards and practices among the various parties.

VIRTUAL DEVICE DRIVERS

Virtual device drivers are a variation on traditional device drivers. They are used when a computer needs to emulate a piece of hardware. This often occurs when an OS runs a program that was created for a different OS by emulating that operating environment. One example would be a Windows OS running a DOS program. If the DOS program needed to interface with an attached printer, the computer would use a virtual device driver.

DEVICE MANAGERS

Most OSs now include device managers that make it easier for the user to manage device drivers. They allow the user to diagnose problems with devices, troubleshoot issues, and update or install drivers. Using the graphical interface of a device manager is less intimidating than typing in text commands to perform driver-related tasks.

—*Scott Zimmer, JD*

BIBLIOGRAPHY

Corbet, Jonathan, Alessandro Rubini, and Greg Kroah-Hartman. *Linux Device Drivers.* 3rd ed. Cambridge: O'Reilly, 2005. Print.

McFedries, Paul. *Fixing Your Computer: Absolute Beginner's Guide.* Indianapolis: Que, 2014. Print.

Mueller, Scott. *Upgrading and Repairing PCs.* 22nd ed. Indianapolis: Que, 2015. Print.

Noergaard, Tammy. *Embedded Systems Architecture: A Comprehensive Guide for Engineers and Programmers.* 2nd ed. Boston: Elsevier, 2012. Print.

Orwick, Penny, and Guy Smith. *Developing Drivers with the Windows Driver Foundation.* Redmond: Microsoft P, 2007. Print.

"What Is a Driver?" *Microsoft Developer Network.* Microsoft, n.d. Web. 10 Mar. 2016.

DIGITAL CITIZENSHIP

FIELDS OF STUDY

Information Systems; Digital Media; Privacy

ABSTRACT

Digital citizenship can be defined as the norms of appropriate, legal, and ethical behavior with regard to the use of information technology in a person's civic and social life. Digital citizenship is a unique phenomenon of the digital age, reflecting the growing importance of digital literacy, digital commerce, and information technology in global culture.

PRINICIPAL TERMS

- **butterfly effect:** an effect in which small changes in a system's initial conditions lead to major, unexpected changes as the system develops.
- **digital commerce:** the purchase and sale of goods and services via online vendors or information technology systems.
- **digital literacy:** familiarity with the skills, behaviors, and language specific to using digital devices to access, create, and share content through the Internet.
- **digital native:** an individual born during the digital age or raised using digital technology and communication.
- **IRL relationships:** relationships that occur "in real life," meaning that the relationships are developed or sustained outside of digital communication.
- **piracy:** in the digital context, unauthorized reproduction or use of copyrighted media in digital form.

A New Form of Citizenship

Digital citizenship can be defined as the norms and rules of behavior for persons using digital technology in commerce, political activism, and social communication. A person's digital citizenship begins when they engage with the digital domain, for instance, by beginning to use a smartphone or e-mail. However, digital citizenship exists on a spectrum based on an individual's level of digital literacy. This can be defined as their familiarity with the skills, jargon, and behaviors commonly used to communicate and conduct commerce with digital tools.

An Evolving Paradigm

Educational theorist Marc Prensky suggested that the modern human population can be divided into two groups. Digital natives were raised in the presence of digital technology. They learned how to use it in childhood. They absorbed the basics of digital citizenship during their early development. Digital immigrants were born before the digital age or have limited access to technology. They adapt to digital technology and communication later in life. Given the growing importance of digital technology, educators and social scientists believe that teaching children and adults to use digital technology safely and ethically is among the most important goals facing modern society.

Digital communication enables people to have relationships online or on mobile devices. The degree to which these digital relationships affect IRL relationships, or those that occur "in real life," is an important facet of digital citizenship. For instance, research suggests that people who spend more time communicating through smartphones or who feel they need constant access to digital media have more difficulty forming and maintaining IRL relationships. Good digital citizenship can help people use digital technology in positive ways that do not detract from their IRL relationships and well-being.

Digital technology has had a powerful, democratizing force on culture. Social media, for instance, has enabled small, local social movements to have national and even international impact. Even a simple tweet or viral video can quickly spread to millions of social media users. Small-scale behaviors can thus have larger, often unexpected consequences, both good and bad. This is sometimes called the butterfly effect.

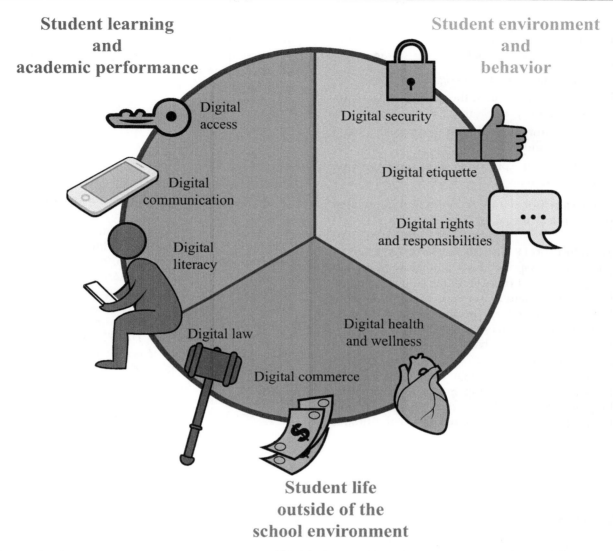

Good digital citizens know and understand the nine themes of digital citizenship. These themes address appropriate ways of interacting with the technology, information, government, companies, and other citizens that comprise the digital world. EBSCO illustration.

The potential impact of even a single user in the digital domain highlights the importance of learning and teaching effective digital behavior and ethics.

ETHICS OF DIGITAL CITIZENSHIP

In *Digital Citizenship in Schools*, educational theorist Mike Ribble outlines nine core themes that characterize digital citizenship. These themes are: digital access, digital commerce, digital communication, digital literacy, digital etiquette, digital law, digital rights and responsibilities, digital health and wellness, and digital

security. They address appropriate ways of interacting with the technology, information, government, companies, and other citizens in the digital world. Ribble writes that well-adjusted digital citizens help others become digitally literate. They strive to make the digital domain harmonious and culturally beneficial.

He also argues that digital citizens are responsible for learning about and following the laws and ethical implications of all digital activities. Piracy is the unauthorized digital reproduction of copyrighted media. It violates laws that protect creative property. It has become one of the most controversial issues of the twenty-first century. Other common digital crimes and forms of misconduct include:

> Plagiarizing content from digital sources,
>
> Hacking (gaining unauthorized access to computer networks or systems),
>
> Sending unwanted communications and spam messages, and
>
> Creating and spreading destructive viruses, worms, and malware.

Such behaviors violate others' rights and so are considered unethical, illegal, or both.

DIGITAL SECURITY AND RESPONSIBILITY

The rights and responsibilities of digital citizenship differ by political environment. In the United States, individuals have the right to free speech and expression. They also have a limited right to digital privacy. The ownership of digital data is an evolving subject in US law.

One responsibility of digital citizenship is to learn about potential dangers, both social and physical, and how to avoid them. These include identity theft, cyberstalking, and cyberbullying. Strong digital security can help protect one's identity, data, equipment, and creative property. Surge protectors, antivirus software,

and data backup systems are some of the tools digital citizens use to enhance their digital security. Though the digital world is a complex, rapidly evolving realm, advocates argue that the rules of digital citizenship can be reduced to a basic concept: respect oneself and others when engaging in digital life.

—*Micah L. Issitt*

BIBLIOGRAPHY

"The Digital Millennium Copyright Act of 1998." *Copyright.* US Copyright Office, 28 Oct 1998. Web. 23 Jan 2016.

McNeill, Erin. "Even 'Digital Natives' Need Digital Training." *Education Week.* Editorial Projects in Education, 20 Oct 2015. Web. 26 Jan. 2016.

Ribble, Mike. *Digital Citizenship in Schools: Nine Elements All Students Should Know.* Eugene: Intl. Soc. for Technology in Education, 2015. Print.

Saltman, Dave. "Tech Talk: Turning Digital Natives into Digital Citizens." *Harvard Education Letter* 27.5 (2011): n. pag. *Harvard Graduate School of Education.* Web. 27 Jan. 2016.

Wells, Chris. *The Civic Organization and the Digital Citizen: Communicating Engagement in a Networked Age.* New York: Oxford UP, 2015. Print.

DIGITAL FORENSICS

FIELDS OF STUDY

Information Technology; System Analysis; Privacy

ABSTRACT

Digital forensics is a branch of science that studies stored digital data. The field emerged in the 1990s but did not develop national standards until the 2000s. Digital forensics techniques are changing rapidly due to the advances in digital technology.

PRINICIPAL TERMS

- **cybercrime:** crime that involves targeting a computer or using a computer or computer network to commit a crime, such as computer hacking, digital piracy, and the use of malware or spyware.

- **Electronic Communications Privacy Act:** a 1986 law that extended restrictions on wiretapping to cover the retrieval or interception of information transmitted electronically between computers or through computer networks.

- **logical copy:** a copy of a hard drive or disk that captures active data and files in a different configuration from the original, usually excluding free space and artifacts such as file remnants; contrasts with a physical copy, which is an exact copy with the same size and configuration as the original.

- **metadata:** data that contains information about other data, such as author information, organizational information, or how and when the data was created.

- **Scientific Working Group on Digital Evidence (SWGDE):** an American association of various academic and professional organizations interested in the development of digital forensics systems, guidelines, techniques, and standards.

AN EVOLVING SCIENCE

Digital forensics is the science of recovering and studying digital data, typically in the course of criminal investigations. Digital forensic science is used to investigate cybercrimes. These crimes target or involve the use of computer systems. Examples include identity theft, digital piracy, hacking, data theft, and cyberattacks. The Scientific Working Group on Digital Evidence (SWGDE), formed in 1998, develops industry guidelines, techniques, and standards.

DIGITAL FORENSICS POLICY

Digital forensics emerged in the mid-1980s in response to the growing importance of digital data in criminal investigations. The first cybercrimes occurred in the early 1970s. This era saw the emergence of "hacking," or gaining unauthorized access to computer systems. Some of the first documented uses of digital forensics data were in hacking investigations.

Prior to the Electronic Communications Privacy Act (ECPA) of 1986, digital data or communications were

Digital forensics encompasses computer forensics, mobile forensics, computer network forensics, social networking forensics, database forensics, and forensic data analysis or the forensic analysis of large-scale data EBSCO illustration.

not protected by law and could be collected or intercepted by law enforcement. The ECPA was amended several times in the 1990s and 2000s to address the growing importance of digital data for private communication. In 2014, the Supreme Court ruled that police must obtain a warrant before searching the cell phone of a suspect arrested for a crime.

DIGITAL FORENSICS TECHNIQUES

Once forensic investigators have access to equipment that has been seized or otherwise legally obtained, they can begin forensic imaging. This process involves making an unaltered copy, or forensic image, of the device's hard drive. A forensic image records the drive's structures, all of its contents, and metadata about the original files.

A forensic image is also known as a "physical copy." There are two main methods of copying computer data, physical copying and logical copying. A physical copy duplicates all of the data on a specific drive, including empty, deleted, or fragmented data, and stores it in its original configuration. A logical copy, by contrast, copies active data but ignores deleted files, fragments, and empty space. This makes the data easier to read and analyze. However, it may not provide a complete picture of the relevant data.

After imaging, forensics examiners analyze the imaged data. They may use specialized tools to recover deleted files using fragments or backup data, which is stored on many digital devices to prevent accidental data loss. Automated programs can be used to search and sort through imaged data to find useful information. (Because searching and sorting are crucial to the forensic process, digital forensics organizations invest in research into better search and sort algorithms). Information of interest to examiners may include e-mails, text messages, chat records, financial files, and various types of computer code. The tools and techniques used for analysis depend largely on the crime. These specialists may also be tasked with interpreting any data collected during an investigation. For instance, they may be called on to explain their findings to police or during a trial.

CHALLENGES FOR THE FUTURE

Digital forensics is an emerging field that lags behind fast-changing digital technology. For instance, cloud computing is a fairly new technology in which data storage and processing is distributed across multiple computers or servers. In 2014, the National Institute of Standards and Technology identified sixty-five challenges that must be addressed regarding cloud computing. These challenges include both technical problems and legal issues.

The SWGDE works to create tools and standards that will allow investigators to effectively retrieve and analyze data while keeping pace with changing technology. It must also work with legal rights organizations to ensure that investigations remain within boundaries set to protect personal rights and privacy. Each forensic investigation may involve accessing personal communications and data that might be protected under laws that guarantee free speech and expression or prohibit unlawful search and seizure. The SWGDE and law enforcement agencies are debating changes to existing privacy and surveillance laws to address these issues while enabling digital forensic science to continue developing.

—*Micah L. Issitt*

BIBLIOGRAPHY

"Digital Evidence and Forensics." *National Institute of Justice.* Office of Justice Programs, 28 Oct. 2015. Web. 12 Feb. 2016.

Gogolin, Greg. *Digital Forensics Explained.* Boca Raton: CRC, 2013. Print.

Holt, Thomas J., Adam M. Bossler, and Kathryn C. Seigfried-Spellar. *Cybercrime and Digital Forensics: An Introduction.* New York: Routledge, 2015. Print.

Pollitt, Mark. "A History of Digital Forensics." *Advances in Digital Forensics VI.* Ed. Kam-Pui Chow and Sujeet Shenoi. Berlin: Springer, 2010. 3–15. Print.

Sammons, John. *The Basics of Digital Forensics: The Primer for Getting Started in Digital Forensics.* Waltham: Syngress, 2012. Print.

Shinder, Deb. "So You Want to Be a Computer Forensics Expert." *TechRepublic.* CBS Interactive, 27 Dec. 2010. Web. 2 Feb. 2016.

DIGITAL SIGNAL PROCESSORS

FIELDS OF STUDY

Computer Science; Information Technology; Network Design

ABSTRACT

Digital signal processors (DSPs) are microprocessors designed for a special function. DSPs are used with analog signals to continuously monitor their output, often performing additional functions such as filtering or measuring the signal. One of the strengths of DSPs is that they can process more than one instruction or piece of data at a time.

PRINICIPAL TERMS

- **fixed-point arithmetic:** a calculation involving numbers that have a defined number of digits before and after the decimal point.
- **floating-point arithmetic:** a calculation involving numbers that have a decimal point that can be placed anywhere through the use of exponents, as is done in scientific notation.
- **Harvard architecture:** a computer design that has physically distinct storage locations and signal routes for data and for instructions.
- **multiplier-accumulator:** a piece of computer hardware that performs the mathematical operation of multiplying two numbers and then adding the result to an accumulator.
- **pipelined architecture:** a computer design where different processing elements are connected in a series, with the output of one operation being the input of the next.
- **semiconductor intellectual property (SIP) block:** a quantity of microchip layout design that is owned by a person or group; also known as an "IP core."

DIGITAL SIGNAL PROCESSING

Digital signal processors (DSPs) are microprocessors designed for a special function. Semiconductor intellectual property (SIP) blocks designed for use as DSPs generally have to work in a very low latency environment. They are constantly processing streams of video or audio. They also need to keep power consumption to a minimum, particularly in the case of mobile devices, which rely heavily on DSPs. To make this possible, DSPs are designed to work efficiently on both fixed-point arithmetic and the more computationally intensive floating-point arithmetic. The latter, however, is not needed in most DSP applications. DSPs tend to use chip architectures that allow them to fetch multiple instructions at once, such as the Harvard architecture. Many DSPs are required to accept analog data as input, convert this to digital data, perform some operation, and then convert the digital signals back to analog for output. This gives DSPs a pipelined architecture. They use a multistep process in which the output of one step is the input needed by the next step.

An example of the type of work performed by a DSP can be seen in a multiplier-accumulator. This is a piece of hardware that performs a two-step operation. First, it receives two values as inputs and multiplies one value by the other. Next, the multiplier-accumulator takes the result of the first step and adds it to the value stored in the accumulator. At the end, the accumulator's value can be passed along as output. Because DSPs rely heavily on multiplier-accumulator operations, these are part of the instruction set hardwired into such chips. DSPs must be able to carry out these operations quickly in order to keep up with the continuous stream of data that they receive.

The processing performed by DSPs can sometimes seem mysterious. However, in reality it often amounts to the performance of fairly straightforward mathematical operations on each value in the stream of data. Each piece of analog input is converted to a digital value. This digital value is then added to, multiplied by, subtracted from, or divided by another value. The result is a modified data stream that can then be passed to another process or generated as output.

APPLICATIONS OF DIGITAL SIGNAL PROCESSORS

There are many applications in which DSPs have become an integral part of daily life. The basic purpose of a DSP is to accept as input some form of analog information from the real world. This could include anything from an audible bird call to a live video-feed broadcast from the scene of a news event.

DSPs are also heavily relied upon in the field of medical imaging. Medical imaging uses ultrasound or other technologies to produce live imagery of what is occurring inside the human body. Ultrasound is often used to examine fetal developmental, from what position the fetus is in to how it is moving to how its heart is beating, and so on. DSPs receive the ultrasonic signals from the ultrasound equipment and convert it into digital data. This digital data can then be used to produce analog output in the form of a video display showing the fetus.

Digital signal processing is also critical to many military applications, particularly those that rely on the use of sonar or radar. As with ultrasound devices, radar and sonar send out analog signals in the form of energy waves. These waves bounce off features in the environment and back to the radar- or sonar-generating device. The device uses DSPs to receive this analog information and convert it into digital data. The data can then be analyzed and converted into graphical displays that humans can easily interpret. DSPs must be able to minimize delays in processing, because a submarine using sonar to navigate underwater cannot afford to wait to find out whether obstacles are in its path.

This is a block diagram for an analog-to-digital processing system. Digital signal processors are responsible for performing specified operations on digital signals. Signals that are initially analog must first be converted to digital signals in order for the programs in the digital processor to work correctly. The output signal may or may not have to be converted back to analog. EBSCO illustration.

BIOMETRIC SCANNING

A type of digital signal processing that many people have encountered at one time or another in their lives is the fingerprint scanner used in many security situations to verify one's identity. These scanners allow a person to place his or her finger on the scanner, and the scanner receives the analog input of the person's fingerprint. This input is then converted to a digital format and compared to the digital data on file for the person to see whether they match. As biometric data becomes increasingly important for security applications, the importance of digital signal processing will likely grow.

—*Scott Zimmer, JD*

BIBLIOGRAPHY

Binh, Le Nguyen. *Digital Processing: Optical Transmission and Coherent Receiving Techniques.* Boca Raton: CRC, 2013. Print.

Iniewski, Krzysztof. *Embedded Systems: Hardware, Design, and Implementation.* Hoboken: Wiley, 2013. Print.

Kuo, Sen M., Bob H. Lee, and Wenshun Tian. *Real-Time Digital Signal Processing: Fundamentals, Implementations and Applications.* 3rd ed. Hoboken: Wiley, 2013. Print.

Snoke, David W. *Electronics: A Physical Approach.* Boston: Addison, 2014. Print.

Sozański , Krzysztof. *Digital Signal Processing in Power Electronics Control Circuits.* New York: Springer, 2013. Print.

Tan, Li, and Jean Jiang. *Digital Signal Processing: Fundamentals and Applications.* 2nd ed. Boston: Academic, 2013. Print.

DIGITAL WATERMARKING

FIELDS OF STUDY

Computer Science; Digital Media; Security

ABSTRACT

Digital watermarking protects shared or distributed intellectual property by placing an additional signal within the file. This signal can be used to inform users of the copyright owner's identity and to authenticate the source of digital data. Digital watermarks may be visible or hidden.

PRINICIPAL TERMS

- **carrier signal:** an electromagnetic frequency that has been modulated to carry analog or digital information.
- **crippleware:** software programs in which key features have been disabled and can only be activated after registration or with the use of a product key.
- **multibit watermarking:** a watermarking process that embeds multiple bits of data in the signal to be transmitted.
- **noise-tolerant signal:** a signal that can be easily distinguished from unwanted signal interruptions or fluctuations (i.e., noise).
- **1-bit watermarking:** a type of digital watermark that embeds one bit of binary data in the signal to be transmitted; also called "0-bit watermarking."
- **reversible data hiding:** techniques used to conceal data that allow the original data to be recovered in its exact form with no loss of quality.

PROTECTING OWNERSHIP AND SECURITY OF DIGITAL DATA

Digital watermarking is a technique that embeds digital media files with a hidden digital code. It was first developed in the late twentieth century. These hidden codes can be used to record copyright data, track copying or alteration of a file, or prevent alteration or unauthorized efforts to copy a copyrighted file. Digital watermarking is therefore commonly used for copyright-protected music, video, and software downloads. Governments and banks also rely on it to ensure that sensitive documents and currency are protected from counterfeiting and fraud.

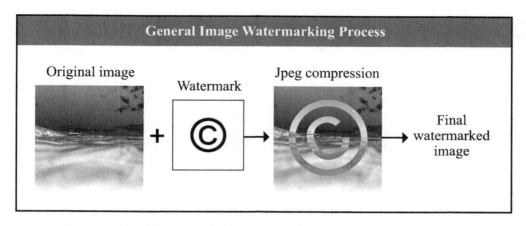

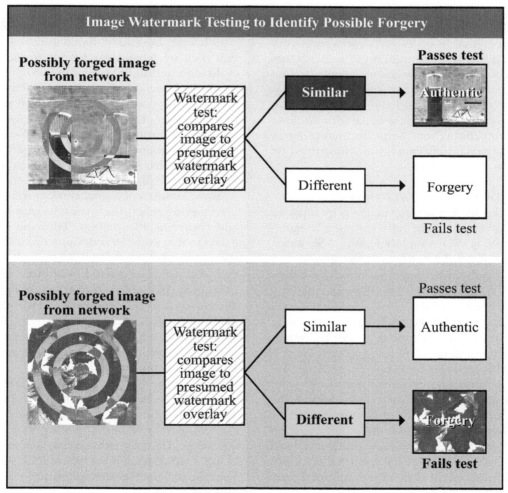

Diagrams of the watermarking process and a process for testing watermarks. EBSCO illustration.

BASICS OF WATERMARKING

A paper watermark is an image embedded within another image or a piece of paper. It can be seen by shining light on the image. Watermarks are used on banknotes, passports, and other types of paper documents to verify their authenticity. Similarly, digital watermarking involves embedding data within a digital signal in order to verify the authenticity of the signal or identify its owners. Digital watermarking was invented in the late 1980s or early 1990s. It uses techniques that are also used in steganography (the concealment of messages, files, or other types of data within images, audio, video, or even text).

Most digital watermarks are not detectable without an algorithm that can search for the signal embedded in the carrier signal. In order for a carrier signal to be watermarked, it must be tolerant of noise. Noise-tolerant signals are generally strong signals that resist degradation or unwanted modulation. Typically, digital watermarks are embedded in data by using an algorithm to encode the original signal with a hidden signal. The embedding may be performed using either public- or private-key encryption, depending on the level of security required.

QUALITIES OF DIGITAL WATERMARKS

One way to classify digital watermarks is by capacity, which measures how long and complex a watermarking signal is. The simplest type is 1-bit watermarking. This is used to encode a simple message that is meant only to be detected or not (a binary result of 1 or 0). In contrast, multibit watermarking embeds multiple bits of data in the original signal. Multibit systems may be more resistant to attack, as an attacker will not know how or where the watermark has been inserted.

Watermarks may also be classified as either robust or fragile. Robust watermarks resist most types of modification and therefore remain within the signal after any alterations, such as compression or cropping. These watermarks are often used to embed copyright information, as any copies of the file will also carry the watermark. Fragile watermarks cannot be detected if the signal is modified and are therefore used to determine if data has been altered.

In some cases, a digital watermark is designed so that users can easily detect it in the file. For instance, a video watermark may be a visible logo or text hovering onscreen during playback. In most cases, however, digital watermarks are hidden signals that can only be detected using an algorithm to retrieve the watermarking code. Reversible data hiding refers to cases in which the embedding of a watermark can be reversed by an algorithm to recover the original file.

APPLICATIONS OF DIGITAL WATERMARKING

A primary function of digital watermarking is to protect copyrighted digital content. Audio and video players may search for a digital watermark contained in a copyrighted file and only play or copy the file if it contains the watermark. This essentially verifies that the content is legally owned.

Certain types of programs, known colloquially as crippleware, use visible digital watermarks to ensure that they are legally purchased after an initial free evaluation period. Programs used to produce digital media files, such as image- or video-editing software, can often be downloaded for free so users can try them out first. To encourage users to purchase the full program, these trial versions will output images or videos containing a visible watermark. Only when the program has been registered or a product key has been entered will this watermark be removed.

Some creators of digital content use digital watermarking to embed their content with their identity and copyright information. They use robust watermarks so that even altered copies of the file will retain them. This allows a content owner to claim their work even if it has been altered by another user. In some cases, watermarked data can be configured so that any copies can be traced back to individual users or distributors. This function can be useful for tracing illegal distribution of copyrighted material. It can also help investigations into the unauthorized leaking of sensitive or proprietary files.

More recently, digital watermarking has been used to create hidden watermarks on product packaging. This is intended to make it easier for point-of-sale equipment to find and scan tracking codes on a product. Digital watermarking is also increasingly being used alongside or instead of regular watermarking to help prevent the counterfeiting of important identification papers, such as driver's licenses and passports.

—*Micah L. Issitt*

BIBLIOGRAPHY

Chao, Loretta. "Tech Partnership Looks beyond the Bar Code with Digital Watermarks." *Wall Street Journal*. Dow Jones, 12 Jan. 2016. Web. 14 Mar. 2016.

"Frequently Asked Questions." *Digital Watermarking Alliance*. DWA, n.d. Web. 11 Mar. 2016.

Gupta, Siddarth, and Vagesh Porwal. "Recent Digital Watermarking Approaches, Protecting Multimedia Data Ownership." *Advances in Computer Science* 4.2 (2015): 21–30. Web. 14 Mar. 2016.

Patel, Ruchika, and Parth Bhatt. "A Review Paper on Digital Watermarking and Its Techniques." *International Journal of Computer Applications* 110.1 (2015): 10–13. Web. 14 Mar. 2016.

Savage, Terry Michael, and Karla E. Vogel. *An Introduction to Digital Multimedia*. 2nd ed. Burlington: Jones, 2014. 256–58. Print.

"Unretouched by Human Hand." *Economist*. Economist Newspaper, 12 Dec. 2002. Web. 14 Mar. 2016.

DIRTY PAPER CODING

FIELD OF STUDY
Computer Science

ABSTRACT

Dirty paper coding is a technique that aims to maximize channel capacity. Communication channels experience a lot of interference. Through this technique, the receiver should receive the signal or message with minimal distortion. In such cases, receivers are unaware of the interference. Adoption and improvement of the technique ensure efficient data transmission with minimal power requirements.

PRINICIPAL TERMS

- **Additive White Gaussian Noise (AWGN):** a model used to represent imperfections in real communication channels.
- **channel capacity:** the upper limit for the rate at which information transfer can occur without error.
- **interference:** anything that disrupts a signal as it moves from source to receiver.
- **noise:** interferences or irregular fluctuations affecting electrical signals during transmission.
- **precoding:** a technique that uses the diversity of a transmission by weighting an information channel.
- **signal-to-noise ratio (SNR):** the power ratio between meaningful information, referred to as "signal," and background noise.

EFFECTIVE DATA TRANSMISSION

Data encounters interference during transmission from source to receiver on a channel. Dirty paper coding (DPC) is used in channels subjected to interference. Using DPC on such channels helps achieve efficient data transmission by ensuring channel capacity. Efficient transmission is possible despite interference because DPC uses precoding. This is a technique that minimizes data's vulnerability to distortion before reaching the receiver. The idea for DPC originated with Max Costa in 1983. Costa compared data transmission to sending a message on paper. The paper gets dirtier along the way before it reaches the intended recipient, who cannot distinguish ink from dirt. Apart from achieving channel capacity, DPC works without additional power requirements and without the receiver being aware of the interference.

ELIMINATING INTERFERENCE

Data transmission via communication systems always faces interference. Through DPC, it is possible to eliminate as much interference as possible. Noise is a type of electrical interference. Applying Additive White Gaussian Noise (AWGN) provides an effective way to develop channels with minimal distortion rates and corrupted signals. In some contexts, AWGN is referred to as a "noise removal algorithm." The model is "additive" because it is added to noise affecting a channel. It is "white" to denote uniform power across

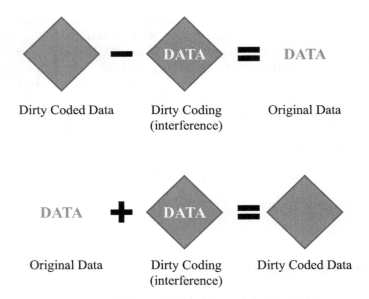

Dirty Coded Data   Dirty Coding (interference)   Original Data

Original Data   Dirty Coding (interference)   Dirty Coded Data

Dirty paper coding is named for the conceptual analogy of deciphering writing on paper that has had known ink splotches added to it. In data transmission, the "ink splotches" are data interference, but someone who knows the interference is there can remove it to see the original data that was covered. EBSCO illustration.

a channel's frequency band. It is Gaussian because of its normal distribution within the time domain.

AWGN is used to simulate distortions facing a channel to make it efficient, a concept that DPC proposes. Models like AWGN have helped developers create multiuser channels with multiple-antenna transmitters. Implementing DPC techniques ensures each user encounters no interference from others in such multiuser channels.

INFORMATION INTEGRITY
In wireless infrastructures, DPC helps improve performance. Improvements have contributed to the development of multicarrier hybrid systems that combine unicast and broadcast connectivity. Implementing architectures that use DPC allows reception of interference-free signals. Thus, cellular, television, and radio signals that use unicast and broadcast connectivity are becoming clearer with each improvement.

DPC has also found its way into information hiding, or "digital watermarking." In that process, an encoded message is sneaked into a waveform using an unknown signal. DPC ensures the receiver can decode the message. The technique also minimizes

distortion to the original message and required power. Attackers with no knowledge of the encoding and signal parameters cannot decode the message. Removing the watermark also requires the hidden parameters.

Military communications apply watermarking. Parameters used in the process become classified information and are only accessible by authorized individuals. Watermarking helps safeguard information integrity.

DIRTY PAPER CODING'S CONTRIBUTION
Despite challenges in the calculations required, DPC has formed the basis for the development of new techniques for achieving efficiency in data transmission and information systems. One example is zero-forcing dirty paper coding. Efficient channels have fewer power demands, minimal message distortion, secure transmissions, and high signal-to-noise ratios (SNR). The combination of such advantages presages the development of affordable wireless infrastructures with fast and reliable data transmission rates. With the growing number of mobile devices accessing wireless networks, such infrastructure improvement might provide one of the solutions to the low battery life of the devices. Models on DPC, such as DPC with phase reshaping (PDC-PR), can help address the problem of resource allocation facing unicast and broadcast systems. However, performance comparisons between different DPC modifications can help establish the best technique to adopt.

—Melvin O

BIBLIOGRAPHY
Cox, Ingemar J., Jessica Fridrich, Matthew L. Miller, Jeffrey A. Bloom, and Ton Kalker. "Practical Dirty-Paper Codes." *Digital Watermarking and Steganography.* 2nd ed. Amsterdam: Elsevier, 2008. 183–212. Digital file.

Devroye, Natasha, Patrick Mitran and Vahid Tarokh. *On Cognitive Graphs: Decomposing Wireless Networks.* New York: Wiley Interscience, 2006. Print.

Devroye, N., P. Mitran, and V. Tarokh. "Limits on Communications in a Cognitive Radio Channel." *IEEE Communication Magazine* 44.6 (2006): 4449. *Inspec.* Web. 9 Mar. 2016.

Habiballah, N., M. Qjani, A. Arbaoui, and J. Dumas. "Effect of a Gaussian White Noise on the Charge Density Wave Dynamics in a One Dimensional Compound." *Journal of Physics and Chemistry of Solids* 75.1 (2014): 153–56. *Inspec.* Web. 9 Mar. 2016.

Kilper, Daniel C., and Tucker, Rodney S. "Energy-Efficient Telecommunications." *Optical Fiber Telecommunications.* 6th ed. N.p.: Elsevier, 2013. 747–91. Digital file.

Savischenko, Nikolay V. *Special Integral Functions Used in Wireless Communications Theory.* N.p.: World Scientific, 2014. Digital file.

SAMPLE PROBLEM

Given a bandwidth of 10 megabytes per second (MBps) and using an acceptable signal-to-noise ratio (SNR) of 25 decibels (dB), calculate the channel capacity (C) in bits per second (bps), using the following formula:

$$C = B \log2(1 + S/N)$$

where B is the bandwidth, S is the signal power, and N is the noise power.

Answer:

First, convert the SNR in decibels to power, using the following equation:

$$SNRdB = 10 \log10(S/N)$$
$$25\ dB = 10 \log10(S/N)$$
$$2.5 = \log10(S/N)$$
$$102.5 = S/N$$
$$316 = S/N$$

Next, convert the bandwidth from megabytes per second (MBps) to bits per second (bps). Recall that 1 byte is equal to 8 bits, and ignore the seconds for now:

$$1\ MB = 16\ B$$
$$= 8\ b/B \times 16\ B$$
$$= 86\ b$$

Then, plug in the found values for the bandwidth and S/N into the given formula, and solve:

$$C = B \log2(1 + S/N)$$
$$C = 86\ bps \cdot \log2(1 + 316)$$
$$C = 86\ (\log2(317))$$
$$C = 86 \times 8.31$$
$$C = 6.648 \times 107\ bps$$

DOS

FIELDS OF STUDY

Computer Science; Information Technology; Operating Systems

ABSTRACT

The term DOS is an acronym for "disk operating system." DOS refers to any of several text-based operating systems that share many similarities. Perhaps the best-known form of DOS is MS-DOS. MS-DOS is an operating system developed by the technology company Microsoft. It is based heavily on an earlier operating system known as QDOS. DOS was largely replaced by operating systems featuring graphical user interfaces by the end of the twentieth century. It continues to be used in certain specialized contexts in the twenty-first century, however.

PRINICIPAL TERMS

- **command line:** a text-based computer interface that allows the user to input simple commands via a keyboard.
- **graphical user interface (GUI):** an interface that allows users to control a computer or other device by interacting with graphical elements such as icons and windows.
- **nongraphical:** not featuring graphical elements.
- **shell:** an interface that allows a user to operate a computer or other device.

BACKGROUND ON DOS

"Disk operating system" (DOS) is a catchall term for a variety of early operating systems designed for personal computers (PCs). Operating systems have existed since the early days of computers. They became more important in the late 1970s and early 1980s. At that time PCs became available to the public. An operating system allows users unfamiliar with computer programming to interface directly with these devices. Among the first companies to offer PCs for sale was the technology company IBM, a long-time leader in the field of computer technology.

In 1980 IBM sought to license an operating system for its new device, the IBM PC. IBM approached the company Digital Research to license its Control Program for Microcomputers (CP/M). However, the two companies were unable to come to an agreement. IBM made a deal with the software company Microsoft instead. Microsoft had been founded in 1975. It initially specialized in creating and licensing computer programs and programming languages. In order to supply IBM with an operating system, Microsoft licensed the Quick and Dirty Operating System (QDOS) from the company Seattle Computer Products. This system, later renamed 86-DOS, was based in part on CP/M and used similar commands but differed in several key ways. Notably, the operating system was designed to be used with computers, such as the IBM PC, that featured sixteen-bit microprocessors, which could process sixteen bits, or

pieces of binary information, at a time. Microsoft also hired QDOS creator Tim Patterson. Microsoft asked Patterson to create a new version of QDOS to be licensed to IBM under the name PC-DOS. Microsoft went on to develop an essentially identical version called MS-DOS, which the company sold and licensed to computer manufacturers itself.

As IBM's PCs became increasingly popular, competing hardware manufacturers created similar computers. Many of these computers, commonly known as "IBM clones," used MS-DOS. In addition to MS-DOS, PC-DOS, and 86-DOS, other DOS or DOS-compatible systems entered the market over the decades. However, MS-DOS dominated the market and was often referred to simply as DOS.

UNDERSTANDING DOS

MS-DOS is the most popular and best-known form of DOS. MS-DOS consists of three key parts: the input/output (I/O) handler, the command processor, and the auxiliary utility programs. The I/O handler manages data input and output and consists of two programs, IO.SYS and MSDOS.SYS. The command processor enables the computer to take commands from the user and carry them out. The most commonly used commands and associated routines are stored in the command processor. Others are stored on the system disk and loaded as needed. Those routines are known as "auxiliary utility programs."

Like all operating systems, DOS functions as a shell. A shell is an interface that allows a user to operate a computer without needing knowledge of computer programming. A DOS or DOS-compatible system does require users to enter text-based commands. These are relatively simple and limited in number, however. As such, the public found PCs featuring the early forms of DOS fairly easy to use.

USING DOS

MS-DOS and later DOS and DOS-compatible systems are in many ways quite similar to those systems that came before, such as QDOS and CP/M. In general, MS-DOS and similar systems are nongraphical systems. Thus, they do not contain graphical elements found in later

```
Welcome to FreeDOS

CuteMouse v1.9.1 alpha 1 [FreeDOS]
Installed at PS/2 port
C:\>ver

FreeCom version 0.82 pl 3 XMS_Swap [Dec 10 2003 06:49:21]

C:\>dir
 Volume in drive C is FREEDOS_C95
 Volume Serial Number is 0E4F-19EB
 Directory of C:\

FDOS                 <DIR>    08-26-04   6:23p
AUTOEXEC BAT           435    08-26-04   6:24p
BOOTSECT BIN           512    08-26-04   6:23p
COMMAND  COM        93,963    08-26-04   6:24p
CONFIG   SYS           801    08-26-04   6:24p
FDOSBOOT BIN           512    08-26-04   6:24p
KERNEL   SYS        45,815    04-17-04   9:19p
         6 file(s)       142,038 bytes
         1 dir(s)   1,064,517,632 bytes free

C:\>_
```

The FreeDOS command line interface is based on the original DOS (disk operating system) command line interface to provide individuals with an alternative to the more prevalent graphical user interface available with most operating systems. Public domain, via Wikimedia Commons.

graphical user interfaces (GUIs), such as clickable icons and windows. Instead, nongraphical systems allow the user to operate the computer by entering commands into the command line. By entering basic commands, the user can instruct the computer to perform a wide range of functions, including running programs and opening or copying files.

Impact of DOS

Although DOS and other nongraphical systems were largely phased out by the mid-1990s as GUIs became more popular, they remained an influential part of the development of PC operating systems. Some graphical operating systems, such as Microsoft's Windows 95, were based in part on DOS, and had the ability to run DOS programs when opened in a specialized mode. Despite advances in computer technology, MS-DOS and similar systems are still used in the twenty-first century for applications. In some cases, companies or institutions continue to use the operating system in order to maintain access to software compatible only with DOS. Some individuals use DOS and DOS-compatible systems to play early games originally designed for those systems. Companies such as Microsoft no longer sell DOS to the public. However, a number of companies and organizations are devoted to providing DOS-compatible systems to users, often in the form of open-source freeware.

—*Joy Crelin*

Bibliography

Doeppner, Thomas W. *Operating Systems in Depth.* Hoboken: Wiley, 2011. Print.

Gallagher, Sean. "Though 'Barely an Operating System,' DOS Still Matters (to Some People)." *Ars Technica.* Condé Nast, 14 July 2014. Web. 31 Jan. 2016.

McCracken, Harry. "Ten Momentous Moments in DOS History." *PCWorld.* IDG Consumer, n.d. Web. 31 Jan. 2016.

Miller, Michael J. "The Rise of DOS: How Microsoft Got the IBM PC OS Contract." *PCMag.com.* PCMag Digital Group, 10 Aug. 2011. Web. 31 Jan. 2016.

"MS-DOS: A Brief Introduction." *Linux Information Project.* Linux Information Project, 30 Sept. 2006. Web. 31 Jan. 2016.

"Part Two: Communicating with Computers—The Operating System." *Computer Programming for Scientists.* Oregon State U, 2006. Web. 31 Jan. 2016.

Shustek, Len. "Microsoft MS-DOS Early Source Code." *Computer History Museum.* Computer History Museum, 2013. Web. 31 Jan. 2016.

DRONES

FIELDS OF STUDY

Computer Engineering; Robotics; Privacy

ABSTRACT

Drones are unmanned aerial vehicles (UAVs). Unlike other UAVs, drones are at least partly automated but may also respond to remote piloting. Drones are used in military operations. They are also used for commercial applications like aerial photography and filmmaking, search and rescue, and environmental research. The concept of unmanned aerial vehicles for combat use emerged in the 1800s. Advancements in computer technology and engineering have made drones commonplace in the 2010s for both military and civilian applications.

PRINICIPAL TERMS

- **actuators:** motors designed to control the movement of a device or machine by transforming potential energy into kinetic energy.
- **communication devices:** devices that allow drones to communicate with users or engineers in remote locations.
- **field programmable gate array:** an integrated circuit that can be programmed in the field and can therefore allow engineers or users to alter a machine's programming without returning it to the manufacturer.
- **sensors:** devices capable of detecting, measuring, or reacting to external physical properties.

- **telemetry:** automated communication process that allows a machine to identify its position relative to external environmental cues.
- **unmanned aerial vehicle (UAV):** an aircraft that does not have a pilot onboard but typically operates through remote control, automated flight systems, or preprogrammed computer instructions.

DRONE DEVELOPMENT

The term "drone" refers to any of a large number of semi-independent flying vehicles, or unmanned aerial vehicles (UAVs). Military organizations began using UAVs in the form of balloons and kites to conduct aerial surveillance and air strikes in the nineteenth century. Radio-controlled UAVs were used in World War II. In the 1990s advancements in computer technology, sensor capacity, and micro-engineering led to the development of the first true drones. True drones are remote-controlled vehicles that have some level of automation or self-control. By 2016, nations around the world were using drones for military operations. Drone technology had also become popular for nonmilitary uses, including photography, filmmaking, ecological and wildlife research, and search-and-rescue operations.

HOW DRONES WORK

While remote-controlled helicopters, planes, and other flying vehicles have been available to hobbyists since the mid-twentieth century, remote-controlled vehicles that qualify as drones differ by having some degree of automation. Most are partly automated and still need some control from a pilot. Drones may be piloted using web-based signals, radio waves, or Bluetooth connections. More advanced drones can automatically avoid obstacles, regulate flight path, and even plot new paths.

Drones typically have sensors that allow the drone to detect, record, transmit, and respond to environmental variables. For instance, drones with high-definition cameras can collect visual data about the environment. Some advanced drones can use this data to adjust their flight path, speed, and elevation. They may also navigate toward certain visual targets. Drones also typically have telemetry equipment. Such equipment collects and transmits data about the drone's location relative to other environmental features. With telemetry and sensors, drone pilots can determine the drone's location. They can then direct the drone toward a target location. Drones may also have communications devices. These devices transmit data back to users and allow them to adjust the drone's activities. Some advanced drones have voice-command software that allows the drone to respond to a user's verbal commands.

Most drones are powered by batteries. The batteries send electrical energy to actuators, engines that use the energy to create movement. This movement powers rotors, jet turbines, and flight control mechanisms. Battery life is one of the major challenges in drone engineering. Long-lasting batteries are also typically heavy and so can only be used in larger drones. Engineers are designing a new generation of solar-electric hybrid drones. These hybrid drones can get additional charge from sunlight, thus greatly increasing the duration of drone flights.

Unmanned aerial vehicles, more commonly known as drones, come in sizes ranging from a small toy to a large plane. They are used by the military for surveillance as well as by civilians for recreation. By Gerald L. Nino, public domain, via Wikimedia Commons.

DRONE APPLICATIONS

Many companies producing drones for the commercial market use open-source software. This allows the same basic software to be altered for a variety of applications. Many companies also outfit drones with adjustable hardware, such as field programmable gate arrays. These arrays allow users to adjust hardware programming in the field without returning the drone to the manufacturer. This flexibility in hardware and software design means that drones can be altered for applications ranging from scientific research to art projects.

Drones have been used by law enforcement agencies for aerial search and surveillance. They have also been used for search-and-rescue operations. Drones with high-definition video equipment onboard have become a popular tool in journalism and filmmaking. Drones can take aerial shots that once would have been impossible without helicopters. Wildlife and ecological researchers have also used UAVs to take aerial footage. Drones with specialized equipment can be used to collect detailed topographical and ecological data or to monitor weather patterns. Some engineers think that the development of solar-electric hybrid drones may be the key to future weather monitoring. Such hybrid drones may one day replace the use of low-Earth orbit satellites for signal transmission and other applications.

CONTROVERSY AND REGULATION

The military use of drones, especially for lethal operations, is among the most controversial issues of the twenty-first century. Drone strikes have been linked to civilian casualties and accidental deaths. Opponents of drone strikes argue that automating military operations is a violation of moral and ethical principles. Another controversy surrounding UAVs is the potential for the technology to be misused for unauthorized or illegal surveillance. Such surveillance would be a threat to personal privacy. In the United States, the Federal Aviation Authority (FAA) is investigating potential regulations for commercial and recreational drone use. State legislators have passed laws that prohibit the use of drones over private property. Governmental regulation of private drone use is still in its infancy. The debate over the legality and potential for misuse has spread around the world, however.

—*Micah Issitt*

BIBLIOGRAPHY

Couts, Andrew. "Drones 101: A Beginner's Guide to Taking Flight, No License Needed." *Digital Trends.* Designtechnica, 16 Nov. 2013. Web. 27 Jan. 2015.

Fowler, Geoffrey, A. "The Drones on Autopilot That Follow Your Lead (Usually)." *Wall Street Journal.* Dow Jones, 23 Dec. 2014. Web. 20 Jan. 2016.

Moynihan, Tim. "Things Will Get Messy If We Don't Start Wrangling Drones Now." *Wired.* Condé Nast, 30 Jan. 2015. Web. 30 Jan. 2016.

Pullen, John Patrick. "This Is How Drones Work." *Time.* Time, 3 Apr. 2015. Web. 27 Jan. 2016.

Stanley, Jay. "'Drones' vs 'UAVs'—What's behind a Name?" *ACLU.* ACLU, 20 May 2013. Web. 27 Jan. 2016.

"Unmanned Aircraft Systems (UAS) Frequently Asked Questions." *Federal Aviation Administration.* US Dept. of Transportation, 18 Dec. 2015. Web. 11 Feb. 2016.

E

ELECTRONIC CIRCUITS

FIELD OF STUDY

Computer Engineering

ABSTRACT

Electronic circuits actively manipulate electric currents. For many years these circuits have been part of computer systems and important home appliances such as televisions. Innovation has helped develop effective circuits such as integrated circuits, which are power efficient, small, and powerfully capable. Integrated circuits form the basic operational units of countless everyday gadgets.

PRINICIPAL TERMS

- **BCD-to-seven-segment decoder/driver:** a logic gate that converts a four-bit binary-coded decimal (BCD) input to decimal numerals that can be output to a seven-segment digital display.
- **counter:** a digital sequential logic gate that records how many times a certain event occurs in a given amount of time.
- **inverter:** a logic gate whose output is the inverse of the input; also called a NOT gate.
- **negative-AND (NAND) gate:** a logic gate that produces a false output only when both inputs are true
- **programmable oscillator:** an electronic device that fluctuates between two states that allows user modifications to determine mode of operation.
- **retriggerable single shot:** a monostable multivibrator (MMV) electronic circuit that outputs a single pulse when triggered but can identify a new trigger during an output pulse, thus restarting its pulse time and extending its output.

Electric versus Electronic Circuits

Electrical circuits have developed over the years since the discovery of the Leyden jar in 1745. An electrical circuit is simply a path through which electric current can travel. Its primary components include resistors, inductors, and capacitors. The resistor controls the amount of current that flows through the circuit. It is so called because it provides electrical resistance. The inductor and the capacitor both store energy. Inductors store energy in a magnetic field, while capacitors store it in the electric field. The Leyden jar was the first capacitor, designed to store static electricity.

Electronic circuits are a type of electrical circuit. However, they are distinct from basic electrical circuits in one important respect. Electrical circuits passively conduct electric current, while electronic circuits actively manipulate it. In addition to the passive components of an electrical circuit, electronic circuits also contain active components such as transistors and diodes. A transistor is a semiconductor that works like a switch. It can amplify an electronic signal or switch it on or off. A diode is a conductor with very low electrical resistance in one direction and very high resistance in the other. It is used to direct the flow of current.

Integrated Circuits

The most important advance in electronic circuits was the development of the integrated circuit (IC) in the mid-twentieth century. An IC is simply a semiconductor chip containing multiple electronic circuits. Its development was enabled by the invention of the transistor in 1947. Previously, electric current was switched or amplified through a vacuum tube. Vacuum tubes are much slower, bulkier, and less efficient than transistors. The first digital computer,

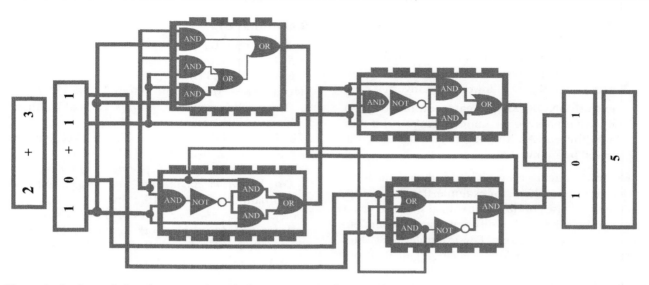

Electronic circuits are designed to use a series of logic gates to send a charge through the circuit in a particular manner. These logic gates control the charge output and thus the output of the circuits. In this example, the circuit is designed to add two binary numbers together using a series of AND, OR, and NOT commands to determine the route of the charge and the resulting output from each circuit component. EBSCO illustration.

ENIAC (Electronic Numerical Integrator and Computer), contained about eighteen thousand vacuum tubes and weighed more than thirty tons. Once the transistor replaced the vacuum tube, electronic circuits could be made much smaller.

By 1958, several scientists had already proposed ideas for constructing an IC. That year, Jack Kilby, a scientist at Texas Instruments, was the first to put the idea into practice. He designed a chip constructed entirely from a single block of semiconductor material. Because there were no individual components, the circuits did not have to be large enough to assemble manually. The number of circuits in the chip was limited only by the number of transistors that could fit in it. Early ICs contained only a few transistors each. By the twenty-first century, the maximum possible number of transistors per IC was in the billions.

LOGIC GATES

Active manipulation of electric current is accomplished through logic gates. A logic gate is an electronic circuit that implements a Boolean function. Broadly speaking, a Boolean function is a function that produces one of two potential outputs—either 0 or 1—based on a given rule. Because transistors work as switches, which can

take one of two values (e.g., "on" or "off"), they are ideal for implementing logic gates. Most logic gates take in two inputs and produce a single output.

There are seven basic types of logic gates: AND, OR, NOT, XOR, NAND, NOR, and XNOR. These logic gates only accept two input values, 0 and 1, which represent "false" and "true" respectively. They are distinguished from one another based on what output is produced by each combination of inputs:

AND gate: output is only true (1) if both inputs are true; otherwise, output is false (0).

OR gate: output is false (0) only if both outputs are false; otherwise, output is true (1).

NOT gate: output is true (1) if input is false (0), and vice versa. A NOT gate is also called an **inverter**, because it takes in only one input and outputs the inverse.

exclusive-OR (XOR) gate: output is true (1) if the inputs are different, that is, if only one input is true; if both inputs are the same, output is false (0).

negative-AND (NAND) gate: output is false (0) if all inputs are true (1); otherwise output is true. A NAND gate is essentially an AND gate followed by an inverter

NOR gate: output is true (1) only if both inputs are false (0); otherwise, output is false. A NOR gate is an OR gate followed by an inverter.

exclusive-NOR (XNOR) gate: output is true (1) if both inputs are the same and false (0) if they are different. An XNOR gate is a XOR gate followed by an inverter.

Electronic circuits transmit binary data in the form of electric pulses, where, for example, 0 and 1 are represented by pulses of different voltages. These seven gates can be combined in different ways to complete more complex operations. For example, a BCD-to-seven-segment decoder/driver is a logic gate that converts binary data from a counter to a decimal number display. The "seven segment" refers to the number display system common in digital clocks and other devices, where each numeral is represented by up to seven short parallel or perpendicular line segments. Another complex circuit is a retriggerable single shot. This is a type of time-delay relay circuit that can generate an output pulse of a predetermined length and then extend the output indefinitely if the input is repeated. The purpose of this is to change the timing of another circuit, such as a programmable oscillator.

LIFE WITHOUT INTEGRATED CIRCUITS

Whether in home appliances, computer systems, or mobile devices, electronic circuits make modern life possible. Without advanced electronic circuits such as ICs, personal computers and small, portable electronic devices could not exist. Despite improvements over the years, ICs have maintained their silicon-based design. Scientists predict that the only thing to replace ICs would be a new kind of biologically based circuit technology.

—Melvin O

BIBLIOGRAPHY

Frenzel, Louis E., Jr. *Electronics Explained: The New Systems Approach to Learning Electronics.* Burlington: Elsevier, 2010. Print.

Harris, David Money, and Sarah L. Harris. *Digital Design and Computer Architecture.* 2nd ed. Waltham: Morgan, 2013. Print.

"The History of the Integrated Circuit." *Nobelprize.org.* Nobel Media, 2014. Web. 31 Mar. 2016.

Kosky, Philip, et al. *Exploring Engineering: An Introduction to Engineering and Design.* 4th ed. Waltham: Academic, 2016. Print.

Tooley, Mike. *Electronic Circuits: Fundamentals and Applications.* 4th ed. New York: Routledge, 2015. Print.

Wilson, Peter. *The Circuit Designer's Companion.* 3rd ed. Waltham: Newnes, 2012. Print.

SAMPLE PROBLEM

Determine the output of a NAND logic gate for all possible combinations of two input values (0 and 1).

Answer:
The combination of an AND gate and a NOT gate forms a NAND logic gate. The output of each input combination is the inverse of the AND gate output. The NAND gate accepts four possible combinations of two inputs and produces outputs as shown:

$$0,0 = 1$$
$$0,1 = 1$$
$$1,0 = 1$$
$$1,1 = 0$$

ELECTRONIC COMMUNICATION SOFTWARE

FIELDS OF STUDY

Information Systems; Information Technology

ABSTRACT

Electronic communication software is used to transfer information via the Internet or other transmission-and-reception technology. As technology has evolved, electronic communication software has taken on many new forms, from text-based instant messaging using computers to SMS messages sent between cell phones on opposite sides of the world. Electronic communication software allows people to communicate in real time using audio and video and to exchange digital files containing text, photos, and other data.

PRINICIPAL TERMS

- **Electronic Communications Privacy Act (ECPA):** a regulation enacted in 1986 to limit the ability of the US government to intrude upon private communications between computers.
- **multicast:** a network communications protocol in which a transmission is broadcast to multiple recipients rather than to a single receiver.
- **push technology:** a communication protocol in which a messaging server notifies the recipient as soon as the server receives a message, instead of waiting for the user to check for new messages.
- **Short Message Service (SMS):** the technology underlying text messaging used on cell phones.
- **voice over Internet Protocol (VoIP):** a set of parameters that make it possible for telephone calls to be transmitted digitally over the Internet, rather than as analog signals through telephone wires.

ASYNCHRONOUS COMMUNICATION

Many types of electronic communication software are asynchronous. This means that the message sender and the recipient communicate with one another at different times. The classic example of this type of electronic communication software is e-mail. E-mail is asynchronous because when a person sense a message, it travels first to the server and then to the recipient. The server may use push technology to notify the recipient that a message is waiting. It is then up to the recipient to decide when to retrieve the message from the server and read it.

E-mail evolved from an earlier form of asynchronous electronic communication: the bulletin-board system. In the 1980s and earlier, before the Internet was widely available to users in their homes, most people went online using a dial-up modem. A dial-up modem is a device that allows a computer to connect to another computer through a telephone line. To connect, the first computer dials the phone number assigned to the other computer's phone line. Connecting in this way was slow and cumbersome compared to the broadband Internet access common today. This was in part because phone lines could only be used for one purpose at a time, so a user could not receive phone calls while online. Because users tended to be online only in short bursts, they would leave messages for each other on online bulletin-board systems (BBSs). Similar to e-mail, the message would stay on the BBS until its recipient logged on and saw that it was waiting.

Another popular method of asynchronous communication is text messaging. Text messaging allows short messages to be sent from one mobile phone to another. The communications protocol technology behind text messages is called Short Message Service (SMS). SMS messages are limited to 160 characters. They are widely used because they can be sent and received using any kind of cell phone.

SYNCHRONOUS COMMUNICATION

Other forms of electronic communication software allow for synchronous communication. This means that both the recipient and the sender interact through a communications medium at the same time. The most familiar example of synchronous communication is the telephone, and more recently the cell phone. Using either analog protocols or voice over Internet Protocol (VoIP), users speak into a device. Their speech is translated into electronic signals by the device's communication software and then transmitted to the recipient. There are sometimes minor delays due to network latency. However, most of the conversation happens in the same way it would if the parties were face to face.

Electronic communication software has many forms to satisfy many uses. The most popular social media and communication companies implement programming that provides users with attributes they deem important for electronic communication, such as private and/or public sharing and posting to a network, saving, contributing, subscribing, and commenting across a number of formats. EBSCO illustration.

Another form of electronic synchronous communication is instant messaging or chat. Instant messaging occurs when multiple users use computers or mobile devices to type messages to one another. Each time a user sends a message, the recipient or recipients see it pop up on their screens. Chat can occur between two users, or it can take the form of a multicast in which one person types a message and multiple others receive it.

Multicast can also be delivered asynchronously. An example of this type of electronic communication is a performer who records a video of themselves using a digital camera and then posts the video on an online platform such as YouTube. The video would then stay online, available for others to watch at any time, until its creator decided to take it down. This type of electronic communication is extremely popular because viewers do not have to be online at a particular time in order to view the performance, as was the case with television broadcasts in the past.

PRIVACY CONCERNS
In some respects the rapid growth of electronic communication software caught regulators off guard. There were many protections in place to prevent the

government from eavesdropping on private communications using the telephone. However similar protections for electronic communications were lacking until the passage of the Electronic Communications Privacy Act (ECPA) in the late 1980s. This act extended many traditional communication protections to VoIP calls, e-mails, SMS messages, chat and instant messaging logs, and other types of communications.

CUTTING EDGE
Some of the newest forms of electronic communication software are pushing the boundaries of what is possible. One example of this is video calling using cell phones. This technology is available in many consumer devices, but in reality its utility is often limited by the amount of bandwidth available in some locations. This causes poor video quality and noticeable delays in responses between users.

—*Scott Zimmer, JD*

BIBLIOGRAPHY
Bucchi, Massimiano, and Brian Trench, eds. *Routledge Handbook of Public Communication of Science and Technology.* 2nd ed. New York: Routledge, 2014. Print.

Cline, Hugh F. *Information Communication Technology and Social Transformation: A Social and Historical Perspective.* New York: Routledge, 2014. Print.

Gibson, Jerry D., ed. *Mobile Communications Handbook.* 3rd ed. Boca Raton: CRC, 2012. Print.

Gillespie, Tarleton, Pablo J. Boczkowski, and Kirsten A. Foot, eds. *Media Technologies: Essays on Communication, Materiality, and Society.* Cambridge: MIT P, 2014. Print.

Hart, Archibald D., and Sylvia Hart Frejd. *The Digital Invasion: How Technology Is Shaping You and Your Relationships.* Grand Rapids: Baker, 2013. Print.

Livingston, Steven, and Gregor Walter-Drop, eds. *Bits and Atoms: Information and Communication Technology in Areas of Limited Statehood.* New York: Oxford UP, 2014. Print.

ELECTRONIC WASTE

FIELDS OF STUDY

Computer Science; Computer Engineering; Information Technology

ABSTRACT

This article discusses the growing problem of waste in society and goes into the specific issues and challenges surrounding the problem of discarded electronic devices. Electronic waste poses a significant threat to the environment, and various efforts to remediate the issue have been proposed.

PRINICIPAL TERMS

- **cathode ray tube (CRT):** a vacuum tube used to create images in devices such as older television and computer monitors.
- **commodities:** consumer products, physical articles of trade or commerce.
- **Environmental Protection Agency (EPA):** US government agency tasked with combating environmental pollution.
- **heavy metal:** one of several toxic natural substances often used as components in electronic devices.
- **Phonebloks:** a concept devised by Dutch designer Dave Hakkens for a modular mobile phone intended to reduce electronic waste.
- **planned obsolescence:** a design concept in which consumer products are given an artificially limited lifespan, therefore creating a perpetual market.

TECHNOLOGY AND WASTE

One of the defining aspects of the twentieth century was technological advancement. Telephones, televisions, computers, and other electronic devices evolved rapidly and became common throughout the world. This growth continued and even accelerated into the twenty-first century, allowing for easier communication and enabling humanity to see itself in a larger, more global context.

However, these advancements have come at a cost. New communication technologies produce an ever-expanding mass of electronic waste, or e-waste, as electronics break or become obsolete. The immense volume of electronic devices compounds traditional waste-disposal challenges. Many such products present serious pollution problems, especially in the developing world. Though many groups have noted and taken action on the issue, e-waste is a global problem that requires widespread cooperation from the individual to the government level.

WHAT IS E-WASTE?

Simply put, e-waste is anything electronic that has been discarded. This includes common consumer items such as televisions, computers, microwave ovens, and mobile phones. Some experts further categorize e-waste into three groups: unwanted items that still work or can be repaired, items that can be harvested for scrap materials, and true waste to be dumped. The first two types are considered commodities that may be bought and sold for recycling purposes.

E-waste is a major problem for several reasons. Many electronics use parts made from toxic substances, especially heavy metals. If items are disposed of improperly, these substances can leak into the environment, causing widespread damage. For example, lead used in cathode ray tubes (CRTs) can contaminate water and poison people and animals. Even the recycling process itself creates pollution, as electronics are exposed to chemicals or burned in order to separate valuable elements such as gold and copper. For this reason, various environmental organizations consider some electronics, such as CRTs, to be household hazardous waste. The US Environmental Protection Agency (EPA) has special regulations for disposing of such materials. However, most e-waste is shipped to recycling centers and dumps in developing nations. Many of these countries lack strong environmental regulations, and e-waste pollution has become a major problem for them.

One reason for the huge amount of e-waste is the popularity of planned obsolescence in the technology industry. In sales-driven industries, product durability eventually becomes a liability. Companies can earn greater profits by using the cheapest possible materials and enticing consumers to upgrade as often as possible, thus generating more and more e-waste. The rapid evolution and improvement of computer

As technology speeds forward, older products become obsolete more quickly. Electronic waste (e-waste) materials are not easily degradable, and the amount of waste accumulating in landfills is increasing. By Volker Thies, CC-BY-SA-3.0 (http://creativecommons.org/licenses/by-sa/3.0/), via Wikimedia Commons

of a new device. Both production and disposal of cell phones cause significant environmental problems.

Various methods have been proposed to address the issue. Dutch designer Dave Hakkens created the Phonebloks concept to combat planned obsolescence. Phonebloks features a modular design for mobile phones, with interlocking parts connected on a frame. If, for example, the camera breaks, a new camera block can be installed, rather than replacing the whole phone. Consumers can upgrade and customize their phones as they choose. The theory behind Phonebloks attracted corporate and public interest, bringing attention to the problem of e-waste. However, some critics suggested the idea would be very difficult to implement. It could also potentially end up increasing e-waste through constant upgrading.

Public awareness is critical to reducing waste. Many people do not realize their electronics are such a problem after being thrown away. The United Nations (UN) founded the Step (Solving the E-waste Problem) Initiative to coordinate anti-waste efforts at all levels. Others have looked into ways to tackle e-waste already in the environment. Bioremediation encompasses several methods of using organisms to metabolize waste and transform it into less hazardous forms. One method, vermiremediation, uses earthworms to produce enzymes that can absorb and biodegrade hazardous materials. Phytoremediation uses plants to help break down toxins in various ways. Bioremediation can be effective, but it has limitations, such as the amount of time required.

IMPACT OF ELECTRONIC WASTE
The most obvious impact of e-waste is due to its sheer volume. In 2014 alone roughly 41.8 million metric tons of e-waste were generated globally. The EPA estimates that the amount of e-waste grows by 5 to 10 percent each year. This volume has turned e-waste disposal into a major industry, especially in the developing world. The city of Guiyu in southern China is a focal point for used mobile phones, computers,

technology also drives this process. Laptops are a good example of planned obsolescence. Laptops are not built as durably as is possible, and it is often easier to buy a new one than to fix a broken component. Even if an older laptop works fine, it soon becomes incompatible with the latest software.

POTENTIAL SOLUTIONS
Cell phones are a notable example of e-waste. They are increasingly vital to everyday life in most societies, yet models have relatively short lifespans. Like laptops, they have many parts that can easily break and are not easily replaced, requiring the purchase

and anything else that contains semiconductors and microchips. Africa also has its share of digital landfills. One of these is on the outskirts of Accra, Ghana. The concentrated pollution in these areas is a global health threat.

One other complication that arises from e-waste is the problem of identity theft. Many discarded computers and phones still contain personal information. This can be collected by criminals working in recycling centers. According to the US State Department, e-waste is a significant factor in the increase in cybercrime.

—*Andrew Farrell*

BIBLIOGRAPHY

Baldé, C. P., et al. *E-waste Statistics: Guidelines on Classification, Reporting and Indicators, 2015.* Bonn: United Nations U, 2015. *United Nations University.* Web. 9 Feb. 2016.

"Ghana: Digital Dumping Ground." *Frontline.* PBS, 23 June 2009. Web. 29 Jan. 2016.

Glaubitz, John Paul Adrian. "Modern Consumerism and the Waste Problem." *ArXiv.org.* Cornell U, 4 June 2012. Web. 9 Feb. 2016.

McNicoll, Arion. "Phonebloks: The Smartphone for the Rest of Your Life." *CNN.* Cable News Network, 19 Sept. 2013. Web. 29 Jan. 2016.

Mooallem, Jon. "The Afterlife of Cellphones." *New York Times Magazine.* New York Times, 13 Jan. 2008. Web. 9 Feb. 2016.

Patel, Shuchi, and Avani Kasture. "E (Electronic) Waste Management Using Biological Systems—Overview." *International Journal of Current Microbiology and Applied Sciences* 3.7 (2014): 495–504. Web. 9 Feb. 2016.

"Planned Obsolescence: A Weapon of Mass Discarding, or a Catalyst for Progress?" *ParisTech Review.* ParisTech Rev., 27 Sept. 2013. Web. 9 Feb. 2016.

"What Is E-waste?" *Step: Solving the E-waste Problem.* United Nations U/Step Initiative, 2016. Web. 9 Feb. 2016.

ENCRYPTION

FIELDS OF STUDY

Security; Privacy; Algorithms

ABSTRACT

Encryption is the encoding of information so that only those who have access to a password or encryption key can access it. Encryption protects data content, rather than preventing unauthorized interception of or access to data transmissions. It is used by intelligence and security organizations and in personal security software designed to protect user data.

PRINICIPAL TERMS

- **asymmetric-key encryption:** a process in which data is encrypted using a public encryption key but can only be decrypted using a different, private key.
- **authentication:** the process by which the receiver of encrypted data can verify the identity of the sender or the authenticity of the data.
- **hashing algorithm:** a computing function that converts a string of characters into a different, usually smaller string of characters of a given length, which is ideally impossible to replicate without knowing both the original data and the algorithm used.
- **Pretty Good Privacy:** a data encryption program created in 1991 that provides both encryption and authentication.

CRYPTOGRAPHY AND ENCRYPTION

Encryption is a process in which data is translated into code that can only by read by a person with the correct encryption key. It focuses on protecting data content rather than preventing unauthorized interception. Encryption is essential in intelligence and national security and is also common in commercial applications. Various software programs are available that allow users to encrypt personal data and digital messages.

The study of different encryption techniques is called "cryptography." The original, unencrypted data is called the "plaintext." Encryption uses an algorithm called a "cipher" to convert plaintext into ciphertext. The ciphertext can then be deciphered by using another algorithm known as the "decryption key" or "cipher key."

TYPES OF ENCRYPTION

A key is a string of characters applied to the plaintext to convert it to ciphertext, or vice versa. Depending on the keys used, encryption may be either symmetric or asymmetric. Symmetric-key encryption uses the same key for both encoding and decoding. The key used to encode and decode the data must be kept secret, as anyone with access to the key can translate the ciphertext into plaintext. The oldest known cryptography systems used alphanumeric substitution algorithms, which are a type of symmetric encryption. Symmetric-key algorithms are simple to create but vulnerable to interception.

In asymmetric-key encryption, the sender and receiver use different but related keys. First, the receiver uses an algorithm to generate two keys, one to encrypt the data and another to decrypt it. The encryption key, also called the "public key," is made available to anyone who wishes to send the receiver a message. (For this reason, asymmetric-key encryption is also known as "public-key encryption.") The decryption key, or private key, remains known only to the receiver. It is also possible to encrypt data using the private key and decrypt it using the public key. However, the same key cannot be used to both encrypt and decrypt.

Asymmetric-key encryption works because the mathematical algorithms used to create the public and private keys are so complex that it is computationally impractical determine the private key based on the public key. This complexity also means that asymmetric encryption is slower and requires more processing power. First developed in the 1970s, asymmetric encryption is the standard form of encryption used to protect Internet data transmission.

AUTHENTICATION AND SECURITY

Authentication is the process of verifying the identity of a sender or the authenticity of the data sent. A common method of authentication is a hashing algorithm, which translates a string of data into a fixed-length number sequence known as a "hash value." This value can be reverted to the original data using

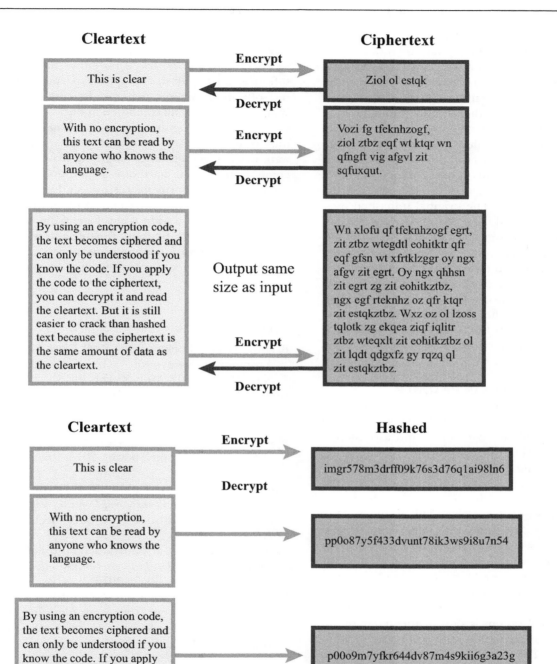

Cleartext

This is clear

With no encryption, this text can be read by anyone who knows the language.

By using an encryption code, the text becomes ciphered and can only be understood if you know the code. If you apply the code to the ciphertext, you can decrypt it and read the cleartext. But it is still easier to crack than hashed text because the ciphertext is the same amount of data as the cleartext.

Ciphertext

Ziol ol estqk

Vozi fg tfeknhzogf, ziol ztbz eqf wt ktqr wn qfngft vig afgvl zit sqfuxqut.

Wn xlofu qf tfeknhzogf egrt, zit ztbz wtegdtl eohitktr qfr eqf gfsn wt xfrtklzggr oy ngx afgv zit egrt. Oy ngx qhhsn zit egrt zg zit eohitkztbz, ngx egf rteknhz oz qfr ktqr zit estqkztbz. Wxz oz ol lzoss tqlotk zg ekqea ziqf iqlitr ztbz wteqxlt zit eohitkztbz ol zit lqdt qdgxfz gy rqzq ql zit estqkztbz.

Output same size as input

Encrypt / Decrypt

Cleartext

This is clear

With no encryption, this text can be read by anyone who knows the language.

By using an encryption code, the text becomes ciphered and can only be understood if you know the code. If you apply the code to the ciphertext, you can decrypt it and read the cleartext. But it is still easier to crack than hashed text because the ciphertext is the same amount of data as the cleartext.

Hashed

imgr578m3drff09k76s3d76q1ai98ln6

pp0o87y5f433dvunt78ik3ws9i8u7n54

p00o9m7yfkr644dv87m4s9kii6g3a23g

This diagram illustrates the output of encrypted content versus hashed content. When text is encrypted, the output will be the same size as the input, and it can be decrypted to show the original input. When text is hashed, input of any size will shrink to an output of a predetermined size, commonly 128 bits. The output cannot be decrypted, only authenticated by comparing it with a known hash value. EBSCO illustration.

131

the same algorithm. The mathematical complexity of hashing algorithms makes it extremely difficult to decrypt hashed data without knowing the exact algorithm used. For example, a 128-bit hashing algorithm can generate 2128 different possible hash values.

For the purpose of authenticating sent data, such as a message, the sender may first convert the data into a hash value. This value, also called a "message digest," may then be encrypted using a private key unique to the sender. This creates a digital signature that verifies the authenticity of the message and the identity of the sender. The original unhashed message is then encrypted using the public key that corresponds to the receiver's private key. Both the privately encrypted digest and the publicly encrypted message are sent to the receiver, who decrypts the original message using their private key and decrypts the message digest using the sender's public key. The receiver then hashes the original message using the same algorithm as the sender. If the message is authentic, the decrypted digest and the new digest should match.

ENCRYPTION SYSTEMS IN PRACTICE

One of the most commonly used encryption programs is Pretty Good Privacy (PGP). It was developed in 1991 and combines symmetric- and asymmetric-key encryption. The original message is encrypted using a unique one-time-only private key called a "session key." The session key is then encrypted using the receiver's public key, so that it can only be decrypted using the receiver's private key. This encrypted key is sent to the receiver along with the encrypted message. The receiver uses their private key to decrypt the session key, which can then can be used to decrypt the message. For added security and authentication, PGP also uses a digital signature system that compares the decrypted message against a message digest. The PGP system is one of the standards in personal and corporate security and is highly resistant to attack. The data security company Symantec acquired PGP in 2010 and has since incorporated the software into many of its encryption programs.

Encryption can be based on either hardware or software. Most modern encryption systems are based on software programs that can be installed on a system to protect data contained in or produced by a variety of other programs. Encryption based on hardware is less vulnerable to outside attack. Some hardware devices, such as self-encrypting drives (SEDs), come with built-in hardware encryption and are useful for high-security data. However, hardware encryption is less flexible and can be prohibitively costly to implement on a wide scale. Essentially, software encryption tends to be more flexible and widely usable, while hardware encryption is more secure and may be more efficient for high-security systems.

—*Micah L. Issitt*

BIBLIOGRAPHY

Bright, Peter. "Locking the Bad Guys Out with Asymmetric Encryption." *Ars Technica*. Condé Nast, 12 Feb. 2013. Web. 23 Feb. 2016.

Delfs, Hans, and Helmut Knebl. *Introduction to Cryptography: Principles and Applications*. 3rd ed. Berlin: Springer, 2015. Print.

History of Cryptography: An Easy to Understand History of Cryptography. N.p.: Thawte, 2013. *Thawte*. Web. 4 Feb. 2016.

"An Introduction to Public Key Cryptography and PGP." *Surveillance Self-Defense*. Electronic Frontier Foundation, 7 Nov. 2014. Web. 4 Feb. 2016.

Lackey, Ella Deon, et al. "Introduction to Public-Key Cryptography." *Mozilla Developer Network*. Mozilla, 21 Mar. 2015. Web. 4 Feb. 2016.

McDonald, Nicholas G. "Past, Present, and Future Methods of Cryptography and Data Encryption." *SpaceStation*. U of Utah, 2009. Web. 4 Feb. 2016.

F

FIREWALLS

FIELDS OF STUDY

Information Systems; Privacy; Security

ABSTRACT

A firewall is a program designed to monitor the traffic entering and leaving a computer network or single device and prevent malicious programs or users from entering the protected system. Firewalls may protect a single device, such as a server or personal computer (PC), or even an entire computer network. They also differ in how they filter data. Firewalls are used alongside other computer security measures to protect sensitive data.

PRINICIPAL TERMS

- **application-level firewalls:** firewalls that serve as proxy servers through which all traffic to and from applications must flow.
- **host-based firewalls:** firewalls that protect a specific device, such as a server or personal computer, rather than the network as a whole.
- **network firewalls:** firewalls that protect an entire network rather than a specific device.
- **packet filters:** filters that allow data packets to enter a network or block them on an individual basis.
- **proxy server:** a computer through which all traffic flows before reaching the user's computer.
- **stateful filters:** filters that assess the state of a connection and allow or disallow data transfers accordingly.

HISTORY OF FIREWALLS

In the early twenty-first century, increasing cybercrime and cyberterrorism made computer security a serious concern for governments, businesses and organizations, and the public. Nearly any computer system connected to the Internet can be accessed by malicious users or infected by harmful programs such as viruses. Both large networks and single PCs face this risk. To prevent such security breaches, organizations and individuals use various security technologies, particularly firewalls. Firewalls are programs or sometimes dedicated devices that monitor the data entering a system and prevent unwanted data from doing so. This protects the computer from both malicious programs and unauthorized access.

The term "firewall" is borrowed from the field of building safety. In that field it refers to a wall specially built to stop the spread of fire within a structure. Computer firewalls fill a similar role, preventing harmful elements from entering the protected area. The idea of computer firewalls originated in the 1980s. At that time, network administrators used routers, devices that transfer data between networks, to separate one network from another. This stopped problems in one network from spreading into others. By the early 1990s, the proliferation of computer viruses and increased risk of hacking made the widespread need for firewalls clear. Some of the advances in that era, such as increased access to the Internet and developments in operating systems, also introduced new vulnerabilities. Early firewalls relied heavily on the use of proxy servers. Proxy servers are servers through which all traffic flows before entering a user's computer or network. In the twenty-first century, firewalls can filter data according to varied criteria and protect a network at multiple points.

TYPES OF FIREWALLS

All firewalls work to prevent unwanted data from entering a computer or network. However, they do so in different ways. Commonly used firewalls can be in various positions relative to the rest of the system.

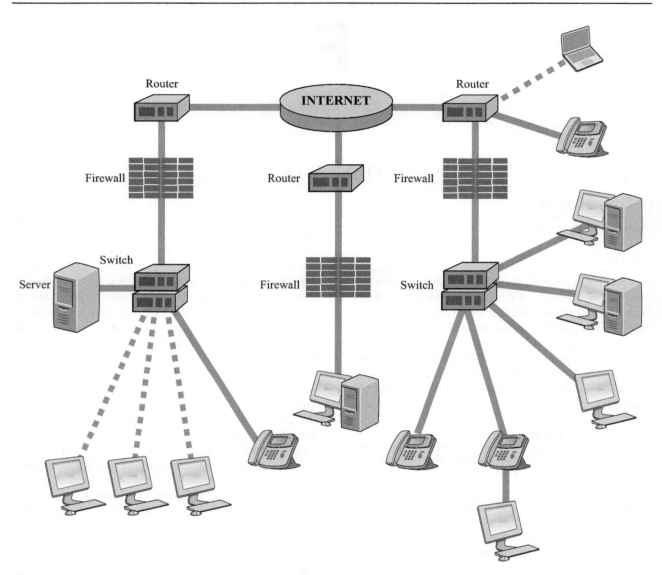

Firewalls are one of many protective measures used to prevent hackers from accessing computers or networks. EBSCO illustration.

An individual computer may have its own personal firewall, such as Windows Firewall or Macintosh OS X's built-in firewall. Other networked devices, such as servers, may also have personal firewalls. These are known as host-based firewalls because they protect a single host rather than the whole network. They protect computers and other devices not only from malicious programs or users on the Internet but also from viruses and other threats that have already infiltrated the internal network, such as a corporate intranet, to which they belong. Network firewalls, on the other hand, are positioned at the entrance to the internal network. All traffic into or out of that network must filter through them. A network firewall may be a single device, such as a router or dedicated computer, which serves as the entrance point for all data. It then blocks any data that is malicious or otherwise unwanted. Application-level firewalls, which monitor and allow or disallow data transfers from and to applications, may be host based or network based.

Firewalls also vary based on how they filter data. Packet filters examine incoming data packets individually and determine whether to block or allow each one to proceed. They decide this based on factors such as the origin and destination of the packets. Stateful filters determine whether to admit or block incoming data based on the state of the connection. Firewalls that use stateful filtering can identify whether data packets trying to enter the computer system are part of an ongoing, active connection and determine whether to let them in based on that context. This allows them to examine and filter incoming data more quickly than their stateless counterparts.

FIREWALLS AND COMPUTER SECURITY

By preventing malicious programs or users from accessing systems, firewalls protect sensitive data stored in or transmitted via computers. They are used to protect personally identifying information, such as Social Security numbers, as well as proprietary trade or government information. Both the technology industry and the public have put increased emphasis on such protections in the early twenty-first century, as identity theft, fraud, and other cybercrimes have become major issues. In light of such threats, firewalls play an essential role in the field of computer security. However, experts caution that a firewall should not be the sole security measure used. Rather, firewalls should be used along with other computer security practices. These practices include using secure passwords, regularly updating software to install patches and eliminate vulnerabilities, and avoiding accessing compromised websites or downloading files from suspicious sources.

—*Joy Crelin*

BIBLIOGRAPHY

"How Firewalls Work." *Boston University Information Services and Technology.* Boston U, n.d. Web. 28 Feb. 2016.

Ingham, Kenneth, and Stephanie Forrest. *A History and Survey of Network Firewalls.* Albuquerque: U of New Mexico, 2002. PDF file.

Morreale, Patricia, and Kornel Terplan, eds. *The CRC Handbook of Modern Telecommunications.* 2nd ed. Boca Raton: CRC, 2009. Print.4 Science Reference Center™ Firewalls

Northrup, Tony. "Firewalls." *TechNet.* Microsoft, n.d. Web. 28 Feb. 2016.

Stallings, William, and Lawrie Brown. *Computer Security: Principles and Practice.* 3rd ed. Boston: Pearson, 2015. Print.

Vacca, John, ed. *Network and System Security.* 2nd ed. Waltham: Elsevier, 2014. Print.

FIRMWARE

FIELDS OF STUDY

Embedded Systems; Software Engineering; System-Level Programming

ABSTRACT

Firmware occupies a position in between hardware, which is fixed and physically unchanging, and software, which has no physical form apart from the media it is stored on. Firmware is stored in nonvolatile memory in a computer or device so that it is always available when the device is powered on. An example can be seen in the firmware of a digital watch, which remains in place even when the battery is removed and later replaced.

PRINICIPAL TERMS

- **embedded systems:** computer systems that are incorporated into larger devices or systems to monitor performance or to regulate system functions.
- **flashing:** a process by which the flash memory on a motherboard or an embedded system is updated with a newer version of software.
- **free software:** software developed by programmers for their own use or for public use and distributed without charge; it usually has conditions attached that prevent others from acquiring it and then selling it for their own profit.
- **homebrew:** software that is developed for a device or platform by individuals not affiliated with the device manufacturer; it is an unofficial or "homemade" version of the software that is developed to provide additional functionality not included or not permitted by the manufacturer.
- **nonvolatile memory:** computer storage that retains its contents after power to the system is cut off, rather than memory that is erased at system shutdown.

OVERVIEW OF FIRMWARE

Many consumer devices have become so complex that they need a basic computer to operate them. However, they do not need a fully featured computer with an operating system (OS) and specially designed software. The answer to this need is to use embedded systems. These systems are installed on microchips inside devices as simple as children's toys and as complex as medical devices such as digital thermometers. The term "embedded" is used because the chips containing firmware are ordinarily not directly accessible to consumers. They are installed within the device or system and expected to work throughout its lifespan.

Computers also use firmware, which is called the "basic input/output system," or BIOS. Even though the computer has its own OS installed and numerous programs to accomplish more specific tasks, there is still a need for firmware. This is because, when the computer is powered on, some part of it must be immediately able to tell the system what to do in order to set itself up. The computer must be told to check the part of the hard drive that contains the start-up sequence, then to load the OS, and so on. The firmware serves this purpose because, as soon as electric current flows into the system, the information stored in the computer's nonvolatile memory is loaded and its instructions are executed. Firmware is usually unaffected even when a different OS is installed. However, the user can also configure the BIOS to some extent and can boot the computer into the BIOS to make changes when necessary. For example, a computer that is configured to boot from the CD-ROM drive first could have this changed in the BIOS so that it would first attempt to read information from an attached USB drive.

MODIFYING AND REPLACING FIRMWARE

Sophisticated users of technology sometimes find that the firmware installed by a manufacturer does not meet all of their needs. When this occurs, it is possible to update the BIOS through a process known as flashing. When the firmware is flashed, it is replaced by a new version, usually with new capabilities. In some cases, the firmware is flashed because the device manufacturer has updated it with a new version. This is rarely done, as firmware functionality is so basic to the operation of the device that it is thoroughly tested prior to release. From time to time, however, security vulnerabilities or other software bugs are found in firmware. Manufacturers helping customers with troubleshooting often recommend using the latest firmware to rule out such defects.

Some devices, especially gaming consoles, have user communities that can create their own versions of firmware. These user-developed firmware versions are referred to as homebrew software, as they are produced by users rather than manufacturers. Homebrew firmware is usually distributed on the Internet as free software, or freeware, so that anyone can download it and flash their device. In the case of gaming consoles, this can open up new capabilities. Manufacturers tend to produce devices only for specialized functions. They exclude other functions because the functions would increase the cost or make it too easy to use the device for illegal or undesirable purposes. Flashing such devices with homebrew software can make these functions available.

AUTOMOBILE SOFTWARE

One of the market segments that has become increasingly reliant on firmware is automobile manufacturing. More and more functions in cars are now controlled by firmware. Not only are the speedometer and fuel gauge computer displays driven by firmware, but cars come with firmware applications for music players, real-time navigation and map display, and interfaces with passengers' cell phones.

FIRMWARE AS VULNERABILITY

Although firmware is not very visible to users, it has still been a topic of concern for computer security professionals. With homebrew firmware distributed over the Internet, the concern is that the firmware may contain "backdoors." A backdoor is a secret means of conveying the user's personal information to unauthorized parties. Even with firmware from official sources, some worry that it would be possible for the government or the device manufacturer to include security vulnerabilities, whether deliberate or inadvertent.

—*Scott Zimmer, JD*

BIBLIOGRAPHY

Bembenik, Robert, Łukasz Skonieczny, Henryk Rybinński, Marzena Kryszkiewicz, and Marek Niezgódka, eds. *Intelligent Tools for Building a Scientific Information Platform: Advanced Architectures and Solutions.* New York: Springer, 2013. Print.

Dice, Pete. *Quick Boot: A Guide for Embedded Firmware Developers.* Hillsboro: Intel, 2012. Print.

Iniewski, Krzysztof. *Embedded Systems: Hardware, Design, and Implementation.* Hoboken: Wiley, 2012. Print.

Khan, Gul N., and Krzysztof Iniewski, eds. *Embedded and Networking Systems: Design, Software, and Implementation.* Boca Raton: CRC, 2014. Print.

Noergaard, Tammy. *Embedded Systems Architecture: A Comprehensive Guide for Engineers and Programmers.* 2nd ed. Boston: Elsevier, 2013. Print.

Sun, Jiming, Vincent Zimmer, Marc Jones, and Stefan Reinauer. *Embedded Firmware Solutions: Development Best Practices for the Internet of Things.* Berkeley: ApressOpen, 2015. Print.

FITBIT

FIELD OF STUDY
Computer Science

ABSTRACT

The term Fitbit refers to activity tracker devices manufactured and sold by Fitbit Incorporated. Designed to help users take control of their health and fitness, Fitbit devices track data such as step counts and calories burned, among other functions. Following their introduction to the market in 2009, Fitbit devices popularized wearable activity trackers among the public, prompting the creation of a wide variety of competing fitness devices.

PRINICIPAL TERMS

- **accelerometer IC:** an integrated circuit that measures acceleration.
- **low-energy connectivity IC:** an integrated circuit that enables wireless Bluetooth connectivity while using little power.
- **microcontroller:** an integrated circuit that contains a very small computer.
- **near-field communications antenna:** an antenna that enables a device to communicate wirelessly with a nearby compatible device.
- **printed circuit board (PCB):** a component of electronic devices that houses and connects many smaller components.
- **vibrator:** an electronic component that vibrates.

HISTORY OF FITBIT

In the first decade of the twenty-first century, growing interest in health and fitness converged with significant advances in consumer technology. This lead to the increasing popularity of wearable electronic devices. The personal activity tracker is essentially an upgraded version of earlier pedometers (step counters). Activity trackers record and measure fitness data. Newer models can sync seamlessly with the user's computer or smartphone. One such family of devices, sold under the brand name Fitbit, was introduced in 2009. Different Fitbit products allow users to track various fitness-related data.

Fitbit Incorporated was founded in 2007 by James Park and Eric Friedman as Healthy Metrics Research Incorporated. The company's initial goal was to incorporate accelerometers into wearable technology. Accelerometers are small sensors that measure acceleration forces. The founders presented a prototype consisting of a circuit board within a small balsawood box to investors in 2008. The company's first device, known as the Fitbit Tracker, was released to the public in 2009. The original Fitbit clipped to the wearer's clothing. The small device could count and track the number of steps taken and estimate released numerous additional Fitbit devices featuring different functions.

TYPES OF FITBIT DEVICES

The Fitbit Tracker could track the number of steps the user took per day and estimate how many calories such activity burned. The device also tracked the user's sleep duration and quality. The Fitbit Tracker could be worn clipped to the user's clothing during the day. It also came with a wristband with which to wear the device at night. The device was paired with a website where users could document their exercise routines and meals and record their weight. It also came with a charging station that enabled users to wirelessly upload the data collected by their Fitbits.

Over the following years, Fitbit introduced more than ten other devices with a variety of different functions. Its products are divided among three product lines: everyday, active, and performance fitness. Devices such as the Fitbit One and the Fitbit Charge belong to the everyday fitness category. They track the user's minutes of activity and number of floors climbed, among other data. Some devices, including the Fitbit Charge and the Fitbit Alta, feature clock displays and can connect to the user's smartphone to display caller information. The active fitness product category includes the Fitbit Charge HR and Blaze. The Charge HR adds heart rate tracking to the Fitbit Charge's features. The Fitbit Blaze features the standard activity-tracking features as well as music controls and phone notifications. The Fitbit Surge, a performance fitness device, can track the user's location via GPS, among other features. Fitbit also manufactures the Aria smart scale. Aria tracks the user's weight, body mass index, and other data and syncs with both the user's activity tracker and the Fitbit website.

Activity trackers include a number of sensors and outputs that allow users to monitor health-related data such as steps taken, heart rate, sleeping patterns, and more. EBSCO illustration.

INSIDE FITBIT DEVICES

As multifunctional wearable devices, Fitbit activity trackers require that numerous components fit into a limited amount of space. The Fitbit Flex wristband-style device, for example, contains a variety of small electronic components, including several integrated circuits (ICs). An IC is a silicon microchip that can perform a specialized task, such as handling memory. One core component of the Fitbit Flex is the three-axis accelerometer IC, which measures acceleration forces to track the user's movement. The device also contains a printed circuit board (PCB). The PCB houses and connects multiple smaller components. It is controlled by a microcontroller, a chip that is essentially a tiny computer, complete with memory and a processor. As the device has a vibrating silent alarm function, the Fitbit Flex likewise has a vibrator component. For connectivity purposes, the device contains a low-energy connectivity IC, which enables wireless Bluetooth connectivity. It also has a near-field communications (NFC) antenna. The NFC antenna enables the device to communicate wirelessly with other NFC-compatible devices such as certain smartphones.

The specific components used in a Fitbit product vary based on the device in question. For example, the Fitbit Surge differs greatly from the clip- or wristband-based devices in terms of composition. However, the components in use generally have a number of common characteristics, including small size and very low power usage.

IMPACT AND CHALLENGES

The popularity of the early Fitbit devices spurred other companies' efforts to create similar devices. Some of these companies had been working on activity trackers prior to the advent of Fitbit. Some other activity-tracking wearable devices include Nike+ products, Jawbone UP bands, and the Apple Watch, which has some activity-tracking capabilities.

However, this popularity has also raised concerns regarding user privacy. The activity data gathered by such devices, and particularly GPS data, could potentially be used for harm if accessed by individuals with malicious intent. The safety and accuracy of Fitbit devices have also been called into question. A 2014 class-action lawsuit alleged that the Fitbit Force's wristband caused users to develop rashes on their wrists. Another lawsuit filed in January 2016 argued that the company had misled consumers about the accuracy of the heart rate data gathered by the Fitbit Surge and Fitbit Charge HR.

—*Joy Crelin*

BIBLIOGRAPHY

Case, Meredith A., et al. "Accuracy of Smartphone Applications and Wearable Devices for Tracking Physical Activity Data." *Journal of the American Medical Association* 313.6 (2015): 625–26. Web. 2 Mar. 2016.

Comstock, Jonah. "Eight Years of Fitbit News Leading Up to Its Planned IPO." *MobiHealthNews.* HIMSS Media, 11 May 2015. Web. 28 Feb. 2016.

"Fitbit Flex Teardown." *iFixit.* iFixit, 2013. Web. 28 Feb. 2016.

Goode, Lauren. "Fitbit Hit with Class-Action Suit over Inaccurate Heart Rate Monitoring." *Verge.* Vox Media, 6 Jan. 2016. Web. 28 Feb. 2016.

Hof, Robert. "How Fitbit Survived as a Hardware Startup." *Forbes.* Forbes.com, 4 Feb. 2014. Web. 28 Feb. 2016.

Mangan, Dan. "There's a Hack for That: Fitbit User Accounts Attacked." *CNBC.* CNBC, 8 Jan. 2016. Web. 28 Feb. 2016.

Marshall, Gary. "The Story of Fitbit: How a Wooden Box Became a $4 Billion Company." *Wareable.* Wareable, 30 Dec. 2015. Web. 28 Feb. 2016.

Stevens, Tim. "Fitbit Review." *Engadget.* AOL, 15 Oct. 2009. Web. 28 Feb. 2016.

FORTRAN

FIELDS OF STUDY

Programming Languages; Software Engineering;

ABSTRACT

Fortran is the oldest programming still in common use, and is used primarily for scientific and engineering computations requiring complex mathematical calculations. Fortran has been continually revised since its development at IBM in the 1950s, and is used presently on "supercomputers" with parallel architecture. It is a robust language that is relatively easy to learn, and Fortran programs adhere to a strictly logical format.

PRINICIPAL TERMS

- **algorithm:** a set of step-by-step instructions for performing computations.
- **character:** a unit of information that represents a single letter, number, punctuation mark, blank space, or other symbol used in written language.
- **function:** instructions read by a computer's processor to execute specific events or operations.
- **object-oriented programming:** a type of programming in which the source code is organized into objects, which are elements with a unique identity that have a defined set of attributes and behaviors.
- **internal-use software:** software developed by a company for its own use to support general and administrative functions, such as payroll, accounting, or personnel management and maintenance.
- **main loop:** the overarching process being carried out by a computer program, which may then invoke subprocesses.
- **programming languages:** sets of terms and rules of syntax used by computer programmers to create instructions for computers to follow. This code is then compiled into binary instructions for a computer to execute.
- **syntax:** rules that describe how to correctly structure the symbols that comprise a language.

FORTRAN HISTORY AND DEVELOPMENT

The FORTRAN programming language was developed by John Backus, working at IBM, between 1954 and 1957. The name is a contraction taken from FORmula TRANslator, a reflection of the development of the language for use in scientific and engineering applications requiring complex mathematical functions. The BASIC programming language, developed in the 1960s, was strongly influenced by the Fortran programming language in regard to program structure, being essentially linear with the ability to use contained subroutines. Fortran has been developed through several versions into the 21st century, including Fortran III, Fortran IV, Fortran 66, Fortran 77, Fortran 90, Fortran 95 and Fortran 2003, and is the oldest of programming languages still in common use. Fortran 90 and later versions are used on present-day advanced parallel computers, which are sometimes referred to as "supercomputers," and it is notable that Fortran 2003 supports object-oriented programming. The version numbers generally refer to the year in which the updated version of Fortran was released for use. Fortran is a versatile and powerful imperative programming language that supports numerical analysis and scientific calculation, as well as structured, array, modular, generic and concurrent programming. Given the specialized fields of science and technology in which the Fortran language is applied, much of the programs written in that language are for internal-use software in which each function may be defined individually within a specific algorithm.

PROGRAM STRUCTURE

Like all programming languages, Fortran has its own syntax, or method in which program commands and statements must be written in order to be "understood" and carried out. A Fortran program is made up of subroutines and function segments that can be compiled independently. Subroutines are termed "external procedures," and it is characteristic of both Fortran and ALGOL that procedures are only invoked as a result of explicit calls from within the main loop of the program or from another procedure within it. Each segment of the program has local data written into the source code, but additional data can be made available to more than one segment by storing it in blocks of code labeled COMMON. Each Fortran program begins with a statement of the name of the program, followed by the statement "implicit none," and ends with an "end program" statement for that program, as

```
program program_name
implicit none
.
.
.
end program program_name
```

Documentation phrases are added into a Fortran program using the ! character at the beginning of a line. Everything after the ! in that line is ignored by the Fortran compiler, allowing the programmer to describe what each program segment or line of code is intended to achieve. Following the "implicit none" statement, Fortran program have a block of type declaration statements to identify the nature of the variables being used in the program. For example, the type declaration statement

```
real :: x, y, z
```

identifies the variables x, y and z as real numbers. A block of executable statements then follows. These are the functions and relations that the program is to calculate. The executable statement block may include, for example, statements such as

```
read (*,*) x
read (*,*) y
z = (x + y)/2
print *, z
```

by which the user enters the two values for x and y from the keyboard. Their average value is calculated as the variable z, which is then printed to the default display device, usually the computer screen or an attached printer.

The heading identifies the name of the program as "average_four." The type declarations identify all of the five variables a, b, c, d and average as real numbers. The executable statements get the four values of a, b, c and d from keyboard input, calculates their

---

SAMPLE PROBLEM

Write a Fortran program that asks for the input of four values and outputs their average.

Answer:

```
program average_four
      implicit none
! this program gets four values and
outputs their average
! type declarations
      real :: a, b, c, d, average
! executable statements
      read (*,*) a
      read (*,*) b
      read (*,*) c
      read (*,*) d
      average = (a + b + c + d)/4
      print *, "The average value is,"
      average
end program average_four
```

average value, and displays the result. The end of the program operation is identified by the "end program average_four" statement.

FORTRAN IN THE PRESENT DAY

The straightforward and strictly logical manner in which Fortran programs are written is very suitable for use in complex applications that call for close attention to detail, such as scientific and engineering calculations. This would account for the longevity of the Fortran programming language in those areas, and it is used extensively on modern supercomputers for numerical analysis of complex systems. The robust character of the Fortran programming language ensures that it will continue to be used well into the future.

—*Richard M. Renneboog M.Sc.*

BIBLIOGRAPHY

Chivers, Ian, and Jane Sleightholme. *Introduction to Programming With Fortran, with Coverage of Fortran 90, 95, 2003, 2008 and 77.* 3rd ed., New York, NY: Springer, 2015. Print.

Counihan, Martin *Fortran 95.* Londion, UK: University College Press. 1996. Print.

Gehrke, Wilhelm. *Fortran 90 Language Guide.* New York, NY: Springer, 1995. Print.

Kupferschmid, Michael *Classical Fortran Programming for Engineering and Scientific Applications.* Boca Raton, FL: CRC Press, 2009. Print.

Metcalf, Michael, and John Reid. *The F Programming Language.* New York, NY: Oxford University Press, 1996. Print.

O'Regan, Gerard. *A Brief History of Computing.* 2nd ed., New York, NY: Springer-Verlag. 2012. Print.

Rajaraman, V. *Computer Programming in Fortran 77.* 4th ed., New Delhi, IND: Prentice-Hall of India Pvt., 2006. Print.

Scott, Michael L. *Programming Language Pragmatics.* 4th ed., Waltham, MA: Morgan Kaufmann, 2016. Print.

FUNCTIONAL DESIGN

FIELDS OF STUDY

Software Engineering; System Analysis

ABSTRACT

Functional design is a paradigm of computer programming. Following functional design principles, computer programs are created using discrete modules that interact with one another only in very specific, limited ways. An individual module can be modified extensively with only minor impacts on other parts of the program.

PRINICIPAL TERMS

- **coupling:** the degree to which different parts of a program are dependent upon one another.
- **imperative programming:** an approach to software development in which the programmer writes a specific sequence of commands for the computer to perform.
- **inheritance:** a technique that reuses and repurposes sections of code.
- **interrupt vector table:** a chart that lists the addresses of interrupt handlers.
- **main loop:** the overarching process being carried out by a computer program, which may then invoke subprocesses.
- **polymorphism:** the ability to maintain the same method name across subclasses even when the method functions differently depending on its class.
- **subtyping:** a relation between data types where one type is based on another, but with some limitations imposed.

BENEFITS OF FUNCTIONAL DESIGN

Functional design is a concept most commonly associated with computer programming. However, it is also relevant to other fields, such as manufacturing and business management. At its core, the concepts underlying functional design are simple. Instead of designing a program in which the individual parts perform many different functions and are highly interconnected, modules should have low coupling. Modules—the individual parts of the program—should be designed to have the simplest possible inputs and outputs. When possible, each module should only perform a single function. Low coupling ensures that each module has a high degree of independence from other parts of the program. This makes the overall program easier to debug and maintain. Most of the labor that goes into programming is related to these two activities. Therefore, it makes sense to use a design approach that makes both of these tasks easier.

Functional design is preferable because when a program has a bug, if the modules perform many different functions and are highly coupled, it is difficult to make adjustments to the malfunctioning part without causing unintended consequences in other parts of the program. Some parts of a program do not easily lend themselves to functional design because their very nature requires that they be connected to multiple parts of the overall program. One example of this is the main loop of the program. By necessity, the main loop interacts with and modifies different modules, variables, and functions. Similarly, the interrupt vector table acts as a directory for interrupt handlers used throughout the program. Still, many other functions of a typical program can be designed in ways that minimize interdependencies.

INTERDEPENDENT COMPLICATIONS

Functional programming is a form of declarative programming. Declarative programs simply specify the end result that a program should achieve. In contrast, imperative programming outlines the particular sequence of operations a program should perform. Programmers must carefully consider how they design variables and functions. While functional design emphasizes modularity and independence between different program components, a number of design techniques rely on interconnections. For example, inheritance draws on the properties of one class to create another. With polymorphism, the same methods and operations can be performed on a variety of elements but will have specific, customized behaviors depending on the elements' class. With subtyping, a type is designed to contain one or more other variables types. While each of these techniques has advantages, their interdependence can cause issues when a change is made to other parts of the program. For example, consider a variable called

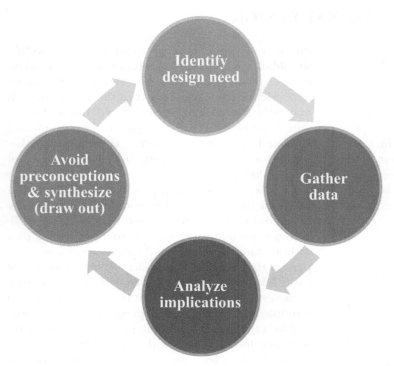

Functional design is a theory of design that focuses foremost on the function of the unit and then on aesthetics and/or economics. Functional design relies on identifying the current functional need, gathering data, and analyzing it to form a cohesive design that meets that function. The design must be tested to determine whether there are any further problems in the function, all while avoiding preconceptions. EBSCO illustration.

into one module. The programmer may then separate these functions into separate, less dependent modules if possible.

MANUFACTURING APPLICATIONS

Functional design also has applications in physical manufacturing. This has been facilitated by the advent of 3-D printing. 3-D printing can make it easier to produce parts, components, and even whole pieces of machinery that would be expensive or impossible to manufacture using traditional methods. In this context, functional design conceptually separates a complex machine into simple sections that can be individually produced and then assembled.

PROCESS AND FUNCTION

Designers who approach a problem from the standpoint of functional design often observe that functional design is not simply an outcome but also a process. Functional design is a way of thinking about problems before any code is written. It asks how a design can be broken down into simpler steps and include only essential processes.

—*Scott Zimmer, JD*

NUM that can take on any numeric value. A second variable is defined as EVENNUM and can take on any NUM value divisible by 2 with no remainder. Because EVENNUM is defined in relation to NUM, if a programmer later makes a change to NUM, it could have unintended consequences on the variable EVENNUM. A program that is highly coupled could have a long chain reaction of bugs due to a change in one of its modules.

Functional design tries to avoid these issues in two ways. First, programmers specifically consider interrelations as they design a program and try to make design choices that will minimize complications. Second, once a program has been written, programmers will proofread the code to find any elements that have unnecessary complications. Some clues are the presence of language such as "and" or "or" in the descriptions of variables, classes, and types. This can suggest that multiple functions are being combined

BIBLIOGRAPHY

Bessière, Pierre, et al. *Bayesian Programming.* Boca Raton: CRC, 2014. Print.

Clarke, Dave, James Noble, and Tobias Wrigstad, eds. *Aliasing in Object-Oriented Programming: Types, Analysis and Verification.* Berlin: Springer, 2013. Print.

Lee, Kent D. *Foundations of Programming Languages.* Cham: Springer, 2014. Print.

Neapolitan, Richard E. *Foundations of Algorithms.* 5th ed. Burlington: Jones, 2015. Print.

Streib, James T., and Takako Soma. *Guide to Java: A Concise Introduction to Programming.* New York: Springer, 2014. Print.

Wang, John, ed. *Optimizing, Innovating, and Capitalizing on Information Systems for Operations.* Hershey: Business Science Reference, 2013. Print.

G

GAME PROGRAMMING

FIELDS OF STUDY

Software Engineering; Programming Language

ABSTRACT

Game programming is a type of software engineering used to develop computer and video games. Depending on the type of game under development, programmers may be required to specialize in areas of software development not normally required for computer programmers. These specialties include artificial intelligence, physics, audio, graphics, input, database management, and network management.

PRINICIPAL TERMS

- **game loop:** the main part of a game program that allows the game's physics, artificial intelligence, and graphics to continue to run with or without user input.
- **homebrew:** a slang term for software made by programmers who create games in their spare time, rather than those who are employed by software companies.
- **object-oriented programming:** a type of programming in which the source code is organized into objects, which are elements with a unique identity that have a defined set of attributes and behaviors.
- **prototype:** an early version of software that is still under development, used to demonstrate what the finished product will look like and what features it will include.
- **pseudocode:** a combination of a programming language and a spoken language, such as English, that is used to outline a program's code.
- **source code:** the set of instructions written in a programming language to create a program.

HOW GAME PROGRAMMING WORKS

While many video games are developed as homebrew projects by individuals working in their spare time, most major games are developed by employees of large companies that specialize in video games. In such a company, the process of creating a video game begins with a basic idea of the game's design. If enough people in the company feel the idea has merit, the company will create a prototype to develop the concept further. The prototype gives form to the game's basic story line, graphics, and programming. It helps developers decide whether the game is likely to do well in the marketplace and what resources are needed for its development.

Games that move past the prototyping stage proceed along many of the same pathways as traditional software development. However, video games differ from most other software applications in that elements of the game must continue to run in the absence of user input. Programmers must therefore create a game loop, which continues to run the program's graphics, artificial intelligence, and audio when the user is not active.

Game programmers often use object-oriented programming (OOP). This is an approach to software development that defines the essential objects of a program and how they can interact with each other. Each object shares attributes and behaviors with other objects in the same class. Because many video games run hundreds of thousands of lines of code, an object-oriented approach to game development can help make the code more manageable. OOP makes programs easier to modify, maintain, and extend. This approach can make it easier to plan out and repurpose elements of the game.

Large numbers of people are involved in writing the source code of a game intended for wide distribution. They are typically organized in teams, with each

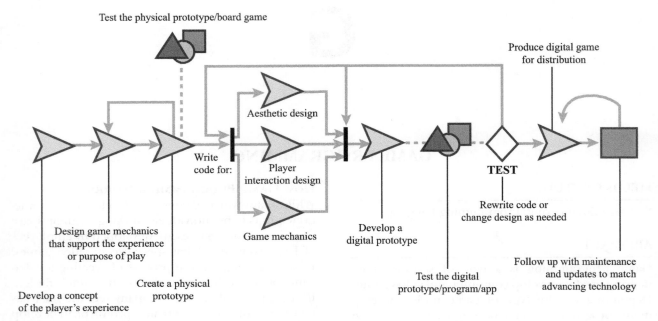

Test the physical prototype/board game

Produce digital game
for distribution

Aesthetic design

Write
code for:

Player
interaction design

TEST

Rewrite code or
change design as needed

Design game mechanics
that support the experience
or purpose of play

Game mechanics

Develop a
digital prototype

Create a physical
prototype

Develop a concept
of the player's experience

Test the digital
prototype/program/app

Follow up with maintenance
and updates to match
advancing technology

The development of a quality game begins with defining the player's experience and the game mechanics. Aesthetics and the digital proto-types of game pieces and environments are actually determined much later in the process. EBSCO illustration.

team specializing in certain aspects of the game's mechanics or certain types of programming. To facilitate collaboration, detailed comments and pseudocode are used to document features of the program and explain choices about coding strategies. Pseudocode is a combination of a programming language and a natural language such as English. It is not executable, but it is much easier to read and understand than source code.

GAME PROGRAMMING TOOLS

Game programmers use the same basic tools as most software developers. These tools include text editors, debuggers, and compilers. The tools used are often determined by the type of device the game is intended to run on. Games can be developed for a particular computer platform, such as Windows, Apple, or Linux. They can also be developed for a gaming console, such as the Wii, Xbox, or PlayStation. Finally, games may be developed for mobile platforms such as Apple's iOS or Google's Android operating system.

A wide variety of programming languages are used to create video games. Some of the most commonly used programming languages in game development are C++, C#, and Java. Many video games are programmed with a combination of several different languages. Assembly code or C may be used to write lower-level modules. The graphics of many games are programmed using a shading language such as Cg or HLSL.

Game developers often use application programming interfaces (APIs) such as OpenGL and DirectX. They also rely heavily on the use of libraries and APIs to repurpose frequently used functions. Many of these resources are associated with animation, graphics, and the manipulation of 3-D elements. One driving force in game development has been the use of increasingly sophisticated visual elements. The video game industry has come to rival motion pictures in their use of complex plotlines, dazzling special effects, and fully developed characters.

A POPULAR BUT DEMANDING CAREER

Game programming has been a popular career choice among young people. Due to the popularity of video games, it is not surprising that large numbers of people are interested in a career that will allow them to get paid to work on games. Becoming a game programmer requires the ability to code and a familiarity with several different programming languages. Game programmers can specialize in developing game engines, artificial intelligence, audio, graphics, user interfaces, inputs, or gameplay.

Game programmers regularly report high levels of stress as they are pressured to produce increasingly impressive games every year. There is intense competition to produce better special effects, more exciting stories, and more realistic action with each new title. Some game companies have been criticized for the high demands and heavy workloads placed upon game developers.

—*Scott Zimmer, JD*

BIBLIOGRAPHY

Harbour, Jonathan S. *Beginning Game Programming.* 4th ed. Boston: Cengage, 2015. Print.

Kim, Chang-Hun, et al. *Real-Time Visual Effects for Game Programming.* Singapore: Springer, 2015. Print.

Madhav, Sanjay. *Game Programming Algorithms and Techniques: A Platform-Agnostic Approach.* Upper Saddle River: Addison, 2014. Print.

Marchant, Ben. "Game Programming in C and C++." *Cprogramming.com.* Cprogramming.com, 2011. Web. 16 Mar. 2016.

Nystrom, Robert. *Game Programming Patterns.* N.p.: Author, 2009–14. Web. 16 Mar. 2016.

Yamamoto, Jazon. *The Black Art of Multiplatform Game Programming.* Boston: Cengage, 2015. Print.

GRAPHICAL USER INTERFACE

FIELDS OF STUDY

Computer Science; Applications; System-Level Programming

ABSTRACT

Graphical user interfaces (GUIs) are human-computer interaction systems. In these systems, users interact with the computer by manipulating visual representations of objects or commands. GUIs are part of common operating systems like Windows and Mac OS. They are also used in other applications.

PRINICIPAL TERMS

- **application-specific GUI:** a graphical interface designed to be used for a specific application.
- **command line:** an interface that accepts text-based commands to navigate the computer system and access files and folders.
- **direct manipulation interfaces:** computer interaction format that allows users to directly manipulate graphical objects or physical shapes that are automatically translated into coding.
- **interface metaphors:** linking computer commands, actions, and processes with real-world actions, processes, or objects that have functional similarities.

- **object-oriented user interface:** an interface that allows users to interact with onscreen objects as they would in real-world situations, rather than selecting objects that are changed through a separate control panel interface.
- **user-centered design:** design based on a perceived understanding of user preferences, needs, tendencies, and capabilities.

GRAPHICS AND INTERFACE BASICS

A user interface is a system for human-computer interaction. The interface determines the way that a user can access and work with data stored on a computer or within a computer network. Interfaces can be either text based or graphics based. Text-based systems allow users to input commands. These commands may be text strings or specific words that activate functions. By contrast, graphical user interfaces (GUIs) are designed so that computer functions are tied to graphic icons (like folders, files, and drives). Manipulating an icon causes the computer to perform certain functions.

HISTORY OF INTERFACE DESIGN

The earliest computers used a text-based interface. Users entered text instructions into a command line. For instance, typing "run" in the command line would

Graphical user interfaces (GUIs) became popular in the early 1980s. Early uses of a GUI included analog clocks, simple icons, charts, and menus. EBSCO illustration.

tell the computer to activate a program or process. One of the earliest text-based interfaces for consumer computer technology was known as a "disk operating system" (DOS). Using DOS-based systems required users to learn specific text commands, such as "del" for deleting or erasing files or "dir" for listing the contents of a directory. The first GUIs were created in the 1970s as a visual "shell" built over DOS system.

GUIs transform the computer screen into a physical map on which graphics represent functions, programs, files, and directories. In GUIs, users control an onscreen pointer, usually an arrow or hand symbol, to navigate the computer screen. Users activate computing functions by directing the pointer over an icon and "clicking" on it. For instance, GUI users can cause the computer to display the contents of a directory (the "dir" command in DOS) by clicking on a folder or directory icon on the screen. Modern GUIs combine text-based icons, such as those found in menu bars and movable windows, with linked text icons that can be used to access programs and directories.

ELEMENTS OF GUIs AND OTHER OBJECT INTERFACES

Computer programs are built using coded instructions that tell the computer how to behave when given inputs from a user. Many different programming languages can be used to create GUIs. These include C++, C#, JavaFX, XAML, XUL, among others. Each language offers different advantages and disadvantages when used to create and modify GUIs.

User-centered design focuses on understanding and addressing user preferences, needs, capabilities, and tendencies. According to these design principles, interface metaphors help make GUIs user friendly. Interface metaphors are models that represent real-world objects or concepts to enhance user understanding of computer functions. For example, the desktop structure of a GUI is designed using the metaphor of a desk. Computer desktops, like actual desktops, might have stacks of documents (windows) and objects or tools for performing various functions. Computer folders, trash cans, and recycle bins are icons whose functions mirror those of their real-world counterparts.

Object-oriented user interfaces (OOUIs) allow a user to manipulate objects onscreen in intuitive ways based on the function that the user hopes to achieve. Most modern GUIs have some object-oriented functionality. Icons that can be dragged, dropped, slid, toggled, pushed, and clicked are "objects." Objects include folders, program shortcuts, drive icons, and trash or recycle bins. Interfaces that use icons can also be direct manipulation interfaces (DMI). These interfaces allow the user to adjust onscreen objects as though they were physical objects to get certain results. Resizing a window by dragging its corner is one example of direct manipulation used in many GUIs.

CURRENT AND FUTURE OF INTERFACE DESIGN

GUIs have long been based on a model known as WIMP. WIMP stands for "windows, icons, menus, and pointer objects," which describes the ways that users can interact with the interface. Modern GUIs are a blend of graphics-based and text-based functions, but this system is more difficult to implement on modern handheld computers, which have less space to hold icons and menus. Touch-screen interfaces represent the post-WIMP age of interface design. With touch screens, users more often interact directly with objects on the screen, rather than using menus and text-based instructions. Touch-screen design is important in a many application-specific GUIs. These interfaces are designed to handle a single process or application, such as self-checkout kiosks in grocery stores and point-of-sale retail software.

Current computer interfaces typically require users to navigate through files, folders, and menus to locate functions, data, or programs. However, voice activation of programs or functions is now available on many computing devices. As this technology becomes more common and effective, verbal commands may replace many functions that have been accessed by point-and-click or menu navigation.

—*Micah L. Issitt*

BIBLIOGRAPHY

"Graphical User Interface (GUI)." *Techopedia.* Techopedia, n.d. Web. 5 Feb. 2016.

Johnson, Jeff. *Designing with the Mind in Mind.* 2nd ed. Waltham: Morgan, 2014. Print.

Lohr, Steve. "Humanizing Technology: A History of Human-Computer Interaction." *New York Times: Bits.* New York Times, 7 Sept. 2015. Web. 31 Jan. 2016.

Reimer, Jeremy. "A History of the GUI." *Ars Technica.* Condé Nast, 5 May 2005. Web. 31 Jan. 2016.

"User Interface Design Basics." *Usability.* US Dept. of Health and Human Services, 2 Feb. 2016. Web. 2 Feb. 2016.

Wood, David. *Interface Design: An Introduction to Visual Communication in UI Design.* New York: Fairchild, 2014. Print.

GRAPHICS FORMATS

FIELDS OF STUDY

Information Systems; Digital Media; Graphic Design

ABSTRACT

Graphics formats are standardized forms of computer files used to transfer, display, store, or print reproductions of digital images. Digital image files are divided into two major families, vector and raster files. They can be compressed or uncompressed for storage. Each type of digital file has advantages and disadvantages when used for various applications.

PRINICIPAL TERMS

- **compressed data:** data that has been encoded such that storing or transferring the data requires fewer bits of information.
- **lossless compression:** data compression that allows the original data to be compressed and reconstructed without any loss of accuracy.
- **lossy compression:** data compression that uses approximation to represent content and therefore reduces the accuracy of reconstructed data.
- **LZW compression:** a type of lossless compression that uses a table-based algorithm.
- **RGB:** a color model that uses red, green, and blue to form other colors through various combinations.

DIGITAL IMAGING

A digital image is a mathematical representation of an image that can be displayed, manipulated, and modified with a computer or other digital device. It can also be compressed. Compression uses algorithms to reduce the size of the image file to facilitate sharing, displaying, or storing images. Digital images may be stored and manipulated as raster or vector images. A third type of graphic file family, called "metafiles," uses both raster and vector elements.

The quality and resolution (clarity) of an image depend on the digital file's size and complexity. In raster graphics, images are stored as a set of squares, called "pixels." Each pixel has a color value and a color depth. This is defined by the number of "bits" allocated to each pixel. Pixels can range from 1 bit per pixel, which has a monochrome (two-color) depth, to 32-bit, or "true color." 32-bit color allows for more than four billion colors through various combinations. Raster graphics have the highest level of color detail because each pixel in the image can have its own color depth. For this reason, raster formats are used for photographs and in image programs like Adobe Photoshop. However, the resolution of a raster image depends on size because the image has the same number of pixels at any magnification. For this reason, raster images cannot be magnified past a certain point without losing resolution. Vector

LOSSLESS

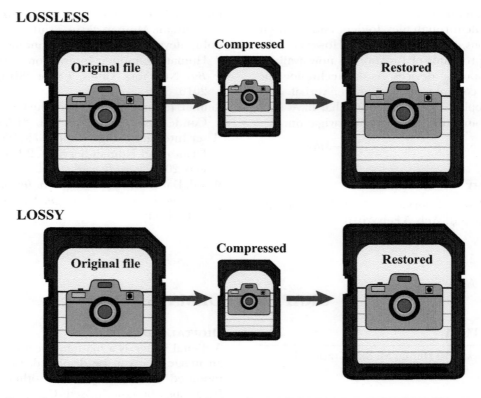

LOSSY

Depending on the image format, data may be lost after compression and restoration. Loss of image data reduces the quality of the image. EBSCO illustration.

graphics store images as sets of polygons that are not size-dependent and look the same at any magnification. For relatively simple graphics, like logos, vector files are smaller and more precise than raster images. However, vector files do not support complex colors or advanced effects, like blurring or drop shadows. Depending on the image format, data may be lost after compression and restoration. Loss of image data reduces the quality of the image.

Two basic color models are used to digitally display various colors. The RGB color model, also called "additive color," combines red, green, and blue to create colors. The CMYK model, also called "subtractive color," combines the subtractive primary colors cyan, magenta, yellow, and black to absorb certain wavelengths of light while reflecting others.

IMAGE COMPRESSION

Image compression reduces the size of an image to enable easier storage and processing. Lossless compression uses a modeling algorithm that identifies repeated or redundant information contained within an image. It stores this information as a set of instructions can be used to reconstruct the image without any loss of data or resolution. One form of lossless compression commonly used is the LZW compression algorithm developed in the 1980s. The LZW algorithm uses a "code table" or "dictionary" for compression. It scans data for repeated sequences and then adds these sequences to a "dictionary" within the compressed file. By replacing repeated data with references to the dictionary file, space is saved but no data is lost. Lossless compression is of benefit when image quality is essential but is less efficient at reducing image size. Lossy compression algorithms reduce file size by removing less "valuable" information. However, images compressed with lossy algorithms continue to lose resolution each time the image is compressed and decompressed. Despite the loss of image quality, lossy compression creates smaller files and is useful when image quality is less important or when computing resources are in high demand.

COMMON GRAPHIC FORMATS

JPEG is a type of lossy image compression format developed in the early 1990s. JPEGs support RGB and CMYK color and are most useful for small images, such as those used for display on websites. JPEGs are automatically compressed using a lossy algorithm. Thus, some image quality is lost each time the image is edited and saved as a new JPEG.

GIF (Graphics Interchange Format) files have a limited color palette and use LZW compression so that they can be compressed without losing quality. Unlike JPEG, GIF supports "transparency" within an image by ignoring certain colors when displaying or printing. GIF files are open source and can be used in a wide variety of programs and applications. However, most GIF formats support only limited color because the embedded LZW compression is most effective when an image contains a limited color palette. PNGs (Portable Network Graphics) are open-source alternatives to GIFs that support transparency and 24-bit color. This makes them better at complex colors than GIFs.

SVGs (Scalable Vector Graphics) are an open-source format used to store and transfer vector images. SVG files lack built-in compression but can be compressed using external programs. In addition, there are "metafile" formats that can be used to share images combining both vector and raster elements. These include PDF (Portable Document Format) files, which are used to store and display documents, and the Encapsulated PostScript (EPS) format, which is typically used to transfer image files between programs.

—*Micah L. Issitt*

BIBLIOGRAPHY

Brown, Adrian. *Graphics File Formats*. Kew: Natl. Archives, 2008. PDF file. Digital Preservation Guidance Note 4.

Celada, Laura. "What Are the Most Common Graphics File Formats." *FESPA*. FESPA, 27 Mar. 2015. Web. 11 Feb. 2016.

Costello, Vic, Susan Youngblood, and Norman E. Youngblood. *Multimedia Foundations: Core Concepts for Digital Design*. New York: Focal, 2012. Print.

Dale, Nell, and John Lewis. *Computer Science Illuminated*. 6th ed. Burlington: Jones, 2016. Print.

"Introduction to Image Files Tutorial." *Boston University Information Services and Technology*. Boston U, n.d. Web. 11 Feb. 2016.

Stuart, Allison. "File Formats Explained: PDF, PNG and More." *99Designs*. 99Designs, 21 May 2015. Web. 11 Feb. 2016.

GREEN COMPUTING

FIELDS OF STUDY

Computer Engineering; Information Technology

ABSTRACT

"Green computing" refers to efforts to manufacture, use, and dispose of computers and digital devices in eco-friendly ways. Concerned with the effects of computer-related pollution and energy use on the environment, various government bodies and industry organizations have sought to promote more sustainable computing practices and the recycling of electronic waste.

PRINICIPAL TERMS

- **e-waste:** short for "electronic waste"; computers and other digital devices that have been discarded by their owners.
- **Green Electronics Council:** a US nonprofit organization dedicated to promoting green electronics.
- **hibernation:** a power-saving state in which a computer shuts down but retains the contents of its random-access memory.
- **TCO certification:** a credential that affirms the sustainability of computers and related devices.
- **undervolting:** reducing the voltage of a computer system's central processing unit to decrease power usage.

BACKGROUND ON GREEN COMPUTING

As pollution and climate change have become more pressing concerns, the need to consider the effects of industry and daily life on the environment has become apparent. As an especially fast-growing field, technology attracted particular notice from lawmakers, nonprofit organizations, and members of the public concerned with sustainability. "Green computing" describes the attempts to address the environmental impacts of widespread digital technology use.

Proponents have highlighted the excessive energy use by large computer systems such as data centers as a particular concern. Centers in the United States consume large amounts of energy, due in part to inefficient operating practices. Data centers used about 2 percent of all electricity in the United States in 2010.

On average, such centers use less than 15 percent of that power for computing data. The bulk of it is used to maintain idle servers in case of a sudden spike in activity. In addition, such data centers often rely on diesel generators for backup power. They thereby emit greenhouse gases that contribute to climate change.

Another concern is the effects of the computer manufacturing process. Digital devices contain minerals such as rare earth elements that must be mined from the earth and processed, which can create toxic waste. Waste is also a matter of concern when an obsolete or otherwise unwanted device is discarded. The resulting electronic waste, or e-waste, often contains materials that may contaminate the surrounding area. Cathode-ray tube (CRT) monitors, for instance, contain lead, which can leach into soil and water from a landfill. Many computer parts are also not biodegradable. This means that they do not break down over time and will thus continue to be a cause of concern for decades.

OVERSIGHT AND CERTIFICATIONS

A number of government and nonprofit organizations promote green computing and address issues such as environmentally responsible energy use and manufacturing. The US Environmental Protection Agency (EPA) launched the Energy Star program in 1992 to reduce air pollution. In the 2000s and 2010s,

Green computing, or sustainable computing, involves practices of environmental sustainability in information technology. Computer technology and components that reach a certain level of recyclability, biodegradability, and energy efficiency are labeled with the Energy Star logo. By United States Environmental Protection Agency, public domain, via Wikimedia Commons.

it has focused on reducing greenhouse-gas emissions through energy efficiency. To earn the Energy Star certification for their products, computer manufacturers must demonstrate that their devices meet the EPA's energy efficiency standards.

TCO Development, a Swedish nonprofit, offers TCO certification. This credential was originally devised in 1992 and modernized in 2012. It is granted to computers, mobile phones, and peripherals that meet certain sustainability requirements in all stages of their life cycle, from manufacture to disposal.

The nonprofit Green Electronics Council manages the Electronic Product Environmental Assessment Tool (EPEAT), created in 2005. Manufacturers provide the council with product information covering an item's life cycle. That information is then compared against other products and listed on the public EPEAT registry, which buyers can access.

GREEN COMPUTING INITIATIVES

Perhaps the most common green computing initiatives have been efforts to educate corporations and the public about the environmental effects of computer use, particularly in regard to energy consumption. Computer users can save energy through techniques such as putting a computer in hibernation mode. When a computer enters this mode, it essentially shuts down and stops using power. It still retains the contents of its random-access memory (RAM), allowing the user to resume their previous activities after exiting hibernation mode.

Another technique, called undervolting, lowers the voltage of the computer system's central processing unit (CPU). The CPU does need a minimum voltage to function, but decreasing the voltage by a small amount can improve the system's energy efficiency. In a mobile device, heat is reduced while battery life is prolonged. Other measures including turning off devices when not in use and enabling power-saving settings when available.

PRESENT AND FUTURE CONCERNS

E-waste is one of the most significant computer-related environmental issues. Advances in computer and mobile technology continue apace, and new devices constantly replace obsolete or outmoded ones. Because of this, the disposal of discarded devices represents a serious and growing environmental problem, in large part because of parts that do not biodegrade and are environmentally hazardous. To prevent unwanted devices from entering landfills, major electronics companies such as Apple, Motorola, and Nintendo offer "take back" programs in which owners may return their devices to the manufacturers. Firms sometimes offer a payment or a discount on a new device in exchange. In some cases, the collected devices are refurbished and resold. In others, they are recycled in an eco-friendly manner.

Another major concern is that as the amount of data being created, stored, and accessed continues to grow, more numerous and powerful data centers will be needed. Unless the issue is addressed, today's energy inefficiency will lead to ever-greater amounts of electricity being wasted in the future.

—*Joy Crelin*

BIBLIOGRAPHY

Dastbaz, Mohammad, Colin Pattinson, and Bakbak Akhgar, eds. *Green Information Technology: A Sustainable Approach*. Waltham: Elsevier, 2015. Print.

Feng, Wu-chun, ed. *The Green Computing Book: Tackling Energy Efficiency at Large Scale*. Boca Raton: CRC, 2014. Print.

Glanz, James. "Power, Pollution and the Internet." *New York Times*. New York Times, 22 Sept. 2012. Web. 28 Feb. 2016.

Ives, Mike. "Boom in Mining Rare Earths Poses Mounting Toxic Risks." *Environment 360*. Yale U, 28 Jan. 2013. Web. 28 Feb. 2016.

Smith, Bud E. *Green Computing: Tools and Techniques for Saving Energy, Money, and Resources*. Boca Raton: CRC, 2014. Print.

"What You Can Do." *Green Computing*. U of California, Berkeley, n.d. Web. 28 Feb. 2016.

I

INFORMATION TECHNOLOGY

FIELDS OF STUDY

Computer Science; Network Design

ABSTRACT

Information technology (IT) is devoted to the creation and manipulation of information using computers and other types of devices. It also involves the installation and use of software, the sets of instructions computers follow in order to function. At one time, most computers were designed with a single purpose in mind, but over time, IT has transformed into a discipline that can be used in almost any context.

PRINICIPAL TERMS

- **device:** equipment designed to perform a specific function when attached to a computer, such as a scanner, printer, or projector.
- **hardware:** the physical parts that make up a computer. These include the motherboard and processor, as well as input and output devices such as monitors, keyboards, and mice.
- **network:** two or more computers being linked in a way that allows them to transmit information back and forth.
- **software:** the sets of instructions that a computer follows in order to carry out tasks. Software may be stored on physical media, but the media is not the software.
- **system:** either a single computer or, more generally, a collection of interconnected elements of technology that operate in conjunction with one another.
- **telecom equipment:** hardware that is intended for use in telecommunications, such as cables, switches, and routers.

HISTORY OF INFORMATION TECHNOLOGY

Information technology (IT) encompasses a wide range of activities, from pure theory to hands-on jobs. At one end of the spectrum are IT professionals who design software and create system-level network designs. These help organizations to maximize their efficiency in handling data and processing information. At the other end are positions in which physical hardware, from telecom equipment to devices such as routers and switches, are connected to one another to form networks. They are then tested to make sure that they are working correctly. This range includes many different types of employees. For example, there are computer technicians, system administrators, programmers at the system and application levels, chief technology officers, and chief information officers.

One way of studying the history of IT is to focus on the ways that information has been stored. IT's history can be divided into different eras based on what type of information storage was available. These eras include prehistoric, before information was written down, and early historical, when information started to be recorded on stone tablets. In the middle historical period, information was recorded on paper and stored in libraries and other archives. In the modern era, information has moved from physical storage to electronic storage. Over time, information storage has become less physical and more Abstract. IT now usually refers to the configuration of computer hardware in business networks. These allow for the manipulation and transfer of electronically stored information.

DOT-COM BUBBLE

IT gained prominence in the 1990s, as the Internet began to grow rapidly and become more user-friendly than it had been in the past. Many companies arose to try to take advantage of the new business models that it made possible. Computer programmers and network technology experts found themselves in high demand. Startup companies tried to build online services quickly and effectively. Investment in technology companies put hundreds of millions of dollars into IT research. Even established companies realized that they needed to invest in their IT infrastructure and personnel if they wanted to stay competitive. As the IT sector of the economy grew rapidly, financial experts began to worry that it was forming an economic bubble. An economic bubble occurs when a market grows rapidly and then that growth declines abruptly. The bubble eventually "pops," and investors pull their money out. This did happen, and many Internet startups shut down.

While the dot-com bubble, as it came to be known, passed quickly, IT remained a central part of life. Simple tasks that used to be done without sophisticated technology, such as banking, shopping, and even reading a book, now involve computers, mobile phones, tablets, or e-readers. This means that the average person must be more familiar with IT in the twenty-first century than in any previous era. Because of this, IT has become a topic of general interest. For example, an average person needs to know a bit about network configuration in order to set up a home system.

DATA PRODUCTION GROWTH

With IT, new information is constantly being created. Once it was possible for a single person to master all of society's knowledge. In the modern world, more data is produced every year than a person could assimilate in lifetime. It is estimated that by 2020, there will be more than five thousand gigabytes (GB) of data for each and every person on earth.

The availability of IT is the factor most responsible for the explosion in data production. Most cell phone plans measure customer data in how many GB per month may be used, for example. This is because of the many photos, videos, and social media status updates people create and share on the Internet every day. It is estimated that every two days, human beings create as much information as existed worldwide before 2003. The pace of this data explosion increases as time goes on.

—*Scott Zimmer, JD*

BIBLIOGRAPHY

Black, Jeremy. *The Power of Knowledge: How Information and Technology Made the Modern World.* New Haven: Yale UP, 2014. Print.

Bwalya, Kelvin J., Nathan M. Mnjama, and Peter M. I. I. M. Sebina. *Concepts and Advances in Information Knowledge Management: Studies from Developing and Emerging Economies.* Boston: Elsevier, 2014. Print.

Campbell-Kelly, Martin, William Aspray, Nathan Ensmenger, and Jeffrey R. Yost. *Computer: A History of the Information Machine.* Boulder: Westview, 2014. Print.

Fox, Richard. *Information Technology: An Introduction for Today's Digital World.* Boca Raton: CRC, 2013. Print.

Lee, Roger Y., ed. *Applied Computing and Information Technology.* New York: Springer, 2014. Print.

Marchewka, Jack T. *Information Technology Project Management.* 5th ed. Hoboken: Wiley, 2015. Print.

INTEGRATED DEVELOPMENT ENVIRONMENTS

FIELDS OF STUDY

Software Engineering; System-Level Programming

ABSTRACT

An integrated development environment (IDE) is a computer program that brings together all of the tools needed for software development. These tools usually include a text editor for writing source code, a compiler for converting code into machine language, and a debugger to check programs for errors. The programmer writes the code, runs the code through the debugger, and then compiles it into a program.

PRINICIPAL TERMS

- **command line:** a prompt where a user can type commands into a text-based interface.
- **graphical user interface (GUI):** an interface that allows a user to interact with pictures instead of requiring the user to type commands into a text-only interface.
- **platform:** the operating system a computer uses and with which all of that computer's software must be compatible. The major platforms are Windows, Mac, and Linux.
- **programming languages:** sets of terms and rules of syntax used by computer programmers to create instructions for computers to follow. This code is then compiled into binary instructions for a computer to execute.
- **visual programming:** a form of programming that allows a programmer to create a program by dragging and dropping visual elements with a mouse instead of having to type in text instructions.

BUILDING AND RUNNING CODE

Computers are not intelligent, even though they can process information much faster than humans. They only know what to do when a human tells them. One way of doing this is by typing instructions using the computer's command line, but this has several drawbacks. First, it requires that a human be present whenever the program has to be run. Second, typing commands into a text-based interface may be uncomfortable or even impossible for some users. They prefer to use a graphical user interface (GUI), which lets them click on icons to run applications (apps). To avoid the need to type commands, computer programmers write programs. These are sets of instructions for a computer to perform, like a recipe used for cooking. Instead of typing in dozens of commands to do each task, the user can simply run the program that contains all of the commands. The computer will follow the instructions just as a cook follows a recipe.

Programs are created using a programming language, such as Java, C++, or Python. A programming language is not as complete as a traditional language such as English or Mandarin, but it contains enough words and rules of syntax to allow programmers to create instructions, or code. Once the source code has been written, the next step is to translate it into machine language that a computer can process. An app called a "compiler" performs this translation. The compiler produces an executable version of the program. This version can be run by any user as long as it is executed on the computer platform it was designed for (such as Windows or Mac).

Programmers often use an integrated development environment (IDE) when coding. An IDE is software that makes programming easier by providing many tools within a single program and automating some simple tasks. An IDE usually includes an editor to write source code, a debugger to fix errors, and build automation tools. Many also have intelligent code completion, which autocompletes commonly used programming expressions.

VISUAL APPLICATION DEVELOPMENT

The first programmers did not use IDEs. They had to create programming languages before they could even write programs. They used many separate programs, including text editors, debuggers, and compilers, to craft their software. This sometimes made the process of programming more complicated. Programmers had to learn how to use several distinct apps that had been designed separately. Thus, it took longer to learn how to use each one. The different tools sometimes conflicted or were incompatible. Tiring of this, some programmers began to develop IDEs made up of multiple programming tools, each designed to work with the others.

The first IDEs were largely text-based, because this is how programming was then done. However,

Integrated development environments pull together the tools for multiple aspects of software development into a single program. Text editors (to write and edit source code), translators (to convert source code into object code), run-time environments (to allow editors to see the output of their code), debugging tools (to fix errors), and auto-documentation tools (to create the necessary software documentation) allow an editor to create software from design to release. EBSCO illustration.

programming gradually took a more visual approach, in the same way that other types of apps went from text-based interfaces to GUIs. Visual programming allows software developers to use a mouse to drag and drop pieces of code in order to design apps. This is often a much more intuitive method for designing an app and can be less time consuming than typing out lines of source code. Today, several IDEs support visual programming. Using these tools, programmers can create blocks of code in one window, then click to debug and compile the code.

ATTITUDES TOWARD IDES

The introduction of IDEs was not greeted with universal praise. Some programmers, even in the twenty-first century, view such tools as limiting their own creativity as programmers. IDEs force programs to be written in certain ways in order to comply with the environment's design. These programmers prefer to write code the old-fashioned way, even if doing so is more labor intensive and time consuming.

One consequence of the emergence of visual IDEs is that younger programmers may never be exposed to the text-based development tools that their predecessors made and used. These contemporary programmers learn in visual environments where there is less need for memorizing programming commands. Veteran programmers have suggested that this is not necessarily a good thing. Not every app or programming task lends itself to visual development, they argue, so programmers should be able to use text-based development tools as well as visual ones.

—*Scott Zimmer, JD*

BIBLIOGRAPHY

Christiano, Marie. "What Are Integrated Development Environments?" *All about Circuits.* EETech Media, 3 Aug. 2015. Web. 23 Feb. 2016.

Freedman, Jeri. *Software Development.* New York: Cavendish Square, 2015. Print.

Mara, Wil. *Software Development: Science, Technology, and Engineering.* New York: Children's, 2016. Print.

Patrizio, Andy. "The History of Visual Development Environments: Imagine There's No IDEs. It's Difficult If You Try." *Mendix.* Mendix, 4 Feb. 2013. Web. 23 Feb. 2016.

Schmidt, Richard F. *Software Engineering: Architecture-driven Software Development.* Waltham: Morgan, 2013. Print.

Tucker, Allen B., Ralph Morelli, and Chamindra de Silva. *Software Development: An Open Source Approach.* Boca Raton: CRC, 2011. Print.

INTELLIGENT TUTORING SYSTEM

FIELD OF STUDY

Tutoring systems

ABSTRACT

An intelligent tutoring system is one that modifies its interaction with the person being tutored bases on the tutee's performance. Just as an experienced teacher will adopt his information presentation to the characteristics of the learner so an intelligent tutoring system with adjust the rate of presentation to the learners prior knowledge, level of interest, and general intelligence.

PRINCIPAL TERMS

- **Andes:** An intelligent tutoring system for a mathematically based course in introductory physics developed for the U. S. Naval Academy.
- **AutoTutor:** a program, which tutors using ordinary language and grades student responses using some form of latent semantic analysis.
- **conceptual physics:** Physics taught with minimum emphasis on mathematical methods. The physics teaching community was surprised by the discovery of Eric Mazur that students could score very well on course exams and yet have very little understanding or ability to explain their understanding of the problems they had just solved.
- **cognitive load (Sweller):** the number of slots in short term memory required to solve problems. When then information presented in a text exceeds the cognitive capacity of the students, little learning will occur.
- **latent semantic analysis (LSA):** a method to summarize the content of a text corpus in terms of correlations between words, allowing for an evaluation of a student's answers to questions in comparison to a subject matter corpus. Generally LSA in involves finding a 300-500 dimensional vector space representation of the text.
- **corpus:** a body of text that the student is expected to be able to answer question about.
- **user modeling:** a mathematical model of student behavior while learning.

The Need for Intelligent Tutoring Systems (ITS's)

While there is no restriction of intelligent tutoring systems to topics of military importance, probably the greatest stimulus to their development comes from the armed forces. Each year several hundred thousand individuals begin advanced individual training as members of the United States' Armed Forces. The majority of these will serve out their initial enlistment period then go on to alternative civilian employment. As a result the will be a tremendous need for basic technical training, probably a greater need than can be met with the available teaching manpower at any time in the near future.

Special Problem with Mathematics-Based Subjects

Though traditional instruction in engineering, physics, and chemistry have focused on students' problem solving ability, students can circumvent the process by memorizing a limited set of problem templates. Indeed, the need to confirm that students are actually developing their understanding of the material is always a background consideration.

Several computer programs have been developed in recent years as intelligent tutoring systems. Among the major sponsors of research in this area are the National Science Foundation, the Institute of Education Sciences and the Office of Naval Research. These researchers are slowly converging upon a general framework for intelligent tutoring systems.

Major Components of an Intelligent Tutoring System

Here we adopt the Generalized Intelligent Framework for Tutoring (GIFT), which has grown out of recent research at the U. S. Army Research Laboratory. It is of course highly desirable that ITS's be modular in design. So that an ITS developed for say organic chemistry could have some interchangeable parts with ITS's for Physical Chemistry or Physics, to minimize the total effort required in in research and development and to maximize transfer of learning from one ITS to another. It is generally accepted that ITS'S will have four major components: (1) a domain model (2) a learner model (3) a pedagogical model and (4) a tutor-user interface. The GIFT model assumes in addition a sensor module, which provides a non-verbal record of the student's psychological state while learning.

The domain model involves the structure of the knowledge to be developed in the student. It normally contains the knowledge that experts bring to the subject along with an awareness of pitfalls and popular misconceptions.

The learner model consists of the succession of psychological states that the learner is expected to go through in the course of learning. It can be viewed as an overlay of the domain model, which will change over the course of learning.

The pedagogical model takes the domain and learner models as input and selects the next move that the tutor will take. In mixed initiative systems, learners may take actions, or ask for help as well.

The tutor – user interface records and interprets the learner's responses obtained through various inputs: mouse clicks. typed-in answers, eye tracking, and possibly postural data and keeps a record as the tutoring progresses.

TRACKING STUDENT PROGRESS

Here are several examples of learner modeling, used in contemporary ITS's:

Knowledge tracing: If the knowledge to be conveyed is contained in a set of production rules then skillometer can be used to visually display the students progress. Step by step knowledge tracing is incorporated in a number of tutoring programs developed in the Pittsburgh Science of Learning Center.

Constraint bases modeling: A relevant satisfied state constraint corresponds to an aspect of the problem solution. Learner modeling is tracked by which constraints are followed by students in solving problems. Successful constraint based tutors include the structured query language (SQL) tutor.

Knowledge space models underlies the very successful Assessment and Learning in Knowledge Spaces (ALEKS) mathematics tutor now widely used on college campuses to deal with students have not studied mathematics in recent semesters. The knowledge space is a large number of possible knowledge states of each mathematical field. As the topics are reviewed the student builds a coherent picture if the field as a whole.

Expectation and misconception tailored dialog. These are particularly important in fields like physics where terms from common usage, like force and energy have specific restricted meanings which students must understand before progress is possible.

Donald Franceschetti, PhD

BIBLIOGRAPHY

Mandl, H., and A. Lesgold, eds. *Learning Issues for Intelligent Tutoring Systems.* New York: Springer, 1988. Print.

Stein, N. L., and S.W. Raudenbush, eds. *Developmental Cognitive Science Goes to School.* New York: Routledge, 2011. Print

Sottilare, R., Graesser, A., Hu, X., and Holden, H. (Eds.). (2013). *Design Recommendations for Intelligent Tutoring Systems: Volume 1 - Learner Modeling.* Orlando, FL: U.S. Army Research Laboratory. ISBN 978-0-9893923-0-3. Available at: https://gift-tutoring.org/documents/42

INTERNET PRIVACY

FIELDS OF STUDY

Computer Science; Privacy; Security

ABSTRACT

Internet privacy is an issue of concern in the early twenty-first century. With increasing Internet use for work, socializing, and daily tasks comes a corresponding increase in potential breaches of privacy. To address this issue, individuals may use a variety of technologies and techniques to ensure the security of their personal information and online communications.

PRINICIPAL TERMS

- **behavioral marketing:** advertising to users based on their habits and previous purchases.
- **cookies:** small data files that allow websites to track users.
- **digital legacy (digital remains):** the online accounts and information left behind by a deceased person.
- **device fingerprinting:** the practice of collecting identifying information about a computer or other web-enabled device.
- **personally identifiable information (PII):** information that can be used to identify a specific individual.
- **Privacy Incorporated Software Agents (PISA):** a project that sought to identify and resolve privacy problems related to intelligent software agents.

INTERNET PRIVACY CONCERNS

The Internet has enabled individuals the world over to communicate with one another and share information. Internet users can socialize, shop, conduct financial transactions, and carry out other tasks online. The privacy of their online actions and the data they share have become increasingly at risk, however. Breaches of online privacy can take many forms. It can be the inadvertent sharing of personally identifiable information (PII) through lax social media security settings. Or it can be the theft of banking or credit card information. Insufficient privacy practices can make it possible for criminals to gain access to individuals' contact details, health data, financial information, or government identification information such as Social Security numbers. Such access can lead to identity theft or fraud.

Cybercrime is one of the major causes of concern in regard to Internet privacy. Privacy breaches by corporations and marketers are also major concerns. Tracking technologies enable advertisers to market to individuals based on their browsing habits and previous purchases. Some users enjoy being served ads relevant to their shopping patterns, but others view that it as an invasion of privacy. Protecting one's online activity from government surveillance is likewise of concern to some Internet users. Their concern increased after the secret US National Security Agency (NSA) Internet surveillance program PRISM was revealed in 2013.

RESEARCH

A number of government and public-sector organizations have researched the privacy needs of Internet users. One such research initiative, the European-run Privacy Incorporated Software Agents (PISA), sought to identify and resolve privacy problems associated with intelligent software agents. These agents are computer systems that act on the behalf of the user in a semiautonomous manner. The researchers also examined existing privacy laws and best practices. They created a prototype interface designed to meet usability requirements of users. The interface was also designed to gain users' trust, so that they would feel comfortable entering their personal information. The researchers found that even when privacy options were provided, as in a settings or control panel interface, many users had difficulty understanding how to use those options to meet their privacy needs. The users also found it hard to understand how the privacy options were connected to the personal information they had entered.

THREATS TO INTERNET PRIVACY

Potential threats to Internet privacy take a number of forms. Perhaps the most high-profile threat is that of hackers who seek to steal individuals' personal data. In some cases, criminals may steal data such as credit card numbers by hacking into the servers of online retailers. In other cases, a criminal may use

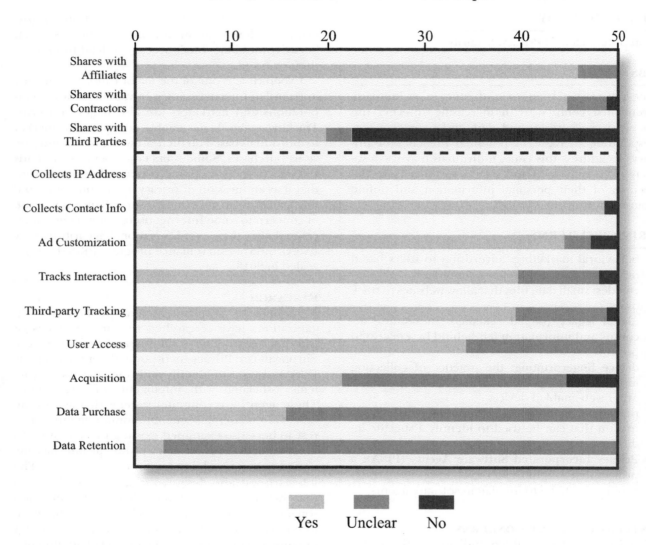

Personal Data Collected from the 50 Most-Popular Websites

Graph showing privacy policy practices of the 50 most visited websites, adapted from KnowPrivacy (knowprivacy.org) research conducted by UC Berkeley School of Information, 2009. Users' personal data is often collected and shared with affiliates or contractors that the site does not consider "third party," making them exempt from privacy policies stating the site does not share users' information. In many cases, it is unclear what information is being collected. EBSCO illustration.

phishing e-mails to trick an individual into revealing personal information. Some criminals use surveillance programs such as keyloggers. Keyloggers can record all keystrokes made on a computer and transmit that data to another user. Most threats to Internet privacy are digital in nature, but the consequences of weak Internet privacy can also extend

past the computer screen. A lack of privacy protections on social media can reveal an individual's address and the dates they will be on vacation to potential burglars, for instance.

Privacy concerns likewise extend to online marketing. Many marketers use cookies, small data files that track individual users on the Internet. Cookies allow marketers to send individuals targeted ads based on prior purchases and browsing habits. For example, an individual may view a pair of shoes online but not immediately buy them. The shoe seller may use cookies to send ads for the shoes to the user as they keep browsing online. This process is called behavioral marketing.

In addition to criminals and ads, the government is sometimes perceived as a threat to online privacy. In the United States, government surveillance of online activity greatly increased following the terrorist attacks of September 11, 2001, and the subsequent passage of the controversial PATRIOT Act. In 2013 the US government confirmed the existence of the Internet surveillance program PRISM. The National Security Agency used PRISM to access data from companies such as Google and Microsoft. The government stated it needed this access to protect national security.

DIGITAL REMAINS

Among the more unusual Internet privacy dilemmas is that of an individual's digital legacy, or digital remains. These are the online information, records of communication, and personal websites or profiles of a deceased person. Digital remains may continue to exist on the Internet long after the person has died. In certain cases, the deceased's next of kin may leave online remains in place as a digital memorial. Or, if allowed, they may delete them to protect the deceased person's privacy. Some online services, such as the social network Facebook, offer users the option to name a contact who can gain access to the account after the user's death. The contact can then determine whether to delete or preserve profile information. Many legal issues remain around who can access and control an individual's digital remains and to what extent this information may be controlled.

PROTECTING INTERNET PRIVACY

Those concerned about Internet privacy can make their personal data more secure. To obtain a basic level of privacy, individuals should observe standard online security procedures. These include using strong, unique passwords and two-factor authentication, in which a password is paired with another form of authentication, when available. Running antivirus or firewall software allows individuals to identify and remove programs such as keyloggers or prevent them from infecting the computer in the first place. Individuals can also protect the privacy of credit card or Social Security numbers by entering them only on trusted websites that encrypt such data. Most standard web browsers offer users the ability to delete cookies, which temporarily prevents such tracking from taking place. However, it is sometimes possible to track devices without the use of cookies through device fingerprinting. Thus, deleting cookies can provide only a limited level of privacy.

—*Joy Crelin*

BIBLIOGRAPHY

Bernal, Paul. *Internet Privacy Rights: Rights to Protect Autonomy.* New York: Cambridge UP, 2014. Print.

"Internet Privacy." *ACLU.* American Civil Liberties Union, 2015. Web. 28 Feb. 2016.

"Online Privacy: Using the Internet Safely." *Privacy Rights Clearinghouse.* Privacy Rights Clearinghouse, Jan. 2016. Web. 28 Feb. 2016.

"Protect Your Privacy on the Internet." *Safety and Security Center.* Microsoft, n. d. Web. 28 Feb. 2016.

"A Short History of US Internet Legislation: Privacy on the Internet." *ServInt.* ServInt, 17 Sept. 2013. Web. 28 Feb. 2016.

"What Is Personally Identifiable Information (PII)?" *U Health.* U of Miami Health System, n.d. Web. 28 Feb. 2016.

IOS

FIELDS OF STUDY

Operating Systems; Software Engineering; Mobile Platforms

ABSTRACT

Apple's iOS is an operating system designed for mobile computing. It is used on the company's iPhone and iPad products. The system, which debuted in 2007, is based on the company's OS X. iOS was the first mobile operating system to incorporate advanced touch-screen controls.

PRINICIPAL TERMS

- **jailbreaking:** the process of removing software restrictions within iOS that prevent a device from running certain kinds of software.
- **multitasking:** in computing, the process of completing multiple operations concurrently.
- **multitouch gestures:** combinations of finger movements used to interact with touch-screen or other touch-sensitive displays in order to accomplish various tasks. Examples include double-tapping and swiping the finger along the screen.
- **platform:** the underlying computer system on which an application is designed to run.
- **3D Touch:** a feature that senses the pressure with which users exert upon Apple touch screens.
- **widgets:** small, self-contained applications that run continuously without being activated like a typical application.

A New Generation of Mobile Operating Systems

Apple's iOS is an operating system (OS) designed for use on Apple's mobile devices, including the iPhone, iPad, Apple TV, and iPod Touch. In 2016, iOS was the world's second most popular mobile OS after the Android OS. Introduced in 2007, iOS was one of the first mobile OSs to incorporate a capacitive touch-screen system. The touch screen allows users to activate functions by touching the screen with their fingers. The Apple iOS was also among the first mobile OSs to give users the ability to download applications (apps) to their mobile devices. The iOS is therefore a platform for hundreds of thousands third-party apps.

The first iOS system and iPhone were unveiled at the 2007 Macworld Conference. The original iOS had a number of limitations. For example, it was unable to run third-party apps, had no copy and paste functions, and could not send e-mail attachments. It was also not designed for multitasking, forcing users to wait for each process to finish before beginning another. However, iOS introduced a sophisticated capacitive touch screen. The iOS touch features allowed users to activate most functions with their fingers rather than needing a stylus or buttons on the device. The original iPhone had only five physical buttons. All other functions, including the keyboard, were integrated into the device's touch screen. In addition, the iOS system supports multitouch gestures. This allows a user to use two or more fingers (pressure points) to activate additional functions. Examples include "pinching" and "stretching" to shrink or expand an image.

JAILBREAKING

Computer hobbyists soon learned to modify the underlying software restrictions built into iOS, a process called jailbreaking. Modified devices allow users greater freedom to download and install apps. It also allows users to install iOS on devices other than Apple devices. Apple has not pursued legal action against those who jailbreak iPhones or other devices. In 2010, the US Copyright Office authorized an exception permitting users to jailbreak their legally owned copies of iOS. However, jailbreaking iOS voids Apple warranties.

VERSION UPDATES

The second version of iOS was launched in July 2008. With iOS 2, Apple introduced the App Store, where users could download third-party apps and games. In 2009, iOS 3 provided support for copy and paste functions and multimedia messaging. A major advancement came with the release of iOS 4 in 2010. This update introduced the ability to multitask, allowing iOS to begin multiple tasks concurrently without waiting for one task to finish before initiating the next task in the queue. The iOS 4 release was also the first to feature a folder system in which similar apps could be grouped together on the device's home screen (called the "springboard"). FaceTime video calls also became available with iOS 4.

iOS integrated Siri, a voice-activated search engine, for the iPhone. Through multiple updates, its capabilities have expanded to work extremely well with a number of popular websites with all Apple apps, as well as with many popular third-party apps. Vasile Cotovanu, CC BY 2.0 (http://creativecommons.org/licenses/by/2.0), via Wikimedia Commons.

The release of iOS 5 in 2011 integrated the voice-activated virtual assistant Siri as a default app. Other iOS 5 updates include the introduction of iMessage, Reminders, and Newsstand. In 2012, iOS 6 replaced Google Maps with Apple Maps and redesigned the App Store, among other updates. Released in 2013, iOS 7 featured a new aesthetic and introduced the Control Center, AirDrop, and iTunes Radio.

NEW INNOVATIONS

With the release of iOS 8, Apple included third-party widget support for the first time in the company's history. Widgets are small programs that do not need to be opened and continuously run on a device. Examples including stock tickers and weather widgets that display current conditions based on data from the web. Widgets had been a feature of Android and Windows mobile OSs for years. However, iOS 8 was the first iOS version to support widgets for Apple. Since their release, Apple has expanded the availability of widgets for users.

The release of iOS 9 in 2015 marked a visual departure for Apple. This update debuted a new typeface for iOS called San Francisco. This specially tailored font replaced the former Helvetica Neue. The release of iOS 9 also improved the battery life of Apple devices. This update introduced a low-power mode that deactivates high-energy programs until the phone is fully charged. Low-power mode can extend battery life by as much as an hour on average.

Coinciding with the release of iOS 9, Apple also debuted the iPhone 6S and iPhone 6S Plus, which introduced 3D Touch. This new feature is built into the hardware of newer Apple devices and can sense how deeply a user is pressing on the touch screen. 3D Touch is incorporated into iOS 9 and enables previews of various functions within apps without needing to fully activate or switch to a new app. For instance, within the camera app, lightly holding a finger over a photo icon will bring up an enlarged preview without needing to open the iPhoto app.

—*Micah L. Issitt*

Bibliography

Heisler, Yoni. "The History and Evolution of iOS, from the Original iPhone to iOS 9." *BGR.* BGR Media, 12 Feb. 2016. Web. 26 Feb. 2016.

"iOS: A Visual History." *Verge.* Vox Media, 16 Sept. 2013. Web. 24 Feb. 2016.

Kelly, Gordon, "Apple iOS 9: 11 Important New Features." *Forbes.* Forbes.com, 16 Sept. 2015. Web. 28 Feb. 2016.

Parker, Jason, "The Continuing Evolution of iOS." *CNET.* CBS Interactive, 7 May 2014. Web. 26 Feb. 2016.

Williams, Rhiannon. "Apple iOS: A Brief History." *Telegraph.* Telegraph Media Group, 17 Sept. 2015. Web. 25 Feb. 2016.

Williams, Rhiannon, "iOS 9: Should You Upgrade?" *Telegraph.* Telegraph Media Group, 16 Sept. 2015. Web. 25 Feb. 2016.

L

LISP

FIELDS OF STUDY

Programming Languages; Software Engineering

ABSTRACT

LISP is the second-oldest programming language still in common use, and is one of the preferred languages for research and development in the field of artificial intelligence. Numerous variations of the language exists, but are becoming standardized in the variations called Common LISP and Scheme. LISP is designed for the manipulation of lists of data objects, rather than for "number crunching," although it is founded on principles of mathematical logic. The language was developed in 1958.

PRINICIPAL TERMS

- **artificial intelligence:** the intelligence exhibited by machines or computers, in contrast to human, organic, or animal intelligence.
- **character:** a unit of information that represents a single letter, number, punctuation mark, blank space, or other symbol used in written language.
- **compliance:** adherence to standards or specifications established by an official body to govern a particular industry, product, or activity.
- **declarative language:** language that specifies the result desired but not the sequence of operations needed to achieve the desired result.
- **floating-point arithmetic:** a calculation involving numbers that have a decimal point that can be placed anywhere through the use of exponents, as is done in scientific notation.
- **programming languages:** sets of terms and rules of syntax used by computer programmers to create instructions for computers to follow. This code is then compiled into binary instructions for a computer to execute.
- **syntax:** rules that describe how to correctly structure the symbols that comprise a language.

HISTORY AND CHARACTERISTICS OF LISP

LISP is the second-oldest of programming languages that is still in common use. LISP, a contraction of the words LISt Processor, was described in 1958, by John McCarthy, at MIT. It has been most commonly used in university laboratories and only recently has it become more generally utilized in specialized fields that rely of the manipulation of data lists. LISP is prefered for research and development in the field of artificial intelligence. It is a high-level declarative language developed for manipulating lists of data objects, rather than for numerical calculation. There are dozens of variations of the LISP language that have not been fully standardized. Efforts to bring LISP into compliance with a standard definition of the language by ANSI and IEEE have focused on the Common LISP and Scheme variations of LISP. Symbols and lists are the essential data types for the theory behind the LISP language. Math data types are not central to LISP programming although LISP does support floating point arithmetic and integers of all sizes. A math calculation carried out in a LISP program will therefore be expected to return a complete numerical value rather than an abbreviated, rounded value.

PROGRAM FEATURES OF LISP

The program structure and syntax of LISP are very unlike those of general purpose programming languages such as C, C++, Java and other languages that rely on statements and subroutines in an object-oriented programming format. LISP programs use primarily expressions and functions stated as a "form." Every LISP expression returns a value, and every LISP procedure returns a data object. The difference in syntax and program execution does allow the translation of source code from another language such as FORTRAN into LISP source code, but the result obtained from such translated code will not be quite the same as that obtained from true LISP code. LISP does not rely on imperative programming, but uses function statements. The simple symbolic structure of such statements is very non-intuitive to humans, but is very amenable to computations carried out by the computer. For example, the operation of multiplying two pairs of variables and then adding their resultant values would normally be written in other languages as

$$(a*b) + (c*d)$$

but in LISP this is written as

$$(+ (* a b) (* c d))$$

When this is understood, the construction of complex mathematical statements using LISP's built-in assortment of functions becomes much easier. For example, writing a LISP expression to evaluate a relation that adds two numbers and multiplies the product by some multiple of a trigonometric function can be written simply in LISP as

$$(* 4.5 (sin 27.5) (+ 6 7))$$

The code for such an expression is both compact and efficient, and conducive to the use of very extensive statements. All LISP procedures have the same value in syntax as a function, and each one returns a data object as its associated value when called. A data object may be anything from a single character to a long string of symbols or a numerical value, depending on the contextual nature of the procedure. A list is a sequence of data objects delimited by parentheses. For example, a list of three data objects (DO_n) would appear as

$$(DO_1 \ DO_2 \ DO_3).$$

Lists can also contain lists. The structure

$$((DO_1 \ DO_2) \ (DO_3 \ DO_4))$$

is a list of two lists. Symbols are strings of letters, digits and special characters, and have different contextual meanings in different statements. They typically serve as identifiers in LISP programs, much like an assigned variable name in other languages. There is a special form in LISP, called the lambda form, that allows mentioning a

procedure without a specified name. For example, to evaluate a function that accepts three numerical values and calculates the solution of the corresponding quadratic equation $y = ax^2 + bx + c$, one could write

<span class="text-node"> (lambda (a b c)
(- (* b b) (* 4.0 a c))) </span>

SAMPLE PROBLEM

Write a lambda form that accepts two numbers b and h, and evaluates the corresponding area of a right triangle.

Answer:

The formula for the area of a right triangle is A = ½ bh. Therefore the corresponding lambda form would be

<span class="text-node">(lambda (b h)
(* 0.5 b h))</span>

VALUE OF LISP PROGRAMMING

The use of data objects and lists in LISP programming and its basis on logic principles have made it a language of choice for artificial intelligence studies and development. These capabilities also make it useful for manipulating the source code of programs written in other languages, and may be an essential feature in developing computer applications that can "learn" as they carry out their functions.

—*Richard M. Renneboog M.Sc.*

BIBLIOGRAPHY

Barski, Conrad. *Land of Lisp. Learn to Program LISP One Game at a Time.* San Francisco, CA: No Starch Press, 2011. Print.

Kramer, Bill. *The Autocadet's Guide to Visual LISP.* Laurence, KS: CMP Books, 2002. Print.

Méndez, Luis Argüelles. *A Practical Introduction to Fuzzy Logic Using LISP.* New York, NY: Springer, 2016. Print.

O'Regan, Gerard. *A Brief History of Computing* 2nd ed., New York, NY: Springer-Verlag, 2012. Print.

Rawls, Rod. R., Paul F. Richard and Mark A. Hagen. *Visual LISP Programming: Principles and Techniques* Tinley Park, IL: Goodheart-Willcox, 2007. Print.

Scott, Michael L. *Programming Language Pragmatics* 4th ed., Waltham, MA: Morgan Kaufmann. Print.

Seibel, Peter. *Practical Common LISP* New York, NY: Apress/Springer-Verlag. Print.

Touretzky, David S. *Common LISP. A Gentle Introduction to Symbolic Computation* Mineola, NY: Dover Publications. Print.

M

MALWARE

FIELDS OF STUDY

Software Engineering; Security

ABSTRACT

Malware, or malicious software, is a form of software designed to disrupt a computer or to take advantage of computer users. Creating and distributing malware is a form of cybercrime. Criminals have frequently used malware to conduct digital extortion.

PRINICIPAL TERMS

- **adware:** software that generates advertisements to present to a computer user.
- **ransomware:** malware that encrypts or blocks access to certain files or programs and then asks users to pay to have the encryption or other restrictions removed.
- **scareware:** malware that attempts to trick users into downloading or purchasing software or applications to address a computer problem.
- **spyware:** software installed on a computer that allows a third party to gain information about the computer user's activity or the contents of the user's hard drive.
- **worm:** a type of malware that can replicate itself and spread to other computers independently; unlike a computer virus, it does not have to be attached to a specific program.
- **zombie computer:** a computer that is connected to the Internet or a local network and has been compromised such that it can be used to launch malware or virus attacks against other computers on the same network.

MALICIOUS PROGRAMMING

Malware, or malicious software, is a name given to any software program or computer code that is used for malicious, criminal, or unauthorized purposes. While there are many different types of malware, all malware acts against the interests of the computer user, either by damaging the user's computer or extorting payment from the user. Most malware is made and spread for the purposes of extortion. Other malware programs destroy or compromise a user's data. In some cases, government defense agencies have developed and used malware. One example is the 2010 STUXNET virus, which attacked digital systems and damaged physical equipment operated by enemy states or organizations. The earliest forms of malware were viruses and worms. A virus is a self-replicating computer program that attaches itself to another program or file. It is transferred between computers when the infected file is sent to another computer. A worm is similar to a virus, but it can replicate itself and send itself to another networked computer without being attached to another file. The first viruses and worms were experimental programs created by computer hobbyists in the 1980s. As soon as they were created, computer engineers began working on the first antivirus programs to remove viruses and worms from infected computers.

Public knowledge about malware expanded rapidly in the late 1990s and early 2000s due to several well-publicized computer viruses. These included the Happy99 worm in 1999 and the ILOVEYOU worm in May 2000, the latter of which infected nearly 50 million computers within ten days. According to research from the antivirus company Symantec in 2015, more than 317 million new malware programs were created in 2014. Yet despite public awareness of malware, many large organizations are less careful than they should be. In a 2015 study of seventy major companies worldwide, Verizon reported that almost 90 percent of data breaches in 2014 exploited known vulnerabilities that were reported in 2002 but had not yet been patched.

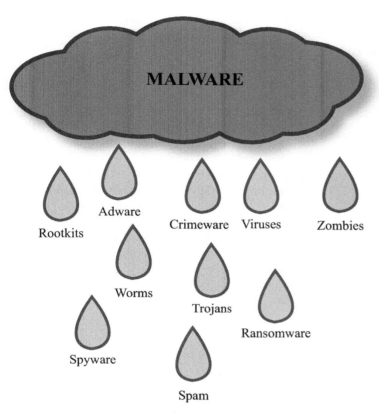

Malware consists of any software designed to cause harm to a device, steal information, corrupt data, confiscate or overwhelm the processor, or delete files. Examples of malware include adware, viruses, worms, Trojans, spam, and zombies. EBSCO illustration.

TYPES OF MALWARE

One of the most familiar types of malware is adware. This refers to programs that create and display unwanted advertisements to users, often in pop-ups or unclosable windows. Adware may be legal or illegal, depending on how the programs are used. Some Internet browsers use adware programs that analyze a user's shopping or web browsing history in order to present targeted advertisements. A 2014 survey by Google and the University of California, Berkeley, showed that more than five million computers in the United States were infected by adware.

Another type of malware is known as spyware. This is a program that is installed on a user's computer to track the user's activity or provide a third party with access to the computer system. Spyware programs can also be legal. Many can be unwittingly downloaded

by users who visit certain sites or attempt to download other files.

One of the more common types of malware is scareware. Scareware tries to convince users that their computer has been infected by a virus or has experienced another technical issue. Users are then prompted to purchase "antivirus" or "computer cleaning" software to fix the problem.

Although ransomware dates back as far as 1989, it gained new popularity in the 2010s. Ransomware is a type of malware that encrypts or blocks access to certain features of a computer or programs. Users with infected computers are then asked to pay a ransom to have the encryption removed.

ADDRESSING THE THREAT

Combating malware is difficult for various reasons. Launching malware attacks internationally makes it difficult for police or national security agencies to target those responsible. Cybercriminals may also use zombie computers to distribute malware. Zombie computers are computers that have been infected with a virus without the owner's knowledge. Cybercriminals may use hundreds of zombie computers simultaneously. Investigators may therefore trace malware to a computer only to find that it is a zombie distributor and that there are no links to the program's originator. While malware is most common on personal computers, there are a number of malware programs that can be distributed through tablets and smartphones.

Often creators of malware try to trick users into downloading their programs. Adware may appear in the form of a message from a user's computer saying that a "driver" or other downloadable "update" is needed. In other cases, malware can be hidden in social media functions, such as the Facebook "like" buttons found on many websites. The ransomware program Locky, which appeared in February 2016, used Microsoft Word to attack users' computers. Users would receive an e-mail containing a document that prompted them to enable "macros" to read the document. If the user followed the instructions, the Locky program would be installed on their computer. Essentially, users infected by Locky made two mistakes. First, they downloaded a Word document attachment from an unknown user. Then they followed a prompt to enable macros within the document—a feature that is automatically turned off in all versions of Microsoft Word. Many malware programs depend on users downloading or installing programs. Therefore, computer security experts warn that the best way to avoid contamination is to avoid opening e-mails, messages, and attachments from unknown or untrusted sources.

—*Micah L. Issitt*

BIBLIOGRAPHY

Bradley, Tony. "Experts Pick the Top 5 Security Threats for 2015." *PCWorld*. IDG Consumer & SMB, 14 Jan. 2015. Web. 12 Mar. 2016.

Brandom, Russell. "Google Survey Finds More than Five Million Users Infected with Adware." *The Verge*. Vox Media, 6 May 2015. Web. 12 Mar. 2016.

Franceschi-Bicchierai, Lorenzo. "Love Bug: The Virus That Hit 50 Million People Turns 15." *Motherboard*. Vice Media, 4 May 2015. Web. 16 Mar. 2016.

Gallagher, Sean. "'Locky' Crypto-Ransomware Rides In on Malicious Word Document Macro." *Ars Technica*. Condé Nast, 17 Feb. 2016. Web. 16 Mar. 2016.

Harrison, Virginia, and Jose Pagliery. "Nearly 1 Million New Malware Threats Released Every Day." *CNNMoney*. Cable News Network, 14 Apr. 2015. Web. 16 Mar. 2016.

Spence, Ewan. "New Android Malware Strikes at Millions of Smartphones." *Forbes*. Forbes.com, 4 Feb. 2015. Web. 11 Mar. 2016.

"Spyware." *Secure Purdue*. Purdue U, 2010. Web. 11 Mar 2016.

MEDICAL TECHNOLOGY

FIELDS OF STUDY

Applications; Information Systems

ABSTRACT

The field of medical technology encompasses a wide range of computerized devices and systems that collect information about a patient's health, process it, and produce reports that can be interpreted by doctors to help them decide what treatment to recommend. Medical technology focuses on the diagnosis and treatment of medical conditions and is considered part of the broader field of health technology.

PRINICIPAL TERMS

- **biosignal processing:** the process of capturing the information the body produces, such as heart rate, blood pressure, or levels of electrolytes, and analyzing it to assess a patient's status and to guide treatment decisions.
- **clinical engineering:** the design of medical devices to assist with the provision of care.
- **Medical Device Innovation Consortium:** a nonprofit organization established to work with the US Food and Drug Administration on behalf of medical device manufacturers to ensure that these devices are both safe and effective.
- **medical imaging:** the use of devices to scan a patient's body and create images of the body's internal structures to aid in diagnosis and treatment planning.
- **telemedicine:** the practice of doctors interacting with and treating patients long distance using telecommunications technology (such as telephones or videoconferencing software) and the Internet.

COMPUTERIZED DIAGNOSTIC TOOLS

Devices that help doctors diagnose health conditions are one very common form of medical technology. Some medical conditions are obvious to even an untrained observer, such as a gunshot wound or a severed limb. These conditions require little diagnosis because it is immediately apparent what is wrong. Other diseases and conditions, however, can be very difficult to detect and identify. Conditions such as low

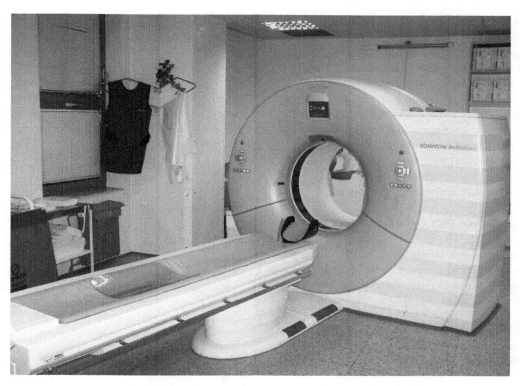

Medical technology is constantly evolving to provide medical practitioners with a better view of what is happening and more information to help determine problems and find solutions. By Tomáš Vendiš, CC BY 3.0 (http://creativecommons.org/licenses/by/3.0), via Wikimedia Commons.

blood sugar, high cholesterol, or thyroid gland problems cannot be diagnosed visually. For such conditions, specially designed devices or procedures must be used to collect information about the patient's bodily processes. This is where medical technology comes in.

For doctors, everything the body does creates information. They must find ways to collect this information, analyze it, and use it to make educated guesses about the patient's condition. Part of medical technology involves clinical engineering, in which devices are designed to collect information about the patient's health. Some of these devices allow doctors to measure the signals created by the body, a process known as biosignal processing. For example, electroencephalograms (EEG) measure the electrical activity of the brain. Detecting abnormalities in these signals helps doctors to diagnose epilepsy, sleep disorders, and brain tumors, among other problems. Doctors also use other types of medical imaging,

such as X-rays. A person with a painful hand injury might receive an X-ray to determine whether or not any bones have been broken. Today, these X-ray images may be taken and stored digitally, rather than on film.

TREATMENT TOOLS

Medical technology can also be used to help treat a condition that has been diagnosed. Some types of medical technology are simple to understand and use. Think of bandages, crutches, and wheelchairs. Other types of medical treatment devices rely on more advanced technology. For example, tiny monitors can be implanted in the body to monitor heart rhythm.

The US Food and Drug Administration (FDA) must approve medical devices before they can be used in general practice. This requires extensive testing to show that the technology is unlikely to cause harm. Other bodies are also involved in this process, such as the Medical Device Innovation Consortium (MDIC). The nonprofit MDIC supports medical technology inventors and guides them through the FDA approval process. Groups such as this must balance the interests of doctors and the FDA with those of the medical technology industry in creating and marketing new devices. It is common for approval of medical devices to take several years. During this time, doctors and even some patients may be waiting for approval to use groundbreaking treatments. This places great pressure on the FDA to move the approval process forward both quickly and safely.

MEDICAL TECHNOLOGY IN THE TWENTY-FIRST CENTURY

The Internet has opened up new avenues of treatment for medical technology. The field of telemedicine, in which patients consult with doctors long distance, has grown greatly. It is particularly useful in bringing more treatment options to rural areas that may not have well-equipped hospitals or many medical specialists. In the past, telemedicine was restricted to telephone consultations. Diagnostic data can now be collected from a patient and transmitted over the Internet to a specialist. The specialist then uses it to recommend a treatment for the patient.

Advances in nonmedical technology have also raised the possibility of devices "printing" new organs or tissue structures that are designed for a particular patient. Today, computer models tell 3-D printers how to build a 3-D object using layers of materials such as plastic or ceramic. Devices could one day use 3-D printing to grow cellular structures according to dimensions defined by medical imaging scans. If a patient needed a healthy kidney, it would be possible to scan them to find out exactly what size and shape of kidney was needed and then 3-D print the kidney using tissue from the patient. This would minimize the risk of tissue rejection that comes along with traditional organ transplant.

—*Scott Zimmer, JD*

BIBLIOGRAPHY

Cassell, Eric J. *The Nature of Healing: The Modern Practice of Medicine.* New York: Oxford UP, 2013. Print.

Liang, Hualou, Joseph D. Bronzino, and Donald R. Peterson, eds. *Biosignal Processing: Principles and Practices.* Boca Raton: CRC, 2012. Print.

Rasmussen, Nicolas. *Gene Jockeys: Life Science and the Rise of Biotech Enterprise.* Baltimore: Johns Hopkins UP, 2014. Print.

Topol, Eric J. *The Creative Destruction of Medicine: How the Digital Revolution Will Create Better Health Care.* New York: Basic, 2013. Print.

Tulchinsky, Theodore H., Elena Varavikova, Joan D. Bickford, and Jonathan E. Fielding. *The New Public Health.* New York: Academic, 2014. Print.

Wachter, Robert M. *The Digital Doctor: Hope, Hype, and Harm at the Dawn of Medicine's Computer Age.* New York: McGraw, 2015. Print.

MESH NETWORKING

FIELDS OF STUDY

Information Technology; Network Design

ABSTRACT

Mesh networking is one way of organizing a network. It was designed for military applications requiring high performance and reliability. All of the nodes in a mesh network transmit data across the network. In some mesh networks, each node has a connection to every other node. These types of mesh networks are highly resilient, but they can become costly as the network grows in size.

PRINICIPAL TERMS

- **flooding:** sending information to every other node in a network to get the data to its appropriate destination.
- **hopping:** the jumping of a data packet from one device or node to another as it moves across the network. Most transmissions require each packet to make multiple hops.
- **node:** a point on a network where two or more links meet, or where one link terminates.
- **peer-to-peer (P2P) network:** a network in which all computers participate equally and share coordination of network operations, as opposed to a client-server network, in which only certain computers coordinate network operations.
- **routing:** selecting the best path for a data packet to take in order to reach its destination on the network.
- **topology:** the way a network is organized, including nodes and the links that connect nodes.

BENEFITS OF MESH NETWORKING

Mesh networking is a type of network topology. It was developed because of the limitations of other types of network topologies, such as star networks, ring networks, and bus networks. While these designs have some advantages, experience showed that they are vulnerable to device failure. In some scenarios, the failure of a single node could prevent the entire network from functioning. Furthermore, adding more nodes to these networks could be challenging, as it sometimes required changing the existing network to accommodate new equipment.

Mesh networks are designed to keep functioning even when one or more nodes fail. When this occurs, unaffected nodes are able to compensate for the damage the network has suffered. The ability of mesh networks to respond to broken nodes has led some to describe them as "self-healing," because they adapt their layouts in order to keep functioning.

Most mesh networks are wireless rather than wired. Wireless technology has advanced enough that different types of wireless antennas can be incorporated in a single device or node on the network. This allows for greater versatility in the type and number of connections that nodes can make with one another. In addition, wireless mesh networks are easier to set up than wired ones, because there is no need to install physical cables connecting all of the nodes on the network.

HOW A MESH NETWORK WORKS

There are two main methods that a mesh network uses to convey information from one point to another. The first of these methods is routing. To route data, the network first calculates the most efficient route for data packets to take. This may be done through the use of a routing table, which is a list of routes that can be used to reach various points on the network. This is analogous to looking up a person's address before driving to their house for a visit. Using the information from the routing table, packets travel by hopping from node to node until they finally reach their destination. This is similar to a person jumping from stone to stone in order to cross a stream, where each stone represents a node on the network.

Another approach to network navigation is flooding. With this method, the network does not need to determine the best path a data packet to take. Instead, the node simply sends a copy of the packet to every other node it is attached to. Each of these nodes then does the same thing, and the process is repeated. This is called flooding because the network is quickly flooded with copies of the packet being transmitted. While this may seem redundant, the redundancy creates greater certainty that the packet will reach its destination.

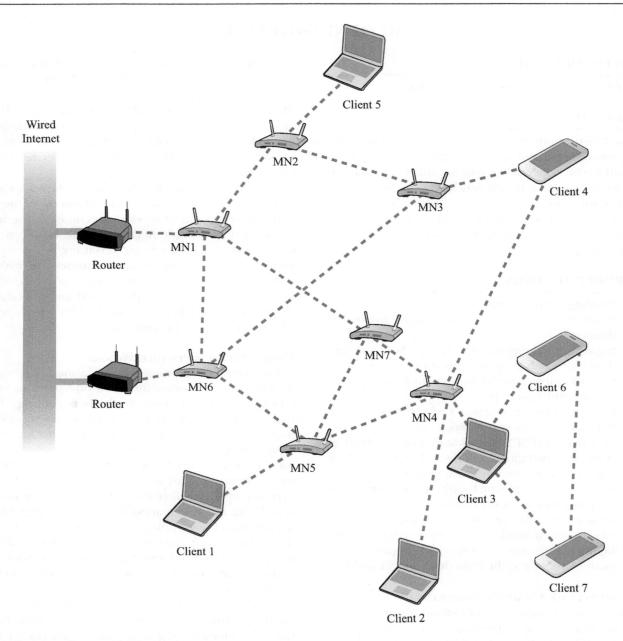

Wireless mesh networks make use of a spiderweb-like design that allows multiple pathways for data to travel from one client to another. This allows data to move quickly by taking the path of least resistance at any given instant. Adapted from the WMN Topology diagram, Communications Network Research Institute.

PEER-TO-PEER NETWORKS

A peer-to-peer (P2P) network is a type of mesh network. P2P networks are used to share files. Hundreds of thousands of computers are joining and leaving the mesh network at any given moment. Each peer computer is able to share its locally stored files with other peers on the network.

P2P networks are legal, but they may be used for legal or illegal purposes. While software developers use P2P networks to share software with users legally, online pirates often use P2P networks to illegally share copyrighted music, movies, and other digital products without permission.

INCREASING INTERNET ACCESS

Mesh networks have useful applications in many different contexts. For example, they have often been used in underdeveloped regions to provide affordable and reliable Internet access to large numbers of people without having to run cables to each household. Instead, homes are provided with wireless Internet access points configured to connect to one another, so that an entire city can quickly be blanketed in Internet access.

—*Scott Zimmer, JD*

BIBLIOGRAPHY

Basagni, Stefano, et al., eds. *Mobile Ad Hoc Networking: Cutting Edge Directions.* 2nd ed. Hoboken: Wiley, 2013. Print.

Gibson, Jerry D., ed. *Mobile Communications Handbook.* 3rd ed. Boca Raton: CRC, 2013. Print.

Khan, Shafiullah, and Al-Sakib Khan Pathan, eds. *Wireless Networks and Security: Issues, Challenges and Research Trends.* Berlin: Springer, 2013. Print.

Pathak, Parth H., and Rudra Dutta. *Designing for Network and Service Continuity in Wireless Mesh Networks.* New York: Springer, 2013. Print.

Wei, Hung-Yu, Jarogniew Rykowski, and Sudhir Dixit. *WiFi, WiMAX, and LTE Multi- Hop Mesh Networks: Basic Communication Protocols and Application Areas.* Hoboken: Wiley, 2013. Print.

Yu, F. Richard, Xi Zhang, and Victor C. M. Leung, eds. *Green Communications and Networking.* Boca Raton: CRC, 2013. Print.

METACOMPUTING

FIELDS OF STUDY

Computer Science; Information Systems

ABSTRACT

Metacomputing is the use of computing to study and design solutions to complex problems. These problems can range from how best to design large-scale computer networking systems to how to determine the most efficient method for performing mathematical operations involving very large numbers. In essence, metacomputing is computing about computing. Metacomputing makes it possible to perform operations that individual computers, and even some supercomputers, could not handle alone.

PRINICIPAL TERMS

- **domain-dependent complexity:** a complexity that results from factors specific to the context in which the computational problem is set.
- **meta-complexity:** a complexity that arises when the computational analysis of a problem is compounded by the complex nature of the problem itself.
- **middle computing:** computing that occurs at the application tier and involves intensive processing of data that will subsequently be presented to the user or another, intervening application.
- **networking:** the use of physical or wireless connections to link together different computers and computer networks so that they can communicate with one another and collaborate on computationally intensive tasks.
- **supercomputer:** an extremely powerful computer that far outpaces conventional desktop computers.
- **ubiquitous computing:** an approach to computing in which computing activity is not isolated in a desktop, laptop, or server, but can occur everywhere and at any time through the use of microprocessors embedded in everyday devices.

REASONS FOR METACOMPUTING

The field of metacomputing arose during the 1980s. Researchers began to realize that the rapid growth in networked computer systems would soon make it difficult to take advantage of all interconnected computing resources. This could lead to wasted resources unless an additional layer of computing power were developed. This layer would not work on a computational problem itself; instead, it would determine the most efficient method of addressing the problem. In other words, researchers saw the potential for using computers in a manner so complicated that only a computer could manage it. This new metacomputing layer would rest atop the middle computing layer that works on research tasks. It would ensure that the middle layer makes the best use of its resources and that it approaches calculations as efficiently as possible.

One reason an application may need a metacomputing layer is the presence of complexities. Complexities are elements of a computational problem that make it more difficult to solve. Domain-dependent complexities arise due to the context of the computation. For example, when calculating the force and direction necessary for an arrow to strike its target, the effects of wind speed and direction would be a domain-dependent complexity. Meta-complexities are those that arise due to the nature of the computing problem rather than its context. An example of a meta-complexity is a function that has more than one possible solution.

METACOMPUTING AND THE INTERNET

Metacomputing is frequently used to make it possible to solve complex calculations by networking between many different computers. The networked computers can combine their resources so that each one works on part of the problem. In this way, they become a virtual supercomputer with greater capabilities than any individual machine. One successful example of this is a project carried out by biochemists studying the way proteins fold and attach to one another. This subject is usually studied using computer programs that model the proteins' behavior. However, these programs consume a lot of time and computing power. Metacomputing allowed the scientists to create a game that users all over the world can play that generates data about protein folding at the same time. Users try to fit shapes together in different

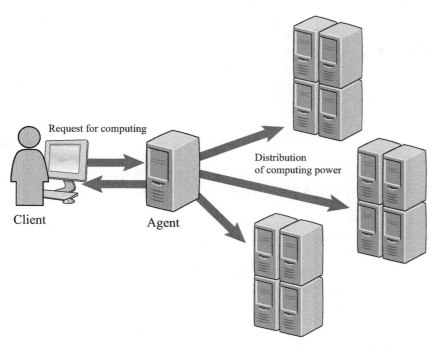

Metacomputing is a system design concept that allows multiple computers to share the responsibility of computation to complete a request efficiently. EBSCO illustration.

ways, contributing their time and computing power to the project in a fun and easy way.

Ubiquitous Metacomputing

Another trend with great potential for metacomputing is ubiquitous computing, meaning computing that is everywhere and in everything. As more and more mundane devices are equipped with Internet-connected microprocessors, from coffee makers to cars to clothing, there is the potential to harness this computing power and data. For example, a person might use multiple devices to monitor different aspects of their health, such as activity level, water intake, and calories burned. Metacomputing could correlate all of this independently collected information and analyze it. This data could then be used to diagnose potential diseases, recommend lifestyle changes, and so forth.

One form of ubiquitous metacomputing that already exists is the way that various smartphone applications use location data to describe and predict traffic patterns. One person's data cannot reveal much about traffic. However, when many people's location, speed, and direction are reported

simultaneously, the data can be used to predict how long one person's commute will be on a given morning.

Mimicking the Brain

Metacomputing is often used when there is a need for a computer system that can "learn." A system that can learn is one that can analyze its own performance to make adjustments to its processes and even its architecture. These systems bear a strong resemblance to the operation of the human brain. In some cases they are intentionally designed to imitate the way the brain approaches problems and learns from its past performance. Metacomputing, in this sense, is not too dissimilar from metacognition.

Metacomputing sometimes conjures up fears of the dangers posed by artificial intelligence in science fiction. In reality, metacomputing is just another category of computer problem to be solved, not the beginning of world domination by machines. Humans can conceive of computer problems so complex that it is nearly impossible to solve them without the aid of another computer. Metacomputing is simply the solution to this dilemma.

—*Scott Zimmer, JD*

Bibliography

Loo, Alfred Waising, ed. *Distributed Computing Innovations for Business, Engineering, and Science.* Hershey: Information Science Reference, 2013. Print.

Mallick, Pradeep Kumar, ed. *Research Advances in the Integration of Big Data and Smart Computing.* Hershey: Information Science Reference, 2016. Print.

Mason, Paul. *Understanding Computer Search and Research.* Chicago: Heinemann, 2015. Print.

Nayeem, Sk. Md. Abu, Jyotirmoy Mukhopadhyay, and S. B. Rao, eds. *Mathematics and Computing: Current Research and Developments.* New Delhi: Narosa, 2013. Print.

Segall, Richard S., Jeffrey S. Cook, and Qingyu Zhang, eds. *Research and Applications in Global*

Supercomputing. Hershey: Information Science Reference, 2015. Print.

Tripathy, B. K., and D. P. Acharjya, eds. *Global Trends in Intelligent Computing Research and Development.* Hershey: Information Science Reference, 2014. Print.

MICROPROCESSORS

FIELDS OF STUDY

Computer Engineering; System-Level Programming

ABSTRACT

Microprocessors are part of the hardware of a computer. They consist of electronic circuitry that stores instructions for the basic operation of the computer and processes data from applications and programs. Microprocessor technology debuted in the 1970s and has advanced rapidly into the 2010s. Most modern microprocessors use multiple processing "cores." These cores divide processing tasks between them, which allows the computer to handle multiple tasks at a time.

PRINICIPAL TERMS

- **central processing unit (CPU):** electronic circuitry that provides instructions for how a computer handles processes and manages data from applications and programs.
- **clock speed:** the speed at which a microprocessor can execute instructions; also called "clock rate."
- **data width:** a measure of the amount of data that can be transmitted at one time through the computer bus, the specific circuits and wires that carry data from one part of a computer to another.
- **micron:** a unit of measurement equaling one millionth of a meter; typically used to measure the width of a core in an optical figure or the line width on a microchip.
- **million instructions per second (MIPS):** a unit of measurement used to evaluate computer performance or the cost of computing resources.
- **transistor:** a computing component generally made of silicon that can amplify electronic signals

or work as a switch to direct electronic signals within a computer system.

BASICS OF MICROPROCESSING

Microprocessors are computer chips that contain instructions and circuitry needed to power all the basic functions of a computer. Most modern microprocessors consist of a single integrated circuit, which is a set of conducting materials (usually silicon) arranged on a plate. Microprocessors are designed to receive electronic signals and to perform processes on incoming data according to instructions programmed into the central processing unit (CPU) and contained in computer memory. They then to produce output that can direct other computing functions. In the 2010s, microprocessors are the standard for all computing, from handheld devices to supercomputers. Among the modern advancements has been development of integrated circuits with more than one "core." A core is the circuitry responsible for calculations and moving data. As of 2016, microprocessors may have as many as eighteen cores. The technology for adding cores and for integrating data shared by cores is a key area of development in microprocessor engineering.

MICROPROCESSOR HISTORY AND CAPACITY

Before the 1970s, and the invention of the first microprocessor, computer processing was handled by a set of individual computer chips and transistors. Transistors are electronic components that either amplify or help to direct electronic signals. The first microprocessor for the home computer market was the Intel 8080, an 8-bit microprocessor introduced in 1974. The number of bits refers to the storage size of each unit of the computer's memory. From the 1970s

CHARACTERISTICS OF MICROPROCESSORS

The small components within modern microprocessors are often measured in microns or micrometers, a unit equaling one-millionth of a meter. Microprocessors are usually measured in line width, which measures the width of individual circuits. The earliest microprocessor, the 1971 Intel 4004, had a minimum line width of 10 microns. Modern microprocessors can have line width measurements as low as 0.022 microns.

All microprocessors are created with a basic instruction set. This defines the various instructions that can be processed within the unit. The Intel 4004 chip, which was installed in a basic calculator, provided instructions for basic addition and subtraction. Modern microprocessors can handle a wide variety of calculations.

Different brands and models of microprocessors differ in bandwidth, which measures how many bits of data a processor can handle per second. Microprocessors also differ in data width, which measures the amount of data that can be transferred between two or more components per second. A computer's bus describes the parts that link the processor to other parts and to the computer's main memory. The size of a computer's bus is known as the width. It determines how much data can be transferred each second. A computer's bus has a clock speed, measured in megahertz (MHz) or gigahertz (GHz). All other factors being equal, computers with larger data width and a faster clock speed can transfer data faster and thus run faster when completing basic processes.

MICROPROCESSOR DEVELOPMENT

Intel cofounder Gordon Moore noted that the capacity of computing hardware has doubled every two years since the 1970s, an observation now known as Moore's law. Microprocessor advancement is complicated by several factors, however. These factors include the rising cost of producing

Microprocessors contain all the components of a CPU on a single chip; this allows new devices to have higher computing power in a smaller unit. By M.ollivander, CC BY-SA 3.0 (http://creativecommons.org/licenses/by-sa/3.0), via Wikimedia Commons.

to the 2010s, microprocessors have followed the same basic design and concept but have increased processing speed and capability. The standard for computing in the 1990s and 2000s was the 32-bit microprocessor. The first 64-bit processors were introduced in the 1990s. They have been slow to spread, however, because most basic computing functions do not require 64-bit processing.

Computer performance can be measured in million instructions per second (MIPS). The MIPS measurement has been largely replaced by measurements using floating-point operations per second (FLOPS) or millions of FLOPS (MFLOPS). Floating-point operations are specific operations, such as performing a complete basic calculation. A processor with a 1 gigaFLOP (GFLOP) performance rating can perform one billion FLOPS each second. Most modern microprocessors can perform 10 GFLOPS per second. Specialized computers can perform in the quadrillions of operations per second (petaFLOPS) scale.

microprocessors and the fact that the ability to reduce power needs has not grown at the same pace as processor capacity. Therefore, unless engineers can reduce power usage, there is a limit to the size and processing speed of microprocessor technology. Data centers in the United States, for instance, used about 91 billion kilowatt-hours of electricity in 2013. This is equal to the amount of electricity generated each year by thirty-four large coal-burning power plants. Computer engineers are exploring ways to address these issues, including alternatives for silicon in the form of carbon nanotubes, bioinformatics, and quantum computing processors.

—*Micah L. Issitt*

BIBLIOGRAPHY

Ambinder, Marc. "What's Really Limiting Advances in Computer Tech." *Week.* The Week, 2 Sept. 2014. Web. 4 Mar. 2016.

Borkar, Shekhar, and Andrew A. Chien. "The Future of Microprocessors." *Communications of the ACM.* ACM, May 2011. Web. 3 Mar. 2016.

Delforge, Pierre. "America's Data Centers Consuming and Wasting Growing Amounts of Energy." *NRDC.* Natural Resources Defense Council, 6 Feb. 2015. Web. 17 Mar. 2016.

McMillan, Robert. "IBM Bets $3B That the Silicon Microchip Is Becoming Obsolete." *Wired.* Condé Nast, 9 July 2014. Web. 10 Mar. 2016.

"Microprocessors: Explore the Curriculum." *Intel.* Intel Corp., 2015. Web. 11 Mar. 2016.

"Microprocessors." *MIT Technology Review.* MIT Technology Review, 2016. Web. 11 Mar. 2016.

Wood, Lamont. "The 8080 Chip at 40: What's Next for the Mighty Microprocessor?" *Computerworld.* Computerworld, 8 Jan. 2015. Web. 12 Mar. 2016.

MICROSCALE 3-D PRINTING

FIELDS OF STUDY

Information Technology; Computer Engineering

ABSTRACT

Microscale 3-D printing is a type of 3-D printing that makes it possible to construct objects at an extremely small scale. Some processes can create objects as small as 100 micrometers. 3-D printing at this scale has a number of applications for computing and medicine. It makes it possible to produce microscopic structures out of organic materials for biomedical applications.

PRINICIPAL TERMS

- **binder jetting:** the use of a liquid binding agent to fuse layers of powder together.
- **directed energy deposition:** a process that deposits wire or powdered material onto an object and then melts it using a laser, electron beam, or plasma arc.
- **material extrusion:** a process in which heated filament is extruded through a nozzle and deposited in layers, usually around a removable support.
- **material jetting:** a process in which drops of liquid photopolymer are deposited through a printer head and heated to form a stable solid.
- **powder bed fusion:** the use of a laser to heat layers of powdered material in a movable powder bed.
- **sheet lamination:** a process in which thin layered sheets of material are adhered or fused together and then extra material is removed with cutting implements or lasers.
- **vat photopolymerization:** a process in which a laser hardens layers of light-sensitive material in a vat.

A REVOLUTIONARY TECHNOLOGY

3-D printing is a relatively new technology. However, it has already revolutionized manufacturing. It takes its name from traditional computer printers that produce pages of printed images. Regular printers operate by depositing small amounts of ink at precise locations on a piece of paper. Instead of ink, a 3-D printer uses a material, such as a polymer, metallic powder, or even organic material. It builds an object by depositing small amounts of that material in successive layers. In some cases, 3-D printing fastens materials to a substrate using heat, adhesives, or other methods. 3-D printing can produce incredibly intricate objects that would be difficult or impossible to create through traditional manufacturing methods.

Microscale 3-D printing advances the innovation of standard 3-D printing to create microscopic structures. The potential applications for microscale 3-D printing are still being explored. However, microscale 3-D printing presents the possibility of creating tissue for transplant. For example, full-scale 3-D printing has already produced some types of tissue, such as muscle, cartilage, and bones. One problem is the printed tissue sometimes did not survive because it had no circulatory system to bring blood and nutrients to the new tissue. Microscale 3-D printing makes it possible to create the tiny blood vessels needed in living tissue, among other potential applications.

3-D PRINTING METHODS

The basic approach used by 3-D printing is to build an object by attaching tiny amounts of material to each

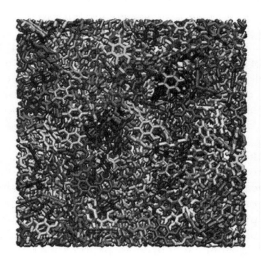

Material engineers can use microscale 3-D printing to develop unique materials for use in a wide range of fields, including bioengineering, architecture, and electronics. The combination and arrangement of particular molecules allows engineers to develop materials with the necessary characteristics to fulfill specific needs, such as the polymers used to build the entry heat shields for NASA. By Alexander Thompson and John Lawson, NASA Ames Research Center.

other at precise locations. Some materials are melted before they are deposited. Material extrusion heats polymer filament. The melted plastic material is then extruded through nozzles and deposited in a layer. The materials then harden into place, and another layer is added. With vat photopolymerization, a light-sensitive liquid polymer is printed onto a platform within a vat of liquid resin. An ultraviolet (UV) laser then hardens a layer of resin. More liquid polymer is then added, and the process is repeated. With material jetting, a printer head deposits liquefied plastic or other light-sensitive material onto a platform. The material is then hardened with UV light.

Other methods melt or fuse materials after they have been deposited. In powder bed fusion, the printer heats a bed of powdered glass, metal, ceramic, or plastic until the materials fuse together in the desired locations. Another layer of powder is then added and fused onto the first. Binder jetting uses a printer head to deposit drops of glue-like liquid into a powdered medium. The liquid soaks into and solidifies the medium. Sheet lamination fuses together thin sheets of paper, metal, or plastic with an adhesive. The layers are then cut with a laser into the desired shape. In directed energy deposition, a metal wire or powder is deposited in thin layers before being melted with a laser.

The method used depends on the physical properties of the material being printed. Metal alloys, for example, cannot easily be liquefied for vat polymerization, material jetting, or material extrusion. Instead, they are printed using binder jetting, powder bed fusion, or sheet lamination.

MICROSCALE METHODS
Microscale 3-D printing requires more exact methods to create objects that are just a few micrometers wide. Microscopic objects require tiny droplets of materials and precise locations of deposition. One microscale 3-D printing technique is optical lithography. This technique uses light to create patterns in a photosensitive resist, where material is then deposited. Optical transient liquid molding (TLM) uses UV light patterns and a custom flow of liquid polymer to create objects that are smaller than the width of a human hair. Optical TLM combines a liquid polymer, which will form the structure of the printed object, with

a liquid mold in a series of tiny pillars. The pillars are arranged based on software that determines the shape of the liquids' flow. Patterned UV light then cuts into the liquids to further shape the stream. The combination of the liquid mold and the UV light pattern allow the creation of highly complex structures that are just 100 micrometers in size.

Microscale 3-D printing makes it possible to create extremely small circuits. This will enable the creation of new devices, such as "smart clothing" that can sense the wearer's body temperature and adjust its properties based on this information. Microscale 3-D printing may also revolutionize the creation of new medicines. Because drug uptake by cells is shape-dependent, the precision of microscale 3-D printing may allow researchers to design custom drugs for specific brain receptors.

A NEW TYPE OF "INK"
Some refer to the build materials used in 3-D printing as "inks" because they take the place of ink as it is used in regular document printers. This can stretch the definition of "ink." Most people think of ink as either the liquid in a pen or the toner of a computer printer. In microscale 3-D printing, the ink might be human cells used to create an organ or metallic powder that will be fused into tiny circuits.

—*Scott Zimmer, JD*

BIBLIOGRAPHY
Bernier, Samuel N., Bertier Luyt, Tatiana Reinhard, and Carl Bass. *Design for 3D Printing: Scanning, Creating, Editing, Remixing, and Making in Three Dimensions*. San Francisco: Maker Media, 2014. Print.
France, Anna Kaziunas, comp. *Make: 3D Printing—The Essential Guide to 3D Printers*. Sebastopol: Maker Media, 2013. Print.
Horvath, Joan. *Mastering 3D Printing: Modeling, Printing, and Prototyping with Reprap- Style 3D Printers*. Berkeley: Apress, 2014. Print.
Hoskins, Stephen. *3D Printing for Artists, Designers and Makers*. London: Bloomsbury, 2013. Print.
Lipson, Hod, and Melba Kurman. *Fabricated: The New World of 3D Printing*. Indianapolis: Wiley, 2013. Print.

MOBILE APPS

FIELDS OF STUDY

Applications; Mobile Platforms

ABSTRACT

Mobile apps are programs designed to run on smartphones, tablets, and other mobile devices. These apps usually perform a specific task, such as reporting the weather forecast or displaying maps for navigation. Mobile devices have special requirements because of their small screens and limited input options. Furthermore, a touch screen is typically the only way to enter information into a mobile device. Programmers need special knowledge to understand the mobile platform for which they wish to create apps.

PRINICIPAL TERMS

- **applications:** programs that perform specific functions that are not essential to the operation of the computer or mobile device; often called "apps."
- **emulators:** programs that mimic the functionality of a mobile device and are used to test apps under development.
- **mobile website:** a website that has been optimized for use on mobile devices, typically featuring a simplified interface designed for touch screens.
- **platform:** the hardware and system software of a mobile device on which apps run.
- **system software:** the operating system that allows programs on a mobile device to function.
- **utility programs:** apps that perform basic functions on a computer or mobile device such as displaying the time or checking for available network connections.

TYPES OF MOBILE SOFTWARE

Mobile applications, or apps, are computer programs designed specifically to run on smartphones, tablets, and other mobile devices. Apps must be designed for a specific platform. A mobile platform is the hardware and system software on which a mobile device operates. Some of the most widely used mobile platforms include Google's Android, Apple's iOS, and Microsoft's Windows. Mobile devices support a variety of software. At the most basic level, a platform's system software includes its operating system (OS). The OS supports all other programs that run on that device, including apps. In addition to the OS, smartphones and tablets come with a variety of preinstalled utility programs, or utilities, that manage basic functions. Examples of utilities include a clock and a calendar, photo storage, security programs, and clipboard managers. While utilities are not essential to the functionality of the OS, they perform key tasks that support other programs.

However, the real power of mobile devices has come from the huge assortment of mobile apps they can run. The app stores for various mobile devices contain hundreds of thousands of different apps to download. Each app has been designed with different user needs in mind. Many are games of one sort or another. There are also vast numbers of apps for every pursuit imaginable, including video chat, navigation, social networking, file sharing and storage, and banking.

Developers of mobile apps use various approaches to design their software. In some cases, an app is little more than a mobile website that has been optimized for use with small touch screens. For example, the Facebook app provides essentially the same functionality as its website, although it can be integrated with the device's photo storage for easier uploading. Other mobile apps are developed specifically for the mobile devices they run on. Programmers must program their apps for a specific mobile platform. App developers usually use a special software development kit and an emulator for testing the app on a virtual version of a mobile device. Emulators provide a way of easily testing mobile apps. Emulators generate detailed output as the app runs, so the developer can use this data to diagnose problems with the app.

THE APP ECONOMY

Mobile apps have evolved into a multibillion-dollar business. Before the advent of mobile devices, software was developed for use on personal computers and a software package often sold for hundreds of dollars. The mobile app marketplace has adopted a very different model. Instead of creating apps that cost a large amount of money and try to provide a wide range of functions, the goal is to create an app that does one thing well and to charge a small amount of money. Many apps are free to download, and paid

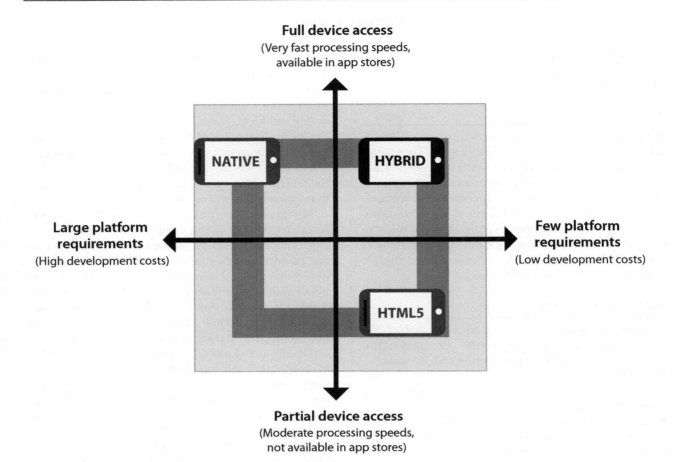

Native mobile apps and hybrid apps offer some features and capabilities unavailable with standard web programming, and they offer faster connectivity than is available with traditional computer software. EBSCO illustration.

apps are typically priced anywhere from ninety-nine cents to a few dollars. Despite such low price points, developers of successful apps can still earn large sums of money. This is in part due to the large numbers of smartphone and tablet users.

Mobile apps also have a low cost of distribution. In the past, software was sold on physical media such as floppy disks or CD-ROMs. These had to be packaged and shipped to retailers at the software developer's expense. With mobile apps, there are no such overhead costs because apps are stored online by the platform's app store and then downloaded by users. Aside from the time and effort of developing an app, the only other financial cost to the developer is an annual registration fee to the platform's app store and a percentage of revenue claimed by the platform. App

stores typically vet apps to make sure they do not contain malware, violate intellectual property, or make false advertising claims. App developers can earn revenue from the download fee, in-app advertisements, and in-app purchases. There have been several developers who have produced an unexpectedly popular app, catapulting them to fame and wealth.

MOBILE APPS AND SOCIAL CHANGE

Mobile apps can do much more than entertain and distract. For example, Twitter is a microblogging app that allows users to post short messages and read updates from others. Twitter has played a significant role in social movements such as the Arab Spring of 2011. Because it relies on telecommunications technology that continues to function even when there are disruptions in other media, the Twitter app has allowed people to communicate even during major disasters and political upheavals. Other apps allow users to report and pinpoint environmental damage or potholes for government agencies to fix.

PLATFORM LOCK

Mobile apps must be designed for a particular mobile platform. Sometimes, a developer will make a version of an app for each major platform. In other cases, however, developers only create an app to run on a single platform. This leaves users of other mobile platforms unable to use the app. In the case of popular apps, this can cause frustration among those who want to be able to use whatever apps they want on their platform of choice.

—*Scott Zimmer, JD*

BIBLIOGRAPHY

Banga, Cameron, and Josh Weinhold. *Essential Mobile Interaction Design: Perfecting Interface Design in Mobile Apps*. Upper Saddle River: Addison-Wesley, 2014. Print.

Glaser, J. D. *Secure Development for Mobile Apps: How to Design and Code Secure Mobile Applications with PHP and JavaScript*. Boca Raton: CRC, 2015. Print.

Iversen, Jakob, and Michael Eierman. *Learning Mobile App Development: A Hands-On Guide to Building Apps with iOS and Android*. Upper Saddle River: Addison-Wesley, 2014. Print.

Miller, Charles, and Aaron Doering. *The New Landscape of Mobile Learning: Redesigning Education in an App-Based World*. New York: Routledge, 2014. Print.

Salz, Peggy Anne, and Jennifer Moranz. *The Everything Guide to Mobile Apps: A Practical Guide to Affordable Mobile App Development for Your Business*. Avon: Adams Media, 2013. Print.

Takahashi, Dean. "The App Economy Could Double to $101 Billion by 2020." *VB*. Venture Beat, 10 Feb. 2016. Web. 11 Mar. 2016.

MOBILE OPERATING SYSTEMS

FIELDS OF STUDY

Mobile Platforms; Operating Systems

ABSTRACT

Mobile operating systems (OSs) are installed on mobile devices such as smartphones, tablets, and portable media players. Mobile OSs differ from ordinary computer OSs in that they must manage cellular connections and are configured to support touch screens and simplified input methods. Mobile OSs tend to have sophisticated power management features as well, since they are usually not connected to a power source during use.

PRINICIPAL TERMS

- **Android Open Source Project:** a project undertaken by a coalition of mobile phone manufacturers and other interested parties, under the leadership of Google. The purpose of the project is to develop the Android platform for mobile devices.
- **iOS:** Apple's proprietary mobile operating system, installed on Apple devices such as the iPhone, iPad, and iPod touch.
- **jailbreak:** the removal of restrictions placed on a mobile operating system to give the user greater control over the mobile device.
- **near-field communication:** a method by which two devices can communicate wirelessly when in close proximity to one another.
- **real-time operating system:** an operating system that is designed to respond to input within a set amount of time without delays caused by buffering or other processing backlogs.

A Brief History of Mobile Operating Systems

The mobile computing market is one of the fastest-growing sectors of the technology field. Its growing popularity began in the late 1990s with the release of the Palm Pilot 1000, a personal digital assistant (PDA). The Pilot 1000 introduced Palm OS, an early mobile operating system (OS). The Palm OS was later extended to smartphones. Smartphones combined the features of PDAs, personal computers (PCs), and cell phones. Prior to smartphones, many people had cell phones but their functionality was extremely limited, and many people owned both a PDA and cell phone. The Ericsson R380, released in 2000, was the first cell phone marketed as a smartphone. The Ericsson R380 ran on Symbian. Symbian was the dominant mobile OS in the early smartphone market.

The arrival of the iPhone in 2007 changed this. The sleek, simple design of Apple's mobile OS, called iOS, gave the device an intuitive user interface. In addition to the iOS, the iPhone has a baseband processor that runs a real-time operating system (RTOS). All smartphones have an RTOS in addition to the manufacturer's mobile OS. In a smartphone, the RTOS communicates with the cellular network, enabling the phone to exchange data with the network. Despite the success of the iPhone, some felt that Apple devices and its App Store were too locked down. Apple does not allow users to make certain changes to the iOS or to install apps from unofficial sources. In response, a community of iPhone hackers began releasing software that could be used to jailbreak Apple devices. Jailbreaking removes some of the restrictions that are built into the

Operating systems on mobile devices have multiple limitations; however, the data input hardware and software can vary greatly, even within one device. One requirement of a good mobile operating system is that it be able to efficiently respond to a variety of data input methods. EBSCO illustration.

iOS in order to give users root access to the iOS. Jailbreaking gives users greater control over their mobile devices.

Google released a mobile OS called Android in 2008. Unlike Apple's iOS, Android's source code is open source. Open-source software is created using publicly available source code. The Android Open Source Project (AOSP) develops modified versions of the Android OS using the open-source code. Android's status as an open system means there is no need to jailbreak. Android quickly became the dominant OS worldwide. However, Apple's iOS accounts for slightly more than half of mobile OS market in the United States and Canada.

Mobile Features

Mobile OSs share many similarities with desktop and laptop OSs. However, mobile OSs are more closely integrated with touch-screen technology. Mobile OSs also typically feature Bluetooth and Wi-Fi connectivity, global positioning system (GPS) navigation, and speech recognition. Furthermore, many new smartphones are equipped with hardware that supports near-field communication (NFC). NFC allows two devices to exchange information when they are placed close to one another. NFC uses radio frequency identification (RFID) technology to enable wireless data transfers. A major benefit of NFC is its low power usage, which is particularly critical to mobile devices. Mobile technology such as NFC facilitates numerous business transactions, including mobile payment systems at the point of sale.

Although there remain some significant differences between desktop OSs and mobile OSs, they are rapidly converging. Cloud computing enables users to share and sync data across devices, further narrowing the differences between smartphones and PCs.

Security Concerns

Features such as NFC and Bluetooth come with privacy and security concerns. Many of the mobile OS features that make smartphones so convenient also make them vulnerable to access by unauthorized users. For example, hackers have been able to sweep up many users' private information by scanning large crowds for mobile devices with unsecured Bluetooth connections. The ability to capture private data so easily creates major vulnerabilities for identity theft. To combat this threat, mobile OSs have incorporated various forms of

security to help make sure that the private data they contain can only be accessed by an authorized user. For example, Apple has integrated fingerprint recognition into its iOS and newer versions of the iPhone.

Both Android and iOS now include features that help a user to locate and recover a mobile device that has been lost or stolen. Mobile owners can go online to geographically locate the device using its GPS data. The owner can also remotely lock the device to prevent its use by anyone else. It can even cause the device to emit a loud alarm to alert those nearby to its presence.

Impact

Mobile OSs and the apps that run on them have revolutionized the way in which people conduct their daily lives. Thanks to mobile OSs, users can track personal health data, transfer funds, connect with social media, receive GPS and weather data, and even produce and edit photo and audiovisual files, among other activities. As the popularity of mobile devices (and, by extension, mobile OSs) has increased, many have questioned the future of the PC. For now, PCs and desktop OSs continue to dominant the business sector, however.

—*Scott Zimmer, JD*

Bibliography

Collins, Lauren, and Scott Ellis. *Mobile Devices: Tools and Technologies*. Boca Raton: CRC, 2015. Print.

Drake, Joshua J. *Android Hacker's Handbook*. Indianapolis: Wiley, 2014. Print.

Dutson, Phil. *Responsive Mobile Design: Designing for Every Device*. Upper Saddle River: Addison-Wesley, 2015. Print.

Elenkov, Nikolay. *Android Security Internals: An In-Depth Guide to Android's Security Architecture*. San Francisco: No Starch, 2015. Print.

Firtman, Maximiliano R. *Programming the Mobile Web*. Sebastopol: O'Reilly Media, 2013. Print.

Neuburg, Matt. *Programming iOS 8: Dive Deep into Views, View Controllers, and Frameworks*. Sebastopol: O'Reilly Media, 2014. Print.

Ravindranath, Mohana. "PCs Lumber towards the Technological Graveyard." *Guardian*. Guardian News and Media, 11 Feb. 2014. Web. 10 Mar. 2016.

Tabini, Marco. "Hidden Magic: A Look at the Secret Operating System inside the iPhone." *MacWorld*. IDG Consumer & SMB, 20 Dec. 2013. Web. 9 Mar. 2016.

MOLECULAR COMPUTERS

FIELDS OF STUDY

Computer Science; Computer Engineering; Information Technology

SUMMARY

The DNA structure is highly amenable to information storage. A single strand of normally helical DNA, an oligomer, containing 20 base pairs, can exist in 420 or roughly 1.2 trillion, distinctly different arrangements. Synthesis of a given oligomer is not difficult and has been automated. While other molecules had been used as computational substrates, molecular biologist Leonard Adelman gained considerable notoriety by demonstrating how a classic mathematical problem could be formulated as a bonding problem for a small set of DNA oligomers.

PRINCIPAL TERMS

- **Hamiltonian path problem:** one of the classic NP-complete problems for which the time of solution grows faster than any power of the number if points involved.
- **ligase:** an enzyme that facilitates bond formation between two molecular fragments. Also called a polymerase.
- **NP-complete problem:** The class of problems for which the solution time grows faster than a polynomial of the number of points involved.
- **Oligomer:** a modest length polymer of DNA, about 20 base pairs long.
- **polymerase chain reaction:** the Nobel-prize winning discovery by Cary Mullis that thermal cycling can produce numerous copies of a single DNA oligomer; used in DNA fingerprinting as well as DNA computing.
- **SAT problem:** determining whether a single assignment of true and false values to the k Boolean values in a one or more statements yields a unique solution.
- **thermal cycling:** bringing a sample repeatedly above the dissociation temperature so that the strands dissociate, and then below that temperature so that replicates form
- **Watson-Crick base-pairing:** the pairing of adenine and thymine and that of cytosine with guanine in H-bonded complexes in the DNA double helix

ADELMAN'S EXPERIMENT

In 1994, L. M. Adelman used standard molecular biology techniques to find a Hamiltonian path through a map of 7 interconnected vertices. A Hamiltonian path is one, which passes through each vertex only once and only once. While the actual problem solved might have required a few seconds thought by an insightful high school student, the Hamiltonian path problem is one that mathematicians consider NP complete, one for which the solution time grows faster than any polynomial in the number of vertices. Given a map with a hundreds of cities connected by a network of roads, some one way, it could take years to determine if a Hamiltonian path actually exists.

Adelman's solution involved treating the road map of one-way and two-way streets as a directed graph G of n vertices with a designated beginning and end. Adelman represented each vertex of his graph by a 20-mer chosen at random. These could be synthesized by well-established techniques and amplified using the polymerase chain reaction. Possible paths representing an allowed connection between vertex i and vertex j would be represented by the complement to the last 10 sites in i and the first 10 in j.

1. Generate a large number of paths through G at random. This involved synthesizing oligomers to represent all allowed vertices and allowing them to pair with each other
2. Remove all paths that do not begin at the designated beginning or end at the designated end. This was easily done using standard biochemical techniques.
3. Remove all paths that do not involve exactly n vertices. This is easily accomplished using gel electrophoresis for which the mobility is inversely proportional to the molecular mass.
4. For each vertex v, remove all paths that do not include it

If any DNA is to be found remaining it represents a Hamiltonian path.

Adelman's experiment as described in the journal *Science* in November 1994 drew a great deal of attention. A conference followed in April 1994 at which Leonard Adelman was the keynote speaker. One of the most significant immediate extensions of Adelman's approach was proposed by R. J. Lipton at the same conference, who showed that the SAT problem could also be solved by manipulating DNA.

THE SAT PROBLEM

The SAT problem involves finding a set of truth values that satisfies a formula in the propositional calculus, in which Boolean variables appear connected by negation, disjunction or conjunction. For instance.

$$a = (x_1.OR..NOT.x_2.OR.x_3).AND.(x_2.OR.x_3).AND.$$
$$(x_1.OR.x_3).AND..NOT.x_3$$

is only satisfied if $x_1 = x_2 =$ true, and x_3 is false.

For a propositional formula with k variables we construct a graph with 3k +1 points. The v_k (k=0,... ,k) are assigned to points on the x axis, points to the right and below the axis are assigned to a_k^1 and the points above the axis are assigned to the a_k^0. Then a path from v_0 to v_k going above or below the like indicates a truth assignment for all the k variables.

The first step in Lipton's method consists in generating all paths from v_1 to v_k.

In analogy to the previous section, assign each on axis vertex to a single stranded oligomer and to each a_k^0 and a_k^1. Then a path $v_{i-1}a_i^j v_i$ encodes the truth value of x_i. Lipton's method begins with a test tube containing all possible paths, then selectively removing

all the molecules for which the first clause is not net, the selectively removing all molecules for which the second clause is not met, and so on. When the number of molecules remaining becomes two small the polymerase chain reaction can be run. If at the end of the sequence of reactions the tube becomes empty, there is no solution to the SAT problem.

We see thus that a great many graph theoretic questions can be answered, in principle at least by DNA computation. There will of course be practical difficulties as the size of problems increases, associated with the statistics of binding, but the true potential of DNA computing has yet to be reached.

—Donald Franceschetti, PhD

BIBLIOGRAPHY

Adelman, L. M. Molecular computation of solutions to combinatorial problems, *Science* 226, 1021-1024 (1994).

Calude, C. S., and Gheorge Paun, *Computing with Cells and Atoms.* (New York: Taylor & Francis, 2001. Print.

Lipton, R. J., and E. B. Baum, eds., DNA Based Computers, DIMACS Series in Discrete Mathematics, and Theoretical Computer *Science,* 27, American Mathematical Society (1995).

MOTHERBOARDS

FIELDS OF STUDY

Computer Engineering; Information Technology

ABSTRACT

The motherboard is the main printed circuit board inside a computer. It has two main functions: to support other computer components, such as random access memory (RAM), video cards, sound cards, and other devices; and to allow these devices to communicate with other parts of the computer by using the circuits etched into the motherboard, which are linked to the slots holding the various components.

PRINICIPAL TERMS

- **core voltage:** the amount of power delivered to the processing unit of a computer from the power supply.
- **crosstalk:** interference of the signals on one circuit with the signals on another, caused by the two circuits being too close together.
- **printed circuit board:** a flat copper sheet shielded by fiberglass insulation in which numerous lines have been etched and holes have been punched, allowing various electronic components to be connected and to communicate with one another and with external components via the exposed copper traces.
- **trace impedance:** a measure of the inherent resistance to electrical signals passing through the traces etched on a circuit board.
- **tuning:** the process of making minute adjustments to a computer's settings in order to improve its performance.

EVOLUTION OF MOTHERBOARDS

The motherboard of a computer is a multilayered printed circuit board (PCB) that supports all of the computer's other components, which are secondary to its functions. In other words, it is like the "mother" of other, lesser circuit boards. It is connected, either directly or indirectly, to every other part of the computer.

In the early days of computers, motherboards consisted of several PCBs connected either by wires or by being plugged into a backplane (a set of interconnected sockets mounted on a frame). Each necessary computer component, such as the central processing

unit (CPU) and the system memory, required one or more PCBs to house its various parts. With the advent and refinement of microprocessors, computer components rapidly shrank in size. While a CPU in the late 1960s consisted of numerous integrated circuit (IC) chips attached to PCBs, by 1971 Intel had produced a CPU that fit on a single chip. Other essential and peripheral components could also be housed in a single chip each. As a result, the motherboard could support a greater number of components, even as it too was reduced in size. Sockets were added to support more peripheral functions, such as mouse, keyboard, and audio support.

In addition to being more cost effective, this consolidation of functions helped make computers run faster. Sending information from point to point on a computer takes time. It is much faster to send information directly from the motherboard to a peripheral device than it is to send it from the CPU PCB across the backplane to the memory PCB, and then from there to the device.

MOTHERBOARD DESIGN

Designing a motherboard is quite challenging. The main issues arise from the presence of a large number of very small circuits in a relatively small area. One of the first considerations is how best to arrange components on the motherboard's various layers. A typical PCB consists of sheets of copper separated by sheets of fiberglass. The fiberglass insulates the copper layers from each other. Most motherboards consist of six to twelve layers, though more or fewer layers are also possible. Certain layers typically have specific functions. The outer layers are signal layers, while other layers carry voltage, ground returns, or carry memory, processor, and input/output data.

Lines etched in the fiberglass insulating each layer allow the familiar copper lines, or traces, to show through. These traces conduct the electrical signals. Most motherboard designers use computer simulations to determine the optimal length, width, and route of the individual traces. For example, a motherboard will have a target trace impedance value, often fifty or sixty ohms, which must be kept constant. Widening a trace will decrease impedance, while narrowing it will make it greater. Another issue

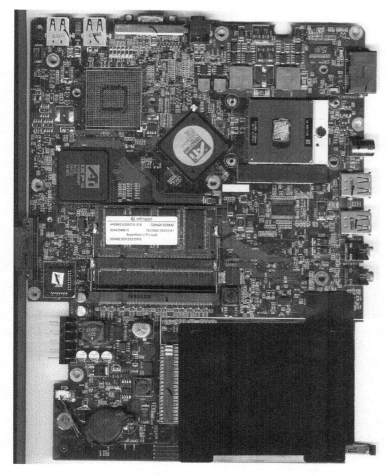

A motherboard is the main printed circuit board (PCB) of a computer. Also known as a logic board or mainboard, it connects the CPU to memory and to peripherals. Often it includes hard drives, sound cards, network cards, video cards, and other components. By .kkursor, public domain, via Wikimedia Commons.

there is usually a buffer zone between the recommended voltage and the maximum safe voltage, setting the voltage too high can still cause processors to overheat or even burn out.

FORM FACTOR

Though somewhat standardized, motherboards still come in different sizes and shapes. The main distinction is between motherboards for laptops and those for desktops. Motherboards designed for one of these categories generally will not fit into the other category. Most large desktop computer cases have enough room inside for just about any model of desktop motherboard, though smaller motherboards leave more space for peripherals to be added later.

BIOS

A motherboard will have some basic software embedded in a read-only memory (ROM) chip. This software is called the BIOS, which stands for "basic input/output system." When the power button is pressed, the BIOS tells the computer what devices to activate in order to locate the operating system and begin running it. If a computer malfunctions, it may be necessary to use the BIOS to change how the motherboard behaves while the system starts up.

—*Scott Zimmer, JD*

is crosstalk resulting from the high level of circuit density. If this happens, traces must be either better insulated or moved farther apart so that interference will diminish.

Some sophisticated computer users may try tuning their systems by adding or removing motherboard components, adjusting power settings, or "overclocking" the CPU to make it run faster. Overclocking can be risky, as it typically requires increasing the core voltage. Most motherboards have a built-in voltage regulator to ensure that the core voltage does not exceed the recommended voltage for the CPU and other processors. However, some regulators allow users to adjust their settings. While

BIBLIOGRAPHY

Andrews, Jean. *A+ Guide to Hardware: Managing, Maintaining, and Troubleshooting.* 6th ed. Boston: Course Tech., 2014. Print.

Andrews, Jean. *A+ Guide to Managing and Maintaining Your PC.* 8th ed. Boston: Course Tech., 2014. Print.

Cooper, Stephen. "Motherboard Design Process." *MBReview.com.* Author, 4 Sept. 2009. Web. 14 Mar. 2016.

Englander, Irv. *The Architecture of Computer Hardware, Systems Software, & Networking: An Information Technology Approach.* 5th ed. Hoboken: Wiley, 2014. Print.

Mueller, Scott. *Upgrading and Repairing PCs*. 22nd ed. Indianapolis: Que, 2015. Print.

Roberts, Richard M. *Computer Service and Repair*. 4th ed. Tinley Park: Goodheart, 2015. Print.

White, Ron. *How Computers Work: The Evolution of Technology*. Illus. Tim Downs. 10th ed. Indianapolis: Que, 2015. Print.

MULTIPROCESSING OPERATING SYSTEMS

FIELDS OF STUDY

Computer Engineering; Operating Systems

ABSTRACT

A multiprocessing operating system (OS) is one in which two or more central processing units (CPUs) control the functions of the computer. Each CPU contains a copy of the OS, and these copies communicate with one another to coordinate operations. The use of multiple processors allows the computer to perform calculations faster, since tasks can be divided up between processors.

PRINICIPAL TERMS

- **central processing unit (CPU):** sometimes described as the "brain" of a computer, the collection of circuits responsible for performing the main operations and calculations of a computer.
- **communication architecture:** the design of computer components and circuitry that facilitates the rapid and efficient transmission of signals between different parts of the computer.
- **parallel processing:** the division of a task among several processors working simultaneously, so that the task is completed more quickly.
- **processor coupling:** the linking of multiple processors within a computer so they can work together to perform calculations more rapidly. This can be characterized as loose or tight, depending on the degree to which processors rely on one another.
- **processor symmetry:** multiple processors sharing access to input and output devices on an equal basis and being controlled by a single operating system.

Multiprocessing versus Single-Processor Operating Systems

Multiprocessing operating systems (OSs) perform the same functions as single-processor OSs. They schedule and monitor operations and calculations in order to complete user-initiated tasks. The difference is that multiprocessing OSs divide the work up into various subtasks and then assign these subtasks to different central processing units (CPUs). Multiprocessing uses a distinct communication architecture to accomplish this. A multiprocessing OS needs a mechanism for the processors to interact with one another as they schedule tasks and coordinate their completion. Because multiprocessing OSs rely on parallel processing, each processor involved in a task must be able to inform the others about how its task is progressing. This allows the work of the processors to be integrated when the calculations are done such that delays and other inefficiencies are minimized. Multiprocessing operating systems can handle tasks more quickly, as each CPU that becomes available can access the shared memory to complete the task at hand so all tasks can be completed the most efficiently.

For example, if a single-processor OS were running an application requiring three tasks to be performed, one taking five milliseconds, another taking eight milliseconds, and the last taking seven milliseconds, the processor would perform each task in order. The entire application would thus require twenty milliseconds. If a multiprocessing OS were running the same application, the three tasks would be assigned to separate processors. The first would complete the first task in five milliseconds, the second would do the second task in eight milliseconds, and the third would finish its task in seven milliseconds. Thus, the multiprocessing OS would complete the entire task in eight milliseconds. From this example, it is clear that multiprocessing OSs offer distinct advantages.

Clustered Symmetric Processing

Multiprocessing operating systems can handle tasks more quickly, as each CPU that becomes available can access the shared memory to complete the task at hand so all tasks can be completed the most efficiently. EBSCO illustration.

COUPLING

Multiprocessing OSs can be designed in a number of different ways. One main difference between designs is the degree to which the processors communicate and coordinate with one another. This is known as processor coupling. Coupling is classified as either "tight" or "loose." Loosely coupled multiprocessors mostly communicate with one another through shared devices rather than direct channels. For the most part, loosely coupled CPUs operate independently. Instead of coordinating their use of devices by directly communicating with other processors, they share access to resources by queueing for them. In tightly coupled systems, each CPU is more closely bound to the others in the system. They

coordinate operations and share a single queue for resources.

One type of tightly coupled multiprocessing system has processors share memory with each other. This is known as symmetric multiprocessing (SMP). Processor symmetry is present when the multiprocessing OS treats all processors equally, rather than prioritizing a particular one for certain operations. Multiprocessing OSs are designed with special features that support SMP, because the OS must be able to take advantage of the presence of more than one processor. The OS has to "know" that it can divide up certain types of tasks among different processors. It must also be able to track the progress of each task so that the results of each operation can be combined once they conclude. In contrast, asymmetric multiprocessing occurs when a computer assigns system maintenance tasks to some types of processors and application tasks to others. Because the type of task assigned to each processor is not the same, they are out of symmetry. SMP has become more commonplace because it is usually more efficient.

MULTITASKING

The advent of multiprocessing OSs has had a major influence on how people perform their work. Multiprocessing OSs can execute more than one program at a time. This enables computers to use more user-friendly interfaces based on graphical representations of input and output. It allows users with relatively little training to perform computing tasks that once were highly complex. They can even perform many such tasks at once.

Multiprocessing OSs, though once a major innovation, have become the norm rather than the exception. As each generation of computers must run more and more complex applications, the processing workload becomes greater and greater. Without the advantages offered by multiple processors and OSs

tailored to take advantage of them, computers would not be able to keep up.

—*Scott Zimmer, JD*

BIBLIOGRAPHY

Garrido, José M., Richard Schlesinger, and Kenneth E. Hoganson. *Principles of Modern Operating Systems.* 2nd ed. Burlington: Jones, 2013. Print.

Gonzalez, Teofilo, and Jorge Díaz-Herrera, eds. *Computing Handbook: Computer Science and Software Engineering.* 3rd ed. Boca Raton: CRC, 2014. Print.

Sandberg, Bobbi. *Networking: The Complete Reference.* 3rd ed. New York: McGraw, 2015. Print.

Silberschatz, Abraham, Peter B. Galvin, and Greg Gagne. *Operating Systems Concepts.* 9th ed. Hoboken: Wiley, 2012. Print.

Stallings, William. *Operating Systems: Internals and Design Principles.* 8th ed. Boston: Pearson, 2014. Print.

Tanenbaum, Andrew S., and Herbert Bos. *Modern Operating Systems.* 4th ed. Boston: Pearson, 2014. Print.

MULTITASKING OPERATING SYSTEMS

FIELDS OF STUDY

Computer Science; Operating Systems

ABSTRACT

A multitasking operating system (OS) is one that can work on more than one task at a time by switching between the tasks very rapidly. The tasks may all pertain to a single user or to multiple users. A multitasking OS can save the current state of each user and task so that it does not lose its place when it comes back to a task to resume its work. This allows the system to switch smoothly between tasks.

PRINICIPAL TERMS

- **context switch:** a multitasking operating system shifting from one task to another; for example, after formatting a print job for one user, the com-

puter might switch to resizing a graphic for another user.
- **cooperative multitasking:** an implementation of multitasking in which the operating system will not initiate a context switch while a process is running in order to allow the process to complete.
- **hardware interruption:** a device attached to a computer sending a message to the operating system to inform it that the device needs attention, thereby "interrupting" the other tasks that the operating system was performing.
- **multiprocessing:** the use of more than one central processing unit to handle system tasks; this requires an operating system capable of dividing tasks between multiple processors.
- **preemptive multitasking:** an implementation of multitasking in which the operating system will initiate a context switch while a process is running,

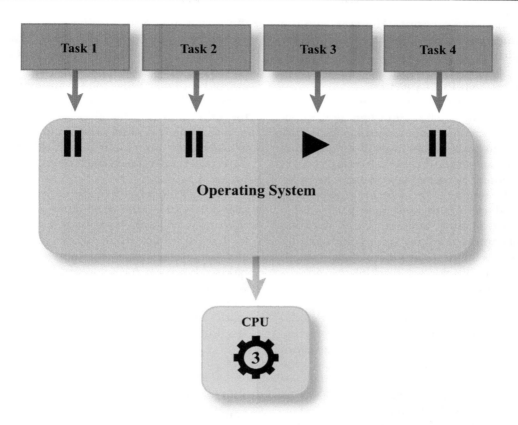

Multitasking operating systems use a single CPU to work on a number of tasks. The operating system determines which task the CPU will work on at any given time, pausing tasks as needed, so that all tasks are completed as efficiently as possible. EBSCO illustration.

usually on a schedule so that switches between tasks occur at precise time intervals.

- **time-sharing:** the use of a single computing resource by multiple users at the same time, made possible by the computer's ability to switch rapidly between users and their needs.

TIME-SHARING AND MULTITASKING

Multitasking operating systems (OSs) are based on the idea of time-sharing. Multitasking OSs do not actually perform multiple tasks at the same exact time. A particular central processing unit (CPU) can only do one thing at any given moment. Instead, they give the appearance of multitasking by switching back and forth between separate tasks for intervals of time too small for humans to perceive. This makes it seem to the user that multiple tasks are being carried out at once.

In the early days of computer science, when computers were huge machines requiring a lot of effort to maintain, researchers observed that it wasted time for a computer to be used by only one person at once. This was because the typical workflow of a computer operator was to submit a task requiring processing time, wait for the process to complete, and then receive the output. After receiving the output, the user evaluated it and prepared the next set of inputs. This time between tasks was wasted as the computer was idle instead of processing. The solution was to allow multiple users to interact with the computer. That way, while the computer was waiting for a response from one user, it could switch over to another user who was ready to proceed. This presented some problems in the early days of computing. Computers were not robust enough to easily store all of a user's context information (a description of the current state of the user's tasks and progress). However, with the passage of time and the use of multiprocessing computers, this changed.

COOPERATION AND PREEMPTION

Multitasking OSs can take a variety of forms. One difference is the way in which they handle the context switch between different users and tasks. A multitasking OS must have some way of deciding when is the appropriate time to make a context switch. The switch requires processing resources in order to be correctly executed, and switching contexts at random would reduce the system's predictability and efficiency over time. One method of deciding the timing of context switching is cooperative multitasking. With this method, the OS will not interrupt a process with a context switch while the process is running but will instead wait for the process to finish.

An alternative approach is preemptive multitasking. Unlike cooperative multitasking, preemptive multitasking does not wait for a process to finish before initiating a context switch. Instead, a multitasking OS using preemption will automatically switch between processes at defined intervals. This allocates resources evenly among all processes. This can be more resource intensive because it requires the system to store information about each process state. That way, the process can smoothly resume when its turn comes round again.

Sometimes, other events can intrude on the OS's time-sharing. An example is hardware interruption. This occurs when a device attached to the computer (such as a printer, scanner, or disk drive) requires the attention of the OS. It sends an interrupt signal to the CPU. This might occur when a printer experiences a paper jam, for instance. It would interrupt the computer's multitasking in order to signal the computer to display an error message to the user.

PRIORITIZING PROCESSES

Multitasking OSs must often assign different priority levels to different processes. One reason for this is that some processes have more of an impact on the user's real-time experience than others do. If a computer were busy keeping track of the time and preparing a system backup when a user closed an application window, the OS would prioritize the tasks associated with closing the window ahead of the others. Doing this gives the user the impression that the computer is operating smoothly. This would not be the case if the window being closed were to freeze in place until the conclusion of the backup process.

Multitasking places heavy demands on a computer. The computer must store enough information about each process to be able to quickly return to the task where it left off. As applications have grown more complex and more reliant on multitasking, hardware has struggled to keep pace. Each year, computers are brought to market with more memory, faster processors, and other hardware upgrades necessary for computers to do more work in less time.

—*Scott Zimmer, JD*

BIBLIOGRAPHY

Ben-Ari, M. *Principles of Concurrent and Distributed Programming*. 2nd ed. New York: Addison, 2006. Print.

Hart, Archibald D., and Sylvia Hart Frejd. *The Digital Invasion: How Technology Is Shaping You and Your Relationships*. Grand Rapids: Baker, 2013. Print.

Kaptelinin, Victor, and Mary P. Czerwinski, eds. *Beyond the Desktop Metaphor: Designing Integrated Digital Work Environments*. Cambridge: MIT P, 2007. Print.

Lee, John D., and Alex Kirlik, eds. *The Oxford Handbook of Cognitive Engineering*. New York: Oxford UP, 2013. Print.

Sinnen, Oliver. *Task Scheduling for Parallel Systems*. Hoboken: Wiley, 2007. Print.

Walker, Henry M. *The Tao of Computing*. 2nd ed. Boca Raton: CRC, 2013. Print.

MULTITHREADING OPERATING SYSTEMS

FIELDS OF STUDY

Computer Science; Operating Systems; Computer Engineering

ABSTRACT

Multithreading operating systems (OSs) are OSs designed to increase the speed of computing by working on multiple threads, which are parts of a programming sequence. There are various types of multithreading processes, based on how various types of threads are handled by the OS. Multithreading is beneficial for increasing computing speed, but makes programming more complicated for software designers and creates additional potential for application errors.

PRINICIPAL TERMS

- **address space:** the amount of memory allocated for a file or process on a computer.
- **context switching:** pausing and recording the progress of a thread or process such that the process or thread can be executed at a later time.
- **fiber:** a small thread of execution using cooperative multitasking.
- **multitasking:** the process of executing multiple tasks concurrently on an operating system (OS).
- **process:** the execution of instructions in a computer program.
- **thread:** the smallest part of a programmed sequence that can be managed by a scheduler in an OS.

THE BASICS OF MULTITHREADING

Multithreading is a way of increasing operating system (OS) efficiency. It does this by dividing the way that the OS handles processes, which occur when a computer executes instructions within a computer program. Multithreading OSs divide processes into subprocesses called threads. Threads are the smallest part of a programmed process that can be managed independently. In multithreading, the OS operates on multiple threads at the same time, or switches rapidly between threads. This essentially allows the OS to work on multiple processes simultaneously or to finish a process more efficiently. Multithreading is a form of multitasking that occurs within applications. Multitasking is the act of handling more than one task or application at the same time at the system level.

MEMORY AND CONTEXT

Processes are the building blocks of applications and all computing functions. All processes within a computer are managed by a scheduler. The scheduler is typically built into the OS and determines when and in what order various tasks should be completed. Each program requires a certain amount of memory, representing the resources available to complete the task. The amount of memory allocated to a process, application, or program is called the address space. To manage ongoing processes, OSs engage in context switching. Context switching means that a process or a thread within a process may be essentially put on hold while the computer switches between tasks.

While each process and each application requires its own address space, threads share the same address space. The way that OSs handle threads depends on the computer's processor. The processor is the essential circuitry that processes tasks through a system of electrical capacitors and transistors. Single-processor computers switch back and forth between threads so rapidly that the threads appear to be processed at the same time. Multicore processors can process more than one thread at a time by dividing threads between processors. Because threads share the same resources and address space, switching between threads is much easier for most computers than switching between processes.

There are two basic types of threads, user threads and kernel threads. Threads can also be classified according to their complexity. The term fiber is used for extremely simple threads that require little processing. User threads are those that are created by user interaction or are built into a program while kernel threads are built into the OS. During processing each user thread is linked to a kernel thread. This provides a link between the computer's central processing and the application. 3 Science Reference Center™ Multithreading Operating Systems

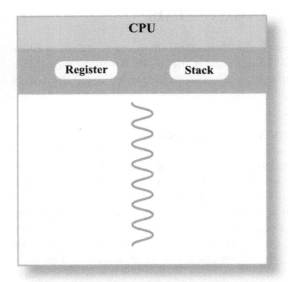

Single-threaded process

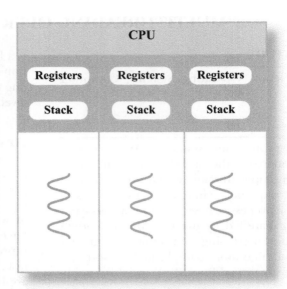

Multi-threaded process

Multithreading operating systems allow more than one part of a process to be conducted simultaneously by a single CPU. Unlike multiprocessing, which uses multiple CPUs working on non- overlapping tasks, multithreading allows full resource sharing so each thread can be dispersed among multiple CPUs in a system EBSCO illustration.

MULTITHREADING OS MODELS

In the one-to-one model, each user thread is linked to a specific kernel thread. Older OSs, like Windows 95 and XP, use a one-to-one model. This model places a limit on the number of threads that can be used to handle a process. In the many-to-one model, all application threads are mapped to a single kernel thread. This model is widely used in single-processor computers because multiple threads cannot be scheduled simultaneously. In a many-to-many model, user threads can be mapped onto an equal or smaller number of kernel threads. The many-to-many OS model is best for multiprocessor computers because it allows the computer to process multiple user threads at once.

BENEFITS AND DRAWBACKS OF MULTITHREADING

Multithreading allows a program to continue running in the background while still accepting new user input or while another task is being completed. This helps to make programs more responsive. Multicore processors have gradually become the norm in personal computing, so designing OSs to handle multithreading provides faster processing overall. Dividing processes into threads is also more efficient than running multiple instances of an entire program. Data from multiple users or different inputs can be integrated more efficiently through threads than through separate processes. This is because threads already share programs, code, and address space, while processes are usually independent of one another.

Creating hardware and software for multithreading is difficult, however, and coding for multithreading applications requires specialized knowledge. In some cases, it can be more difficult to locate and identify errors in multithreading systems. Problems in the code or errors in multithreading systems can also lead to system lag or can crash an entire process. Errors in the execution of any single thread can threaten the progress of an entire process. Therefore, multithreading essentially adds a variety of new variables that can lead to system-wide errors and processing issues.

—*Micah L. Issitt*

BIBLIOGRAPHY

"About Threads and Processes." *MSDN.Microsoft.* Windows, n.d. Web. 15 Mar 2016.

Aravind, Alex A., and Sibsankar Haldar. *Operating Systems.* Upper Saddle River: Pearson, 2010. Print.

Ballew, Joli, and Ann McIver McHoes. *Operating Systems DeMYSTiFieD.* New York: McGraw, 2012. Print.

"Differences between Multithreading and Multitasking for Programmers." *NI.* National Instruments, 20 Jan. 2014. Web. 15 Mar 2016.

Herlihy, Maurice, and Nir Shavit. *The Art of Multiprocessor Programming.* New York: Elsevier, 2012. Print.

Tanenbaum, Andrew S., and Herbert Bos. *Modern Operating Systems.* 4th ed. New York: Pearson, 2014. Print.

MULTITOUCH DISPLAYS

FIELDS OF STUDY

Software Engineering; Information Technology; Mobile Platforms

ABSTRACT

Devices with multitouch displays allow a user to touch a screen to enter commands and data. Traditional touchscreens could only respond to a single touch, as when a user touches a button on the screen. Multitouch displays can interpret touches from more than one conductive surface, such as two or more fingers, allowing for a greater range of inputs and responses.

PRINICIPAL TERMS

- **force-sensing touch technology:** touch display that can sense the location of the touch as well as the amount of pressure the user applies, allowing for a wider variety of system responses to the input.
- **gestures:** combinations of finger movements used to interact with multitouch displays in order to accomplish various tasks. Examples include tapping the finger on the screen, double-tapping, and swiping the finger along the screen.
- **optical touchscreens:** touchscreens that use optical sensors to locate the point where the user touches before physical contact with the screen has been made.
- **projected capacitive touch:** technology that uses layers of glass etched with a grid of conductive material that allows for the distortion of voltage flowing through the grid when the user touches the surface; this distortion is measured and used to determine the location of the touch.
- **resistive touchscreens:** touchscreens that can locate the user's touch because they are made of several layers of conductive material separated by small spaces; when the user touches the screen, the layers touch each other and complete a circuit.
- **surface capacitive technology:** a glass screen coated with an electrically conductive film that draws current across the screen when it is touched; the flow of current is measured in order to determine the location of the touch.

INNOVATIVE TOUCHSCREEN TECHNOLOGY

Multitouch displays are an innovation in computer input technology. Input technology takes several forms. Users can enter information into a computer using a keyboard. They type commands that the computer executes. Users can also move a mouse to control the movement of a cursor on the screen. More recently, touchscreens began to be used. Touchscreens display functions that users can activate by touching graphics on the screen. Multitouch displays take this capability a step further. They allow users to touch the screen with more than one finger at once. Users can combine touch and gestures to issue commands to the computer. For example, if a user wished to move a photo on a multitouch display screen, the user could touch the screen and slide the photo across it, just as if they were sliding a piece of paper across a table. A multitouch display senses the user's touches and movement and determines how the image should appear to move.

Multitouch displays allow users to manipulate things onscreen using multiple fingers at once. Standard gestures cause specific actions for objects onscreen (e.g., touching at two points and turning to rotate), much as specific mouse gestures are expected to cause specific actions (e.g., click and hold to drag an object). Kenneth M Pennington, CC BY 3.0 (http://creativecommons.org/licenses/by/3.0), via Wikimedia Commons.

From an engineering standpoint, there are several ways to design multitouch displays. Each has its own method for the computer to sense multiple touches and movements and then to use software to interpret what these touches and movements mean. Some methods rely primarily on electrical conductivity between the user (a person's fingers, most often) and the display. Surface capacitive technology uses a conductive film to coat the display. When the user touches the film, a circuit is completed and the display can detect where it was touched. Projected capacitive touch uses the same basic idea of detecting completed circuits, but instead of a conducting film coating the display, a grid of conductive material is etched into the display's glass.

OPTICS AND PRESSURE-SENSING

Some multitouch displays rely on optics instead of capacitive touch. Optical touchscreens use visual sensors, which are like tiny cameras. They detect when an object is about to touch the display and at what location. These systems can be preferable when touchscreen input from nonconductive objects may be necessary. For instance, a major drawback of capacitive touchscreens is that one cannot use them while wearing gloves. An optical touchscreen, on the other hand, will still respond to a touch from a gloved hand because it detects the touch without relying on conductivity. This means that

multitouch displays located outdoors in cold climates will be much more user-friendly if they have optical touchscreens.

Another category of multitouch displays uses neither optics nor capacitive touch. Instead, it can sense the amount of force the user employs when touching the display. This category is described as force-sensing touch technology. Within this category, resistive touchscreens are the most common form. Resistive touchscreens can tell how hard a user is pressing because they contain several layers of material separated by tiny spaces. When a user touches the display lightly, only a few layers are compressed enough to come into contact with one another. A touch with greater force behind it will cause more layers to compress. As layers touch one another, they complete a circuit and send a signal to the device about the location and amount of force used.

ADVANTAGES

One main advantage to using multitouch displays is that they allow users with little or no training to input information that would otherwise be time-consuming to enter into a computer or that would require more technical knowledge than the average user has. For example, a user wishing to rotate an image to a more pleasing orientation would probably not know that they want to turn the image precisely 37 degrees to the right. Even if the user did know how many degrees to rotate the image, typing this into a computer or using a mouse to make the adjustment would take much more care than simply touching the corner of the image and sliding a finger in an arc to the right.

PINCH TO ZOOM

Not surprisingly, much of the growth in multitouch displays has been driven by the mobile computing industry. As ever more computing power is packed into ever smaller devices, consumers find themselves navigating websites on ever smaller screens. One of the more frequent multitouch gestures used is the pinch-to-zoom option. This feature lets one magnify a particular part of the information being displayed on the screen, as if one were zooming in with a microscope.

—Scott Zimmer, JD

Bibliography

Horspool, Nigel, and Nikolai Tillmann. *Touchdevelop: Programming on the Go.* New York: Apress, 2013. Print.

Hughes, John F. *Computer Graphics: Principles and Practice.* Upper Saddle River: Addison, 2014. Print.

Moss, Frank. *The Sorcerers and Their Apprentices: How the Digital Magicians of the MIT Media Lab Are Creating the Innovative Technologies That Will Transform Our Lives.* New York: Crown Business, 2011. Print.

Rogers, Scott. *Swipe This!: The Guide to Great Touchscreen Game Design.* Chichester: Wiley, 2012. Print.

Saffer, Dan. *Designing Gestural Interfaces.* Beijing: O'Reilly, 2008. Print.

Soares, Marcelo M., and Francisco Rebelo. *Advances in Usability Evaluation.* Boca Raton: CRC, 2013. Print.

MULTI-USER OPERATING SYSTEMS

FIELDS OF STUDY

Computer Science; Operating Systems

ABSTRACT

A multi-user operating system (OS) is one that can be used by more than one person at a time while running on a single machine. Different users access the machine running the OS through networked terminals. The OS can handle requests from users by taking turns among connected users. This capability is not available in a single-user OS, where one user interacts directly with a machine with a single-user operating system installed on it.

PRINICIPAL TERMS

- **multi-terminal configuration:** a computer configuration in which several terminals are connected to a single computer, allowing more than one person to use the computer.
- **networking:** connecting two or more computers to one another using physical wires or wireless connections.
- **resource allocation:** a system for dividing computing resources among multiple, competing requests so that each request is eventually fulfilled.
- **system:** a computer's combination of hardware and software resources that must be managed by the operating system.
- **terminals:** a set of basic input devices, such as a keyboard, mouse, and monitor, that are used to connect to a computer running a multi-user operating system.
- **time-sharing:** a strategy used by multi-user operating systems to work on multiple user requests by switching between tasks in very small intervals of time.

Multi-User Operating Systems

A computer's operating system (OS) is its most fundamental type of software. It manages the computer system (its hardware and other installed software). An OS is often described as the computer's "traffic cop." It regulates how the different parts of the computer can be used and which users or devices may use them. Many OS functions are invisible to the user of the computer. This is either because they occur automatically or because they happen at such a low level, as with memory management and disk formatting.

A multi-user OS performs the same types of operations as a single-user OS. However, it responds to requests from more than one user at a time. When computers were first developed, they were huge, complex machines that took up a great deal of physical space. Some of the first computers took up whole rooms and required several people to spend hours programming them to solve even simple calculations. These origins shaped the way that people thought about how a computer should work. Computers became more powerful and able to handle more and more complex calculations in shorter time periods. However, computer scientists continued to think of a computer as a centralized machine usable by more than one person at a time through multiple terminals connected by networking. This is why some of the earliest OSs developed, such as UNIX, were designed

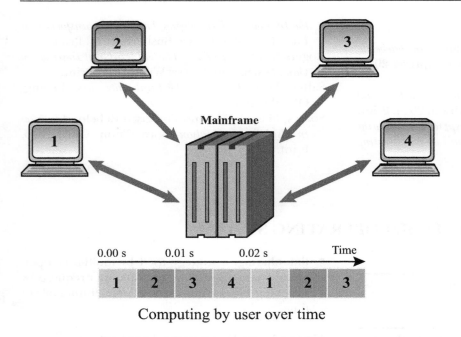

Computing by user over time

Multi-user operating systems are designed to have multiple terminals (monitor, keyboard, mouse, etc.) all connected to a single mainframe (a powerful CPU with many microprocessors) that allocates time for each user's processing demands so that it appears to the users that they are all working simultaneously. EBSCO illustration.

with a multi-terminal configuration in mind. Given the nature of early computers, it made more sense to share access to a single computer. Only years later, when technology advanced and PCs became widely available and affordable, would the focus switch to single-user OSs. Multi-user operating systems are designed to have multiple terminals (monitor, keyboard, mouse, etc.) all connected to a single mainframe (a powerful CPU with many microprocessors) that allocates time for each user's processing demands so that it appears to the users that they are all working simultaneously.

SHARED COMPUTING

In order for an OS to be able to serve multiple users at once, the system needs to have either multiple processors that can be devoted to different users or a mechanism for dividing its time between multiple users. Some multi-user OSs use both strategies, since there comes a point at which it becomes impractical to continue adding processors. A multi-user OS may appear to be responding to many different requests at once. However, what is actually happening inside

the machine is that the computer is spending a very small amount of time on one task and then switching to another task. This is called time-sharing. This task switching continues at a speed that is too fast for the user to detect. It therefore appears that separate tasks are being performed at once. Multi-user OSs function this way because if they handled one task at a time, users would have to wait in line for their requests to be filled. This would be inefficient because users would be idle until their requests had been processed. By alternating between users, a multi-user OS reduces the amount of time spent waiting. Time-sharing is one aspect of resource allocation. This is how a system's resources are divided between the different tasks it must perform.

One example of a multi-user OS is the software used to run the servers that support most webmail applications. These systems have millions or even billions of users who continually log on to check their messages, so they require OSs that can handle large numbers of users at once. A typical webmail application might require hundreds of computers, each running a multi-user OS capable of supporting thousands of users at once.

ECONOMY OF SCALE

Many companies are moving back toward the use of multi-user OSs in an effort to contain costs. For a large organization to purchase and maintain full-featured computers for each employee, there must be a sizable investment in personnel and computer hardware. Companies seeking to avoid such expense find that it can be much more cost effective to deploy minimally equipped terminals for most users. This allows them to connect to servers running multi-user OSs. Backing up user data is also simpler with a multi-user OS because all of the data to be backed up is on the same machine. Therefore, it is less likely that users will lose their data and it saves time and money for the organization.

—*Scott Zimmer, JD*

BIBLIOGRAPHY

Anderson, Thomas, and Michael Dahlin. *Operating Systems: Principles and Practice*. West Lake Hills: Recursive, 2014. Print.

Garrido, José M, Richard Schlesinger, and Kenneth E. Hoganson. *Principles of Modern Operating Systems*. Burlington: Jones, 2013. Print.

Holcombe, Jane, and Charles Holcombe. *Survey of Operating Systems*. New York: McGraw, 2015. Print.

Silberschatz, Abraham, Peter B. Galvin, and Greg Gagne. *Operating System Concepts Essentials*. 2nd ed. Wiley, 2014. Print.

Stallings, William. *Operating Systems: Internals and Design Principles*. Boston: Pearson, 2015. Print.

Tanenbaum, Andrew S. *Modern Operating Systems*. Boston: Pearson, 2015. Print.

N

NATURAL LANGUAGE PROCESSING

FIELDS OF STUDY

Algorithms, Biotechnology, Programming Language

ABSTRACT

Natural language processing, or NLP, is a type of artificial intelligence that deals with analyzing, understanding, and generating natural human languages so that computers can process written and spoken human language without using computer-driven language. Natural language processing, sometimes also called "computational linguistics," uses both semantics and syntax to help computers understand how humans talk or write and how to derive meaning from what they say. This field combines the power of artificial intelligence and computer programming into an understanding so powerful that programs can even translate one language into another reasonably accurately. This field also includes voice recognition, the ability of a computer to understand what you say well enough to respond appropriately.

PRINICIPAL TERMS

- **algorithm:** a set of step-by-step instructions for performing computations.
- **anthropomorphic:** resembling a human in shape or behavior; from the Greek words *anthropos* (human) and *morphe* (form).
- **artificial intelligence:** the intelligence exhibited by machines or computers, in contrast to human, organic, or animal intelligence.
- **neural network:** in computing, a model of information processing based on the structure and function of biological neural networks such as the human brain.
- **process:** the execution of instructions in a computer program.
- **semantics:** a branch of linguistics that studies the meanings of words and phrases
- **syntax:** a branch of linguistics that studies how words and phrases are arranged in sentences to create meaning.

THE BEGINNINGS OF NATURAL LANGUAGE PROCESSING

It has long been a dream of scientists, inventors, and computer programmers to make a robot, computer, or program, such as a voice response program, that can be mistaken for a human. Alan Turing once said, " A computer would deserve to be called intelligent if it could deceive a human into believing it was human." One of the roadblocks to creating a machine like this is that human language has been nearly impossible for machines to understand and respond to appropriately.

That hasn't stopped people from trying. Many early science fiction stories are based on the idea of a robot as a human. Early in the 1950s, programmers attempted to get computers to be able to understand language enough to be able to translate from one language to another. Success was limited. An undocumented (and probably not exactly true) story that is told about early attempts to have computers understand human language correctly goes like this: A programmer typed "The spirit is willing but the flesh is weak" into a computer program that was supposed to translate the sentence into Russian, which it did. Then the programmer asked the computer to translate the Russian sentence back into English. The result was "The vodka is good, but the meat is rotten." As you can see, this makes some sense if you read the original sentence very literally, but the meaning of the sentence was completely lost.

WHY IT'S SO HARD FOR COMPUTERS TO UNDERSTAND US

Humans learn and use language in a way that is difficult for computers to understand. As a simple example, here is a sentence that might mean different things: "Baby swallows fly." Is "baby" a noun or an adjective? Is "swallows" a verb or a noun? Is "fly" a noun or a verb? Depending on the context of the conversation, a human is likely to understand this ambiguous sentence. However, a computer, without any anthropomorphic understanding, is not likely to understand the linguistic structure.

Computer programmers have made great strides in this field. They have combined the linguistic fields of semantics and syntax with powerful computer programs using neural network processes that "learn" how to look for the same kinds of signals humans look for to create meaning. What words are surrounding the words we want to understand? In our example above, if "birds" are mentioned anywhere around the phrase, the computer program can "understand" that this sentence is talking about small birds that fly. If "insect" or "child" is mentioned anywhere around the sentence, the computer can "understand" that this sentence means that a small child ate an insect. This is a very simple example that doesn't really give the scope of the power behind computer programs that perform natural language processing but is easy to understand.

WAYS NATURAL LANGUAGE PROCESSING WORKS

Programmers use a variety of techniques to help machines understand natural language. For example, automatic summarization consists of two techniques, extraction or abstraction. Extraction is a technique that attempts to extract the most important segments of the text and make a summary list of it. Abstraction, which is much more complex, involves writing a summary of the information.

Sentiment analysis tries to identify the emotions conveyed in a text. For example, on a trip review site, a program would try to identify whether a review was positive or negative based on the words used in the review, such as "liked," "enjoyed," "unhappy," or "problem."

Text classification is a way to assign predefined categories to a text. For example, your email spam detector determines whether a message is spam or something you actually want to see, with various degrees of success. Other types of text classification techniques organize, for example, news stories by topics, such as sports or headlines. There are text classification programs, called author attribution, that are so sophisticated that they can determine who wrote the text based on the style of writing, word frequency, vocabulary richness, phrase structure, and sentence length. Some people have used a program like this to determine, for example, whether Shakespeare really wrote all of the plays attributed to him.

Conversational agents are systems that attempt to have a conversation with a human. You may have seen this in customer service situations, where it is often called a "chatbot." You may not recognize right at first that you are not talking to a human, but the program often gives itself away—eventually you realize that you are not chatting with a human based on an unpredictable response in the dialogue.

NATURAL LANGUAGE PROCESSING IN EVERYDAY APPLICATIONS

It is almost impossible to imagine a world without computers translating one language into another. Programs like Google Translate exist for translating nearly any language into nearly any other language. Even FaceBook has a "translate" feature where if your friends type in Spanish, you can read it in English. As you have probably noticed, it's not perfect, but it's generally pretty close. This kind of translation works best in fields or with text where the vocabulary is well known and there are few idioms used. For example, a machine translation of a technical manual could work well, but a translation of a novel or short story is often almost comical.

Natural language processing as voice recognition is also everywhere. The iPhone feature "Siri" takes a question that you ask and makes enough sense of it to be able to answer your question adequately most of the time. You can speak to your phone and ask it to call someone for you or give you directions to get somewhere. Other applications are when your email recognizes an event and suggests you add it to your calendar (called information extraction). These types of features seemed like science fiction at one time but are now part of our everyday lives.

—*Marianne Moss Madsen, MS*

BIBLIOGRAPHY

Bird, Steven, Ewan Klein, and Edward Loper. *Natural Language Processing with Python, 2nd ed.* O'Reilly Media, 2017.

Clark, Alexander, Chris Fox, and Shalom Lappin (eds.) *The Handbook of Computational Linguistics and Natural Language Processing.* Wiley-Blackwell, 2012.

Flach, Peter. *Machine Learning: The Art and Science of Algorithms that Make Sense of Data.* Cambridge University Press, 2012.

Ingersoll, Grant S., Thomas S. Morton, and Drew Farris. *Taming Text: How to Find, Organize, and Manipulate It.* Manning Publications, 2013.

Jurafsky, Daniel, and James H. Martin. *Speech and Language Processing: An Introduction to Natural Language Processing, Computational Linguistics, and Speech Recognition,* PEL, 2008.

Kumar, Ela. *Natural Language Processing.* I K International Publishing House, 2011.

Mihalcea, Rada and Radev Dragomir. *Graph-based Natural Language Processing and Information Retrieval.* Cambridge University Press, 2011.

Watanabe, Shinji, and Jen-Tzung Chien. *Bayesian Speech and Language Processing.* Cambridge University Press, 2015.

NETWORKING: ROUTING AND SWITCHES

FIELDS OF STUDY

Information Technology; Network Design

ABSTRACT

Routing and switches are two vital parts of any computer network. A switch is a piece of hardware that connects devices on a network, directing communications traffic within the network as terminals and hardware devices send data to one another. Routing is the part of the network protocol that directs individual packets of data to the right destination.

PRINICIPAL TERMS

- **bridge:** a connection between two or more networks, or segments of a single network, that allows the computers in each network or segment to communicate with one another.
- **firewall:** a virtual barrier that filters traffic as it enters and leaves the internal network, protecting internal resources from attack by external sources.
- **gateway:** a device capable of joining one network to another that has different protocols.
- **node:** any point on a computer network where communication pathways intersect, are redistributed, or end (i.e., at a computer, terminal, or other device).

- **packet forwarding:** the transfer of a packet, or unit of data, from one network node to another until it reaches its destination.
- **packet switching:** a method of transmitting data over a network by breaking it up into units called packets, which are sent from node to node along the network until they reach their destination and are reassembled.

ROUTING

Routing is the method by which a packet of data is transmitted to its destination. In computer networks, routing is accomplished by packet switching. This term describes how packets of data are moved from one node to another in a network. A network is made up of multiple nodes, which can be devices such as printers or computers. Nodes can also be network hardware such as switches, which direct network traffic. These devices communicate with each other by sending messages over the network. Each message is divided into units of data called "packets," which are sent across the network individually and then reassembled once they all reach their destination. Each packet may take a different path to the destination, because when a packet reaches a node, it is forwarded to another node based on the state of the network at that instant. This is called packet forwarding.

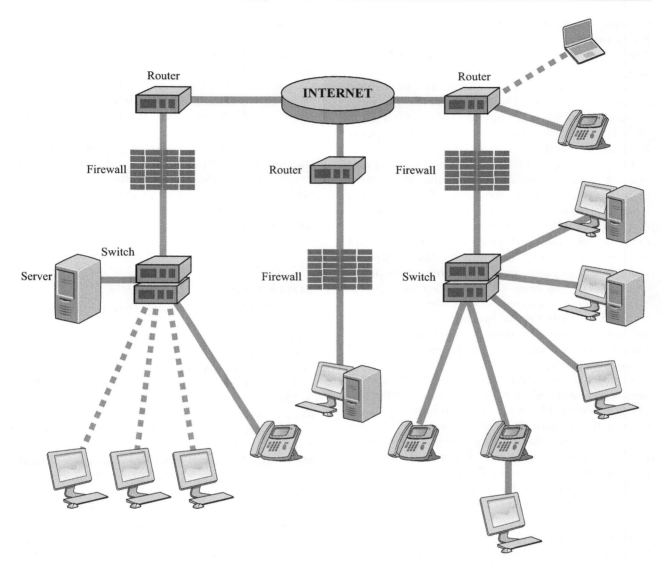

Switches are built into computer circuitry and systems to augment the signal flow throughout the system. The arrangement of switches and the criteria that open and close them will determine what components of the system are used. EBSCO illustration.

HISTORY OF NETWORK ROUTING

The reason that data is transmitted via packet switching is due to the origins of the Internet. The Internet began as a research project of the US Department of Defense, which wanted to design a communication network that could function even if some parts of the network failed. A direct line of communication, such as a telephone wire, could be easily disrupted by cutting the wire. However, a packet-switched network, with many possible pathways from origin to destination, would be less vulnerable. If one part of the network is damaged, the packets can be routed around the affected nodes and reach their destination by another path. This communication method proved so successful that the resulting networks continued to grow and connect with one another, network to network, until they formed the Internet as it is known today.

COMMUNICATION WITHIN AND BETWEEN NETWORKS

The power of the Internet is due in part to the robust performance of packet-switched networks and in part to the fact that it is essentially a network made up of other networks. The fact that networks can be connected to each other has allowed the Internet to grow at a phenomenal rate. Several different types of hardware are used to make this happen.

When an organization decides to build its own network, it uses a switch to connect the different devices that will use the network. In effect, the switch creates the network. In many cases, an organization then divides its internal network into smaller subnetworks, or "subnets." This is done when groups of computers need to communicate often with each other but only rarely with other computers in the organization. It is more efficient to set such groups apart so that the subnet's communications do not compete with other traffic on the network. Different subnets are connected to one another through a network bridge. A bridge can also be used to connect separate networks.

In order to communicate outside of the organization, a local network must be connected to the rest of the Internet. This is done by configuring one or more gateways, typically using a piece of hardware called a "router." Making this connection can leave the local network vulnerable to external attack. To prevent this, a common practice is to set up a firewall on the gateway to filter out suspicious activity, such as hackers probing the network to see if it is vulnerable.

INTERNET ACCESS

When a user on a local network goes online to check the weather forecast, that person uses their computer to send a request to the server containing the weather data. The request is broken down into multiple packets of data. Packets are transmitted between nodes on the local network until they reach the gateway. They then travel out into the Internet, where they eventually reach their destination and are reassembled. The response is sent from the server back to the local user in the same way.

Understanding the nature and function of routing and switches is important for a number of reasons. Among the most important is that it helps people understand the mechanics of protecting their computers from hackers, viruses, and other dangers found on the Internet.

—*Scott Zimmer, JD*

BIBLIOGRAPHY

Agrawal, Manish. *Business Data Communications.* Hoboken: Wiley, 2011. Print.

Comer, Douglas E. *Computer Networks and Internets.* 6th ed. Boston: Pearson, 2015. Print.

Elahi, Ata, and Mehran Elahi. *Data, Network, and Internet Communications Technology.* Clifton Park: Thomson, 2006. Print.

Kurose, James F., and Keith W. Ross. *Computer Networking: A Top-Down Approach.* 6th ed. Boston: Pearson, 2013. Print.

Mir, Nader F. *Computer and Communication Networks.* 2nd ed. Upper Saddle River: Prentice, 2015. Print.

Mueller, Scott. *Upgrading and Repairing PCs.* 22nd ed. Indianapolis: Que, 2015. Print.

Serpanos, Dimitrios N., and Tilman Wolf. *Architecture of Network Systems.* Burlington: Morgan, 2011. Print.

NEURAL NETWORKS

FIELDS OF STUDY

Network Design; Computer Engineering; Robotics Computer Science

ABSTRACT

Neural networks, or artificial neural networks (ANNs) are modeled on the structure and interconnectedness of neurons in biological brains. They have a three-layered structure consisting of an input layer, a hidden layer, and an output layer. Like a biological system, the input layer acquires data and delivers it to the hidden layer for processing. When a suitable solution to the input data has been determined, the hidden layer delivers it to the output layer. ANNs are controlled by the system architecture, the activity function and the learning function, enabling them to learn, or "train," by adjusting to changing input information. Inputs are assigned different weights and activities, and computation is carried out using "fuzzy logic" as the key to artificial intelligence. ANNs learn and compute most commonly through the backpropagation algorithm, which is akin to negative feedback, to minimize error by successive approximation. When the ANN has determined the solution having the least error, this becomes the output of the system. Typically, an ANN of this type gives no indication of how the solution was obtained. Neural networks can be analog or digital, and can be constructed as electronic, optical, hydraulic and software-simulated systems. Parallel processing is the most compatible type of computer system for software-simulated ANNs. Several types of ANN can be defined according to the logic principles by which they function.

PRINICIPAL TERMS

- **artificial intelligence:** the intelligence exhibited by machines or computers, in contrast to human, organic, or animal intelligence.
- **bridge:** a connection between two or more networks, or segments of a single network, that allows the computers in each network or segment to communicate with one another.
- **nervous (neural) system:** the system of nerve pathways by which an organism senses changes in itself and its environment and transmits electrochemical signals describing these changes to the brain so that the brain can respond.
- **neural network:** in computing, a model of information processing based on the structure and function of biological neural networks such as the human brain.
- **processor coupling:** the linking of multiple processors within a computer so that they can work together to perform calculations more rapidly. This can be characterized as loose or tight, depending on the degree to which processors rely on one another.

BASIC CONCEPT

The concept of a neural network is founded in the multiple interconnections of neurons in the brain, which is the ultimate component of the nervous (neural) system. Individual brain cells are called "neurons," and each has the ability to communicate with adjacent cells. The structure of each brain cell consists of a central body, from which a number of "axons" protrude. Each neuron can have several axons, that form connections with other axons. The "synapse" is the interface where axons from two different neurons come into contact with each other, and where the exchange of information takes place. Between actual brain neurons, this process involves the exchange of special compounds called "neurotransmitters." Each connection between neurons that is formed in this way represents a new access route to and from the knowledge that is stored in the brain, and it is this multiplicity of interconnections that is the key to the processing capacity of the brain. This is analogous to processor coupling in parallel computing systems, which use a network of central processing units (CPUs) each linked by a bridge to carry out a large number of computing processes at the same time rather than one after another in a single CPU. Data input to a biological brain comes from an "input layer" composed of the physical senses. This "data" is processed in the "hidden layer" of neurons within the brain. The result of the processing is then sent to an "output layer" of the muscles appropriate to effect the desired physical response. Processing within the brain is not a binary, deterministic process of computation with "on" and "off" signals. It is a probabilistic process that uses many "partial values" of varying importance at the same time. This kind of processing is called "fuzzy logic," a term coined by Lofti Zadeh in 1965. Fuzzy logic considers weighted values of data according to the degree to which they are true rather than as absolute "black or white" truths. Artificial neural networks are constructed on these same principles, and as an integral component of machine learning and parallel processing are believed to be essential to the development of artificial intelligence.

ARTIFICIAL NEURAL NETWORKS

An artificial neural network can be analog or digital, and may be electronic, optical, hydraulic or software-simulated in nature. For the purposes of computer science, however, only electronic and optical constructs are considered. The principal electronic components are digital CPUs. Optical components that use light for data transmission and storage are still in early development, though they are expected to provide greatly enhanced computational and operational abilities in comparison to silicon-based processors. An artificial neural network (ANN) has a functional structure analogous to that of the brain, consisting of an input layer, a hidden or intermediate layer that carries out the computation and processing of data, and an output layer. In an ANN, the interactivity of the processor network is structured to mimic the interactivity of neurons in the brain and each of the processors in the ANN is effectively connected to all of the other processors in the network. The interconnectivity may be electronic or optical in transistor-transistor based communication, or the entire network may be simulated by software. Currently, ANNs are physically limited to just a few thousands of concurrent processors in the largest interconnected systems, in what is effectively a two-dimensional array. By comparison, the human brain contains about 100 billion (1011) neurons, each with multiple interconnections in a three-dimensional array. Miniaturization of integrated circuitry, the development of three-dimensional integrated circuits, and the development of

quantum computers is expected to produce a very large increase in the power of ANNs.

SPECIFICATION OF ANNs

Three factors are typically used to specify an ANN. The first is the architecture of the system. The architecture specifies the variables involved in the composition and functioning of the ANN and typically assigns weights, or relative importance, to the elements in the input layer and assigns their respective activities. The second factor is the "activity rule," which describes how the weight values of the inputs change as they respond to each other. The third factor is the "learning rule," which specifies how the input weights change over time. The learning rule functions on a longer time scale than the activity rule and modifies the weights of the various interconnections according to the input patterns that are received. The most common learning rule used is based on the backpropagation of error, or what might otherwise be termed negative feedback, for the adjustment of the weights of input signals. Through repetition of backpropagated error correction the variance of the computed result from its actual value is minimized, and triggers the output of the result. Effectively, the functioning of an ANN can be thought of as simply a repeated process of approximation to achieve the outcome that has the highest likelihood of being true.

INPUT WEIGHTING

The weighting of inputs to an ANN is based on the logical principle of the "threshold logic unit" (TLU). The TLU is a processing unit for threshold values having n inputs but just one output. The TLU returns an output of 1 when the sum of the proportional, or weighted, values of all inputs is greater than or equal to the threshold value, and an output of 0 otherwise. In effect it is a "truth test" of the input conditions. For example, given the threshold value of 3 and the number of inputs as 6, then the TLU will return a value of 1 if the total input over all six inputs is 3 or more. It will return a value of 0 if the total input over all 6 inputs is less than 3.

USES AND LIMITATIONS OF ANNS

The development of ANNs is currently still very basic, although algorithms for the concept were first produced in 1947. The limiting factor since that time has always been the limitations inherent in available computing capabilities. General applications of the concept could not be produced until computational ability became sufficiently small and agile enough to be feasible. While it is tempting to think of neural networks in terms of supercomputers, the first generally available device to make use of an ANN for the control of its function was a household vacuum cleaner that could self-adjust its operational settings according to the floor-level conditions that it encountered. A similar system has since been incorporated into many other kinds of household appliances. The feature of interest in these, and in all ANNs, is that they have the ability to learn , or "train," to carry out tasks. An exceptional example of this is the "Roomba" floor sweeping device and its competitors. The task the device is required to learn is the most efficient manner of sweeping a mapped space of the physical dimensions of the floor. Neural networks learn their tasks by capturing associations and regularities within patterns, just as a Roomba captures associations and regularities within its map of a floor area. They also are very effective in applications involving a large volume of diverse data that includes numerous variables for input, or when the relationships between variables are poorly understood. Similarly, ANNs are useful when the relationships between input variables are not described well or responsive to conventional computational approaches. The major problem with the functioning of ANNs that rely on backpropagation is that their computation strategies are not transparent to the user. They function ultimately in a "black box" manner to produce an output from specified inputs, without providing any indication of how that output was achieved. In another context, a software-simulated ANN consisting of several processor nodes can be slow to run on a single computer, since that computer consists of just a single processor node. Consequently, it must compute the result from each of the software-simulated nodes before the back-propagation of error can be investigated. This feature renders parallel processing computers more capable for the manipulation of software-simulated ANNs. Progress in the fields of parallel processing and quantum computing to increase the ability of neural networks to perform increasingly complex computations and "learn as they go" will bring with it corresponding advances in the development of artificial intelligence.

—*Richard M. Renneboog M.Sc.*

BIBLIOGRAPHY

Galushkin, Alexander I. *Neural Network Theory.* New York, NY: Springer, 2007. Print.

Graupe, Daniel. *Principles of Artificial Neural Networks.* 2nd ed. Hackensack, NJ: World Scientific, 2007. Print.

Hagan, Martin T., Howard B. Demuth, , Mark H. Beale, and Orlando de Jesús. *Neural Network Design.* 2nd ed. Martin Hagan, 2014. Print.

Pandzu, Abhujit S. and Robert B. Macy. *Pattern Recognition with Neural Networks in C++.* Boca Raton, FL: CRC Press, 1996. Print.

Prasad, Bhanu, and S.R. Mahadeva Prasanna, eds. *Speech, Audio, Image and Biomedical Signal Processing Using Neural Networks.* New York, NY: Springer, 2008. Print.

Priddy, Kevin L. and Keller, Paul E. *Artificial Neural Networks, An Introduction.* Bellingham, WA: SPIE Press, 2005. Print.

Rao, M. Ananda, and J. Srinavas. *Neural Networks. Algorithms and Applications.* Pangbourne, UK: Alpha Science International, 2003. Print.

Rogers, Joey. *Object-Oriented Neural Networks in C++.* New York, NY: Academic Press, 1997. Print.

NEUROMORPHIC CHIPS

FIELDS OF STUDY

Computer Engineering; Information Systems

ABSTRACT

Neuromorphic chips are a new generation of computer processors being designed to emulate the way that the brain works. Instead of being locked into a single architecture of binary signals, neuromorphic chips can form and dissolve connections based on their environment, in effect "learning" from their surroundings. These chips are needed for complex tasks such as image recognition, navigation, and problem solving.

PRINICIPAL TERMS

- **autonomous:** able to operate independently, without external or conscious control.
- **Human Brain Project:** a project launched in 2013 in an effort at modeling a functioning brain by 2023; also known as HBP.
- **memristor:** a memory resistor, a circuit that can change its own electrical resistance based on the resistance it has used in the past and can respond to familiar phenomena in a consistent way.
- **nervous (neural) system:** the system of nerve pathways by which an organism senses changes in itself and its environment and transmits electrochemical signals describing these changes to the brain so that the brain can respond.
- **neuroplasticity:** the capacity of the brain to change as it acquires new information and forms new neural connections.

BRAIN-BASED DESIGN

Neuromorphic chips have much in common with traditional microprocessor chips. Both kinds of chip control how a computer receives an input, processes that information, and then produces output either in the form of information, action, or both. The difference is that traditional chips consist of millions of tiny, integrated circuits. These circuits store and process information by alternating between binary "on" and "off" states. Neuromorphic chips are designed to mimic the way that the human body's nervous (neural) system handles information. They are designed not only to process information but to learn along the way. A system that can learn may be more powerful than a system that must be programmed to respond in every possible situation. Systems that can learn can function autonomously, that is, on their own without guidance. A good example of this is car navigation systems. These systems have to store or access detailed maps as well as satellite data about their current position. With this data, a navigation system can then plot a course. The neuromorphic chips may not need all of that background data. They may be better able to understand information about their immediate surroundings and use it to predict outcomes

Neuromorphic chips are designed to detect and predict patterns in data and processing pathways to improve future computing. They simulate the brain's neuroplasticity, allowing for efficient abstraction and analysis of visual and auditory patterns. Each of the chips on this board has hundreds of millions of connections mimicking the synapses that connect neurons. By DARPA SyNAPSE, public domain, via Wikimedia Commons.

based on past events. They would function much the way someone dropped off a few blocks from home could figure out a way to get there without consulting a map.

In order to design neuromorphic chips, engineers draw upon scientific research about the brain and how it functions. One group doing such research is the Human Brain Project (HBP). HBP is trying to build working models of a rodent brain. Eventually HBP will try to build a fully working model of a human brain. Having models like these will allow scientists to test hypotheses about how the brain works. Their research will aid in the development of computer chips that can mimic such operations.

THE LIMITS OF SILICON

One reason researchers have begun to develop neuromorphic chips is that the designs of traditional chips are approaching the limits of their computational power. For many years, efforts were focused on developing computers capable of the type of learning and insights that the human brain can accomplish. These efforts were made on both the hardware and the software side. Programmers

designed applications and operating systems to use data storage and access algorithms like those found in the neural networks of the brain. Chip designers found ways to make circuits smaller and smaller so they could be ever more densely packed onto conventional chips. Unfortunately, both approaches have failed to produce machines that have either the brain's information processing power or its neuroplasticity. Neuroplasticity is the brain's ability to continuously change itself and improve at tasks through repetition.

Neuromorphic chips try to mimic the way the brain works. In the brain, instead of circuits, there are about 100 billion neurons. Each neuron is connected to other neurons by synapses, which carry electrical impulses. The brain is such a powerful computing device because its huge amount of neurons and synapses allow it to take advantage of parallel processing. Parallel processing is when different parts of the brain work on a problem at the same time. Parallel processing allows the brain to form new connections between neurons when certain pathways have proven especially useful. This forming of new neural pathways is what happens when a person learns something new or how to do a task more efficiently or effectively. The goal of neuromorphic chip designers is to develop chips that approach the brain's level of computational density and neuroplasticity. For example, researchers have proposed the use of a memristor, a kind of learning circuit. To bring such ideas about, a chip architecture completely different from the traditional binary chips is needed.

COMPLEX TASKS

Neuromorphic chips are especially suitable for computing tasks that have proven too intense for traditional chips to handle. These tasks include speech-to-text translation, facial recognition, and so-called smart navigation. All of these applications require a computer with a large amount of processing power and the ability to make guesses about current and future decisions based on past decisions. Because neuromorphic chips are still in the design and experimentation phase, many more uses for them have yet to emerge.

THE WAY OF THE FUTURE

Neuromorphic chips represent an answer to the computing questions that the future poses. Most of the computing applications being developed require more than the ability to process large amounts of data. This capability already exists. Instead, they require a device that can use data in many different forms to draw conclusions about the environment. The computers of the future will be expected to act more like human brains, so they will need to be designed and built more like brains.

—*Scott Zimmer, JD*

BIBLIOGRAPHY

Burger, John R. *Brain Theory from a Circuits and Systems Perspective: How Electrical Science Explains Neuro-Circuits, Neuro-Systems, and Qubits.* New York: Springer, 2013. Print.

Human Brain Project. Human Brain Project, 2013. Web. 16 Feb. 2016.

Lakhtakia, A., and R. J. Martín-Palma. *Engineered Biomimicry.* Amsterdam: Elsevier, 2013. Print.

Liu, Shih-Chii, Tobi Delbruck, Giacomo Indiveri, Adrian Whatley, and Rodney Douglas. *Event-Based Neuromorphic Systems.* Chichester: Wiley, 2015. Print.

Prokopenko, Mikhail. *Advances in Applied Self-Organizing Systems.* London: Springer, 2013. Print.

Quian, Quiroga R., and Stefano Panzeri. *Principles of Neural Coding.* Boca Raton: CRC, 2013. Print.

Rice, Daniel M. *Calculus of Thought: Neuromorphic Logistic Regression in Cognitive Machines.* Waltham: Academic, 2014. Print.

OBJECT-ORIENTED DESIGN

FIELDS OF STUDY

Software Engineering; Programming Language; Computer Science

ABSTRACT

Object-oriented design (OOD) is an approach to software design that uses a process of defining objects and their interactions when planning code to develop a computer program. Programmers use conceptual tools to transform a model into the specifications required to create the system. Programs created through OOD are typically more flexible and easier to write.

PRINICIPAL TERMS

- **attributes:** the specific features that define an object's properties or characteristics.
- **class-based inheritance:** a form of code reuse in which attributes are drawn from a preexisting class to create a new class with additional attributes.
- **class:** a collection of independent objects that share similar properties and behaviors.
- **method:** a procedure that describes the behavior of an object and its interactions with other objects.
- **object:** an element with a unique identity and a defined set of attributes and behaviors.
- **prototypal inheritance:** a form of code reuse in which existing objects are cloned to serve as prototypes.

ADVANCING SOFTWARE DEVELOPMENT

Object-oriented design (OOD) was developed to improve the accuracy of code while reducing software development time. Object-oriented (OO) systems are made up of objects that work together by sending messages to each other to tell a program how to behave. Objects in software design represent real-life objects and concerns. For example, in the code for a company's human resources website, an object might represent an individual employee. An object is independent of all other objects and has its own status. However, all objects share attributes or features with other objects in the same class. A class describes the shared attributes of a group of related objects.

OOD typically follows object-oriented analysis (OOA), which is the first step in the software development process when building an OO system. OOA involves planning out the features of a new program. OOD involves defining the specific objects and classes that will make up that program. The first OO language, Simula, was developed in the 1960s, followed by Smalltalk in 1972. Examples of well-known OO programming languages include Ruby, C++, Java, PHP, and Smalltalk.

OOD consists of two main processes, system design and object design. First, the desired system's architecture is mapped out. This involves defining the system's classes. Using the example of a human resources website, a class might consist of employees, while an object in that class would be a specific employee. Programmers plan out the essential attributes of a class that are shared across all objects within the class. For example, all employees would have a name, a position, and a salary. The class defines these attributes but does not specify their values. Thus, the class would have a field for the employee's name, and the individual object for a specific employee might have the name Joe Smith. All the objects within a class also share methods, or specific behaviors. For example, employee objects could be terminated, given a raise, or promoted to a different position.

Object design is based on the classes that are mapped out in the system design phase. Once software engineers have identified the required

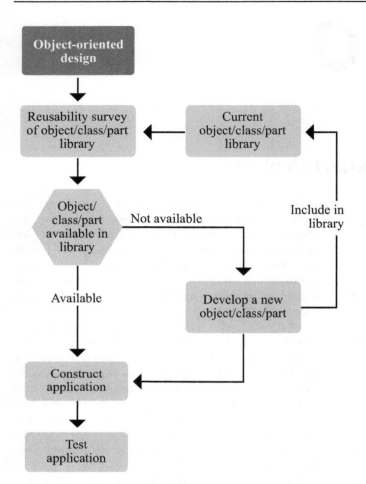

The benefit of object-oriented design lies in the reusability of the object data-base. Because each object has only one job, the code for an object remains applicable for any instance where the object is needed. EBSCO illustration.

classes, they are able to write code to create the attributes and methods that will define the necessary objects. As development progresses, they may decide that new classes or subclasses need to be created. New classes can inherit features from existing classes through class-based inheritance. Inheritance is a form of code reuse. Class-based inheritance draws attributes from an existing class to create a new class. New attributes can then be added to the new class. This is distinct from prototypal inheritance, which involves cloning existing objects, called prototypes, instead of drawing from existing classes.

OOD PRINCIPLES AND APPLICATIONS
Software engineers often follow a set of OOD principles identified by the acronym SOLID. These principles were first compiled by software engineer and author Robert Cecil Martin. The SOLID principles provide guidance for better software development, maintenance, and expansion. These principles are:

Single responsibility principle: Each class should have only one job.

Open-closed principle: A class should be open for extension but closed to modification (that is, it can be easily extended with new code without modifying the class itself).

Liskov substitution principle: Each subclass can be substituted for its parent class.

Interface segregation principle: Do not force clients to implement an interface (a list of methods) they do not use. Interfaces should be as small as possible. Having multiple smaller interfaces is better than having one large one.

Dependency inversion principle: High-level modules must not be dependent on low-level modules. Attempt to minimize dependencies between objects.

OOD is used to create software that solves real-world problems, such as addressing a user's needs when interfacing with an automatic teller machine (ATM). In this example, the software engineer first defines the classes, such as the transaction, screen, keypad, cash dispenser, and bank database. Classes or objects can be thought of as nouns. Some classes, such as transactions, might be further categorized into more specific subclasses, such as inquiry, withdrawal, deposit, or transfer.

Relationships between classes are then defined. For example, a withdrawal is related to the card reader, where the user inserts an ATM card, and the keypad, where the user enters their personal identification number (PIN). Next, classes are assigned attributes. For a withdrawal, these might be the account number, the balance, and the amount of cash to withdraw.

Finally, methods or operations are assigned to each class, based on user needs identified in the OOA phase. Methods can be thought of as verbs, such as execute, display, or withdraw. Detailed diagrams describe changes that happen within the objects, the activities they carry out, and how they interact with each other within the system.

PRECAUTIONS AND DRAWBACKS

Successful development with OOD requires realistic expectations. Possible problems with this approach include insufficient training, which can lead to the belief that OOD will clear up every development delay, resulting in missed deadlines. Further issues arise when timelines are shortened in expectation of OOD's promise of speed. Thorough training is a programmer's best course to understanding the limitations and subtleties of OOD.

OOD has been a mainstay in software programming and will be used for years to come, but advances in technology will require programming languages that can deliver faster search results, more accurate calculations, and better use of limited bandwidth for the applications users increasingly rely upon. In addition, OOD is not the best choice for every requirement. It is appropriate for systems with a large amount of graphics, interfaces, and databases, but for simpler systems, task-oriented design (TOD) might be a better choice.

—*Teresa E. Schmidt*

BIBLIOGRAPHY

Booch, Grady, et al. *Object-Oriented Analysis and Design with Applications.* 3rd ed. Upper Saddle River: Addison, 2007. Print.

Dennis, Alan, Barbara Haley Wixom, and David Tegarden. *Systems Analysis and Design: An Object-Oriented Approach with UML.* 5th ed. Hoboken: Wiley, 2015. Print.

Garza, George. "Working with the Cons of Object Oriented Programming." Ed. Linda Richter. *Bright Hub.* Bright Hub, 19 May 2011. Web. 6 Feb. 2016.

Harel, Jacob. "SynthOS and Task-Oriented Programming." *Embedded Computing Design.* Embedded Computing Design, 2 Feb. 2016. Web. 7 Feb. 2016.

Metz, Sandi. *Practical Object-Oriented Design in Ruby: An Agile Primer.* Upper Saddle River: Addison, 2012. Print.

Oloruntoba, Samuel. "SOLID: The First 5 Principles of Object Oriented Design." *Scotch.* Scotch.io, 18 Mar. 2015. Web. 1 Feb. 2016.

Puryear, Martin. "Programming Trends to Look for This Year." *TechCrunch.* AOL, 13 Jan. 2016. Web. 7 Feb. 2016.

Weisfeld, Matt. *The Object-Oriented Thought Process.* 4th ed. Upper Saddle River: Addison, 2013. Print.

P

PARALLEL PROCESSORS

FIELDS OF STUDY

Computer Engineering

ABSTRACT

A single processor carries out program processes sequentially. With parallel processors, a program can be parallelized such that various independent processes in the program can be computed simultaneously, rather than sequentially. Large arrays of parallel processors, called supercomputers, can generally mitigate against the effects of excess heat generation and electrical noise that can interfere with successful computation. Amdahl's Law describes the limit to which program execution can be sped up by parallelization.

PRINICIPAL TERMS

- **algorithm:** a set of step-by-step instructions for performing computations.
- **bridge:** a connection between two or more networks, or segments of a single network, that allows the computers in each network or segment to communicate with one another.
- **central processing unit (CPU):** electronic circuitry that provides instructions for how a computer handles processes and manages data from applications and programs.
- **channel capacity:** the upper limit for the rate at which information transfer can occur without error.
- **communication architecture:** the design of computer components and circuitry that facilitates the rapid and efficient transmission of signals between different parts of the computer.
- **processor coupling:** the linking of multiple processors within a computer so that they can work together to perform calculations more rapidly. This can be characterized as loose or tight, depending on the degree to which processors rely on one another.
- **processor symmetry:** multiple processors sharing access to input and output devices on an equal basis and being controlled by a single operating system.
- **supercomputer:** an extremely powerful computer that far outpaces conventional desktop computers.

PARALLEL AND CONVENTIONAL PROCESSORS

A conventional computer is constructed around a single central processing unit (CPU) that normally carries out all computational activities, including those needed for active graphic displays such as those in simulation programs and video games, and communicates with a limited set of input/output (I/O) devices. The computing capability of the CPU depends on the number of transistor structures that are etched onto the surface of the silicon chip that makes up the CPU. The architecture of the CPU chip includes registers for manipulating 'bits' of data through various Boolean calculation operations, and cache memory for the temporary storage of data between calculation operations and before writing to an I/O device. That operations are carried out in a sequential manner limits the flow of operations, due to the size of the registers (the number of bits of data that can be held in a register at any one time) and the number of 'clock cycles' required for execution of an operation. This is alleviated by providing a 'graphics processing unit', or GPU, dedicated to the manipulation of graphics. The resultant sharing of computational duties between these separate processing units is the essential feature of parallel processing, although the CPU-GPU combination is not precisely what defines parallel processors and their function. True parallel processors provide a means of by-passing the 'bottleneck' by carrying out sequential processes simultaneously in multiple registers across multiple CPUs.

PROPERTIES OF PARALLEL PROCESSORS

Splitting computational processes across multiple processors not only by-passes the register size bottleneck, it allows a much greater quantity of data to be processed. Just as the computational capacity of a single CPU is determined by the number of transistor structures that exist on that chip, the computational capacity of a parallel processor system is determined by the number of individual CPUs that make up a particular system. The overall capacity of the resulting machine is such that it is often referred to as a supercomputer. Two major problems that have become increasingly important as transistor size has become smaller and CPUs more compact in size are heat and noise. The reason for the difficulties these two factors pose is the close proximity of transistor structures to each other. Heat generated by the flow of electrons through the material of the transistors must be dissipated effectively in order to prevent transistor failure due to overheating. Similarly, the closer in proximity transistor structures are to each other, the more readily do signals from one 'leak' into another, resulting in scrambled data and inaccurate results. Electrical leakage also has the potential to short out and destroy a transistor structure, effectively destroying the entire value of the CPU. Splitting processes across multiple processors mitigates the heat generated by distributing it across the array so that it is rather less localized. In addition, the physical size of the array enables the incorporation of large-scale cooling systems that would not be feasible for smaller or individual CPU systems. For example, supercomputers such as IBM's 'Blue' series consist of CPU arrays housed in cabinetry large enough to occupy several hundred square meters of floor space. The cabinetry for such systems incorporates a refrigeration cooling system that maintains the entire array at its optimum operating temperature. Electrical noise can also be mitigated in supercomputer arrays of parallel processors by enabling use of CPUs having transistor structures etched on a larger scale, unlike the close-packed arrays of transistors on the CPU chips of stand-alone systems.

PARALLELISM AND CONCURRENCE

The structure of a parallel processor array requires a well-designed and implemented bridge system for processor coupling. Additionally, each implementation depends on a proper algorithm for distributing the computation task in the desired manner across the CPU array. This must take in the channel capacity of the system in order to avoid overloading and bottlenecks in data flow. In most applications, the desired function of the communication architecture is to produce process symmetry such that the computational load is evenly distributed among the CPUs in the parallel array. Two methodologies describe process symmetry. One is concurrence, and the other is parallelism. Concurrent processes are different processes being carried out on different processors at the same time. This is the condition achieved when processes 'time-share' on one or more CPUs. In effect, the CPU carries out a number of operations for one process, then switches over to carry out a number of operations for the other process. One might liken concurrence to working out two or more different crossword puzzles at the same time, filling in a few answers of one, then a few of another, until all are finished at roughly the same time. True parallelism, in comparison, can actually be carried out at bit level, but it is most often the case that the operational algorithm divides the overall program into a number of different subtasks and program functions that can be processed independently. Upon completion, the individual results are combined to yield the solution or result of the computation. A third methodology called 'distributed computing' can utilize both concurrency and parallelism, but typically requires that various CPUs communicate with each other for the completion of their respective tasks when the computation of one task relies on the result or a parameter being passed from another task on another CPU.

AMDAHL'S LAW

As a first approximation, increasing the number of processors in a parallel processor array would be expected to have a geometric effect on the performance of a parallel processing algorithm. That is to say, doubling the number of processors would be expected to cut the processing time in half. Doubling the number of processors again would be expected to take up only one quarter of the original processing time. This is not observed, however, and while an increasing rate relationship is observed for parallel processor systems having a fairly small number of processors, most processing rates tend to plateau rather than increase beyond the sixth or seventh processor doubling iteration. The discrepancy is described as the latency of the system, and is described by Amdahl's Law

$$S_{latency}(s) = (1 - p + p/s)^{-1}$$

Latency is due to the fact that a program cannot be made completely parallel in execution; there is always some part of the program code that either does not translate to a parallel counterpart, or that must be run sequentially. Amdahl's Law considers S. the 'speed up' of the entire executable program, relative to s, the speed up of just the part of the program that can be made parallel, in terms of p, the percentage of the execution time of the program due to the part that can be parallelized before the parallelization occurs. This fraction will always be detrimental to the extent to which the execution of program code can be sped up, regardless of the number of additional processors over which the program is spread. For example, a program in which 10% of the code cannot be parallelized has a corresponding value of p = 0.9, and cannot run any more than ten times faster. The addition of more processors than are required to achieve that result will have no further effect on the speed at which the program will run.

—*Richard M. Renneboog M.Sc.*

BIBLIOGRAPHY

Fountain, T J. *Parallel Computing: Principles and Practice.* Cambridge: Cambridge University Press, 1994. Print.

Hughes, Cameron, and Tracey Hughes. *Parallel and Distributed Programming Using C++.* Boston: Addison-Wesley, 2004. Print.

Jadhav, S.S. (2008) *Advanced Computer Architecture & Computing.* Pune, IND: Technical Publishers, 2008. Print.

Kirk, David B, and Wen-mei Hwu. *Programming Massively Parallel Processors: A Hands-on Approach.* Burlington, Massachusetts: Morgan Kaufmann Elsevier, 2013. Print.

Lafferty, Edward L, Marion C. Michaud, and Myra J. Prelle. *Parallel Computing: An Introduction.* Park Ridge: Noyes Data Corporation, 1993. Print.

Openshaw, Stan, and Ian Turton. *High Performance Computing and the Art of Parallel Programming: An Introduction for Geographers, Social Scientists, and Engineers.* London: Routledge, 2005. Print.

Roosta, Seyed H. *Parallel Processing and Parallel Algorithms: Theory and Computation.* New York: Springer, 2013. Print.

Tucker, Allen B, Teofilo F. Gonzalez, and Jorge L. Diaz-Herrera. *Computing Handbook.* Boca Raton, FL: CRC Press, 2014. Print.

PERSONAL HEALTH MONITOR TECHNOLOGY

FIELDS OF STUDY

Biotechnology

ABSTRACT

Modern health care is largely based on the reliable and accurate gathering of patient information. By collecting and analyzing data over a period of time, health care providers can gain a deeper understanding of their patients and make better decisions. New technology has facilitated patients' participation in their own health care in this respect. However, the safeguarding of this information has raised concerns about patient privacy and rights.

PRINICIPAL TERMS

- **biometrics:** measurable physical or behavioral characteristics that can be used to describe, characterize, or identify a given individual.
- **data granularity:** the level of detail with which data is collected and recorded.
- **data integrity:** the degree to which collected data is and will remain accurate and consistent.
- **multimodal monitoring:** the monitoring of several physical parameters at once in order to better evaluate a patient's overall condition, as well as how different parameters affect one another or respond to a given treatment.
- **temporal synchronization:** the alignment of signals from multiple devices to a single time standard, so that, for example, two different devices that record the same event will show the event happening at the exact same time.

PERSONAL HEALTH DATA

Personal health monitoring involves gathering physiological data and evaluating it by comparing it to a set of standard indicators. Digital technology has made this process much more efficient, both for individuals and in medical settings. New mobile applications, or "apps," allow people to monitor different aspects of their health. A growing awareness of the importance of a healthy lifestyle has fueled the creation of new devices and applications (apps) for tracking exercise, nutrition, and other health indicators. Some of these devices are affordable and even free. However, concerns have been raised about patient privacy rights and data accuracy and retention.

TOOLS FOR MONITORING HEALTH

Ever more health-tracking devices are produced and purchased each year. A growing number of these are wearable tracking devices, such as fitness bands. Some companies participate in programs that make such tracking devices available to employees who wish to monitor their activity or fitness levels. These programs often provide incentives for employees to participate, such as prizes or cash rewards.

Other health-monitoring systems include apps that can be downloaded onto mobile devices. Advocates say that health apps encourage self-care and empower users by involving them more with their own health. Health care providers have become more supportive of quality wearable tracking devices because they reduce the need for appointments and make necessary appointments more efficient. In some cases, these devices can be connected to other forms of monitoring technology, providing useful data to medical personnel.

Multimodal monitoring is the monitoring of several physical indicators at once. This typically involves connecting different monitoring tools to a central display. A growing number of health care providers are exploring the possible benefits of multimodal self-monitoring devices, particularly for chronic conditions. These systems may range from apps to remote monitoring and increasingly include biometrics. Developers are creating devices that would connect with others in order to provide greater data granularity.

However, such devices require a greater level of patient involvement than regular trackers. Multimodal monitoring usually requires the support of technicians to perform complex tasks such as temporal synchronization. The accuracy of the results can be skewed by even tiny differences in timekeeping.

ADVANTAGES AND DISADVANTAGES

The health data gathered by personal health-monitoring devices can be highly sensitive. Because data gathering and sharing methods vary from system

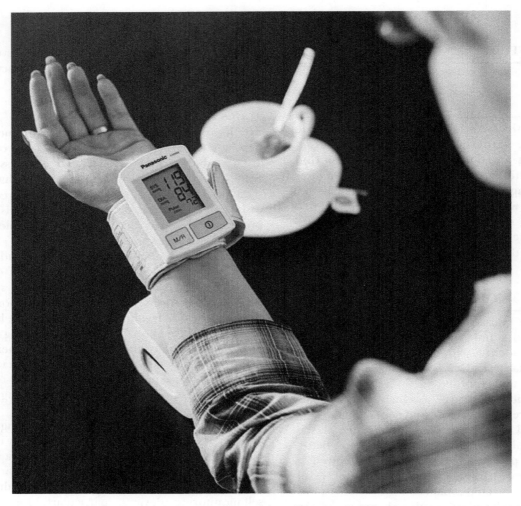

Sensors on health-monitoring devices are able to pick up heart-rate and blood pressure measurements. Advances in technology as a whole have improved the information gathered about a patient by making personal health-monitoring technology more portable and more accurate. By Bogdanradenkovic, CC BY-SA 4.0 (http://creativecommons.org/licenses/by-sa/4.0), via Wikimedia Commons

to system, concerns have been raised about privacy and accuracy. Legally, the situation remains murky. Patient-rights regulations appear not to apply to information gathered by many of these devices. For example, the companies that provide device support are not bound by medical confidentiality requirements. It is important to ensure that these services have safeguards in place to protect patient information and prevent the misuse of health data.

The benefits of innovative methods of self-monitoring are undeniable. Devices have been developed that can warn of an impending heart attack or the possibility of an asthma event. However, some health experts are concerned that self-monitoring devices may also pose grave risks. Some argue that data integrity may be compromised. Others are concerned about privacy issues, such as who has access to the information collected and how it will be used. Mental health experts have argued that excessive monitoring may even be harmful, as it tends to increase anxiety and self-centeredness. Finally, some are worried because there is very little government regulation of these devices, particularly the noninvasive type.

Yet research has shown that self-monitoring can in fact help people who suffer from anxiety to become more aware of their triggers. Moreover, as users track their exercise and diet, they become better aware of how their daily actions affect their health. In other words, personal health monitoring encourages users to become more responsible for their well-being.

—*Trudy Mercadal, PhD*

BIBLIOGRAPHY

Briassouli, Alexia, Jenny Benois-Pineau, and Alexander Hauptmann, eds. *Health Monitoring and Personalized Feedback Using Multimedia Data.* Cham: Springer, 2015. Print.

Cha, Ariana Eunjung. "Health and Data: Can Digital Fitness Monitors Revolutionise Our Lives?" *Guardian.* Guardian News and Media, 19 May 2015. Web. 26 Feb. 2016.

Fasano, Philip. *Transforming Health Care: The Financial Impact of Technology, Electronic Tools and Data Mining.* Hoboken: Wiley, 2013. Print.

Gulchak, Daniel J. "Using a Mobile Handheld Computer to Teach a Student with an Emotional and Behavioral Disorder to Self-Monitor Attention." *Education and Treatment of Children* 31.4 (2008): 567–81. PDF file.

Havens, John C. *Hacking Happiness: Why Your Personal Data Counts and How Tracking It Can Change the World.* New York: Tarcher, 2014. Print.

McNeill, Dwight. *Using Person-Centered Health Analytics to Live Longer: Leveraging Engagement, Behavior Change, and Technology for a Healthy Life.* Upper Saddle River: Pearson, 2015. Print.

Paddock, Catharine. "How Self-Monitoring Is Transforming Health." *Medical News Today.* MediLexicon Intl., 15 Aug. 2013. Web. 26 Feb. 2016.

Schmidt, Silke, and Otto Rienhoff, eds. *Interdisciplinary Assessment of Personal Health Monitoring.* Amsterdam: IOS, 2013. Print.

PERSONALIZED MEDICINE

FIELDS OF STUDY

Applications; Information Systems

ABSTRACT

Personalized medicine is the use of information about a patient's biology—especially his or her personal genome—to provide custom tailored health care. Practitioners seek to use a patient's genetic and biological characteristics to better understand and predict his or her health risks and drug responses.

PRINICIPAL TERMS

- **biomarker:** short for "biological marker"; a measurable quality or quantity (e.g., internal temperature, amount of iron dissolved in blood) that serves as an indicator of an organism's health, or some other biological phenomenon or state.
- **biometrics:** measurements that can be used to distinguish individual humans, such as a person's height, weight, fingerprints, retinal pattern, or genetic makeup.
- **genome-wide association study:** a type of genetic study that compares the complete genomes of individuals within a population to find which genetic markers, if any, are associated with various traits, most often diseases or other health problems.
- **pharmacogenomics:** the study of how an individual's genome influences his or her response to drugs.
- **toxgnostics:** a subfield of personalized medicine and pharmacogenomics that is concerned with whether an individual patient is likely to suffer a toxic reaction to a specific medication.

CUSTOM HEALTH CARE

No two people are identical. Individuals differ from one another in countless ways, some obvious (height, weight, sex) and some less so (hormone production, brain chemistry, genetic makeup). Because of these differences, no two people get sick in exactly the same way. Personalized medicine seeks to use a patient's biometrics to make predictions about his or her health and provide specialized treatment.

Personalized medicine researchers look for connections between certain biomarkers and an individual's health-related traits. Biomarkers are biometrics that correlate to important biological traits or conditions, such as disease risk. For example, a low number of red blood cells in a patient's blood is a biomarker for anemia.

HUMAN GENOMICS

The backbone of personalized medicine is human genomics, the study of the human genome. "Genome" may refer to a species-wide or a personal genome. A species-wide genome is the order in which genes appear in the DNA of all members of one species. The particular alleles, or variants, of those genes differ between individuals. The Human Genome Project was a species-wide genome project, completed in 2003, that produced the first comprehensive map of the human genetic sequence. A personal genome is an individual's unique DNA makeup, including both the genetic sequence and the specific alleles of each gene.

One powerful method of identifying potential genetic biomarkers is a genome-wide association study (GWAS). This process involves studying the genomes of many individuals to search for correlations between certain genes and certain health traits. Once a potential genetic marker is identified, researchers can use more targeted methods to further explore the connection.

Examples of such biomarkers include mutations in the *BRCA1* and *BRCA2* genes, which produce tumor-suppressing proteins in humans. Mutations in these genes are associated with a higher risk of cancer, particularly breast, ovarian, and prostate cancers. (The "BRCA" in the gene names stands for "breast cancer.") This marker was discovered and described prior to GWAS technology, using a mix of patient health histories and more limited genetic analysis to identify the genes in question.

Personal genomics companies, such as 23 and Me, sequence individuals' genomes to provide them with their genetic information. This includes information about a person's ancestry as well as health traits such as disease risk. Doctors use prenatal genetic screening to determine babies' health risks

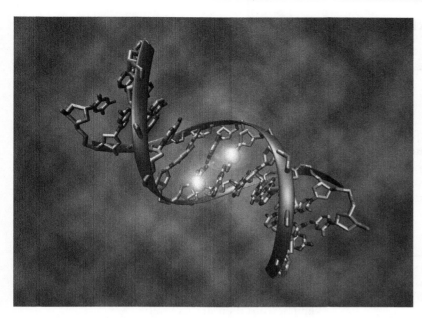

Personalized medicine is developed by analyzing the patient's DNA and determining the ideal treatment based on his or her unique genetic makeup. By Christoph Bock, Max Planck Institute for Informatics, CC BY-SA 3.0 (http://creativecommons.org/licenses/by-sa/3.0), via Wikimedia Commons.

even before birth. Personalized medicine may one day move from screening and preventive treatment to custom post-diagnosis treatment.

THE GENOMICS OF DRUG RESPONSE

The vanguard of personalized medicine and human genomics is a subfield known as pharmacogenomics. This field studies the influence of individuals' genomes on their responses to various drugs. Toxgnostics, a related subfield, focuses on whether a patient will have a toxic response to certain medications based on his or her personal genome. The goal of pharmacogenomics is to replace the "one size fits all" drug model with prescriptions tailored to each patient's biology.

A COMPUTATION-INTENSIVE PROCESS

The amount of data contained in a single human genome is very large. When studying many individuals' genomes, this information quickly grows unwieldy. Personalized medicine, with its emphasis on genomics, is only made possible by extremely powerful computers and specialized software designed to analyze large amounts of data. Fortunately, technology has advanced very quickly in this field, so that functions that once took several years can now be done in days.

—Kenrick Vezina, MS

BIBLIOGRAPHY

"A Catalog of Published Genome-Wide Association Studies." *National Human Genome Research Institute.* Natl. Insts. of Health, 16 Sept. 2015. Web. 23 Dec. 2015.

Hamburg, Margaret A., and Francis S. Collins. "The Path to Personalized Medicine." *New England Journal of Medicine* 363.4 (2010): 301–4. Web. 23 Dec. 2015.

McMullan, Dawn. "What Is Personalized Medicine?" *Genome* Spring 2014: n. pag. Web. 23 Dec. 2015.

"Personalized Medicine and Pharmacogenomics." *Mayo Clinic.* Mayo Foundation for Medical Education and Research, 5 June 2015. Web. 23 Dec. 2015.

"Precision (Personalized) Medicine." *US Food and Drug Administration.* Dept. of Health and Human Services, 18 Nov. 2015. Web. 23 Dec. 2015.

"What Is Pharmacogenomics?" *Genetics Home Reference.* Natl. Insts. of Health, 21 Dec. 2015. Web. 23 Dec. 2015.

"When Healthcare and Computer Science Collide." *Health Informatics and Health Information Management.* Pearson/U of Illinois at Chicago, n.d. Web. 23 Dec. 2015.

PRIVACY REGULATIONS

FIELDS OF STUDY

Privacy; Information Systems

ABSTRACT

Privacy regulations are laws and policies put in place to protect digital privacy and to regulate access to digital data and equipment. While US law has no general consumer privacy protection laws, it does protect certain types of digital data, including medical and financial data.

PRINICIPAL TERMS

- **access level:** in a computer security system, a designation assigned to a user or group of users that allows access a predetermined set of files or functions.
- **Computer Fraud and Abuse Act (CFAA):** a 1986 legislative amendment that made accessing a protected computer without authorization, or exceeding one's authorized level of access, a federal offense.
- **Health Insurance Portability and Accountability Act (HIPAA):** a 1996 law that established national standards for protecting individuals' medical records and other personal health information.
- **PATRIOT Act:** a 2001 law that expanded the powers of federal agencies to conduct surveillance and intercept digital information for the purpose of investigating or preventing terrorism.
- **pen/trap:** short for pen register or trap-and-trace device, devices used to record either all numbers called from a particular telephone (pen register) or all numbers making incoming calls to that phone (trap and trace); also refers to the court order that permits the use of such devices.

PRIVACY PROTECTIONS

As of 2016, the United States has no general laws protecting computer privacy. However, access to computers and certain types of digital data are restricted by various federal and state laws. While the US Constitution has no specific provision protecting the right to privacy, the Supreme Court has repeatedly interpreted several amendments to implicitly guarantee it. For example, the Fourth Amendment protects against unwarranted search and seizure. This has been taken to apply to an individual's personal communications. With advances in digital technology, millions of Americans have begun lobbying for new protections specifically for digital communication and data.

GENERAL FEDERAL PRIVACY LAWS

Within an organization, permission to access digital data may be restricted according to a system of access levels. In such a system, users are grouped into categories with varying levels of computer clearance. Network administrators usually have access to all data and operations. Users at other levels may have more limited access. In corporate and government systems, users are prohibited from accessing computers or data beyond their access level.

The Computer Fraud and Abuse Act (CFAA) of 1986 amended the United States Code statutes on federal crimes and criminal procedures. This act made unauthorized access to computer systems involved in interstate or foreign communications a federal offense. It allows for the prosecution of persons who attempt to gain unlawful computer access. The CFAA was specifically designed to protect government and financial institutions.

Also in 1986, Congress passed the Electronic Communications Privacy Act (ECPA). This law extended wiretap restrictions to apply to electronic data transmissions as well as pen/trap devices. It also specified what information Internet service providers (ISPs) cannot disclose about their users. One category of protected information is electronic communications, such as e-mails. However, the ECPA only protects e-mail stored on an ISP server for 180 days. After that time, the government can compel the ISP to disclose it. The ECPA has been criticized for not keeping pace with Internet technology. When the ECPA was first passed, ISPs only stored a user's e-mail for a short time, until it was downloaded to the user's computer. This changed with the emergence of web-based e-mail services. E-mail ISPs store users' e-mails on their servers indefinitely, often until the users delete them. Under the ECPA, all of these e-mails can be freely accessed after 180 days. Had they been downloaded to a computer and deleted from

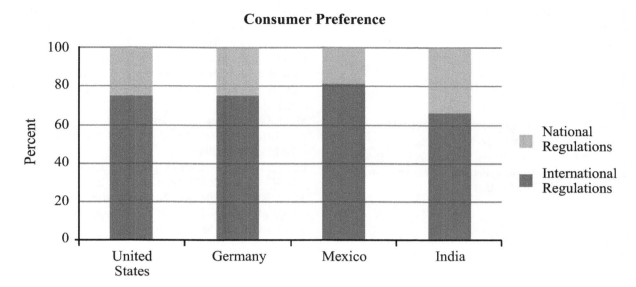

Consumer Preference

The Internet extends beyond national boundaries, making the governance and regulation of Internet privacy challenging. The majority of Internet users in countries around the world believe Internet privacy should be regulated globally, rather than locally. Adapted from data presented in "Market Research in the Mobile World" by Lightspeed, LLC.

the server instead, they could not be accessed without a warrant.

After the September 11, 2001, terrorist attacks, Congress passed the PATRIOT Act. This act gave federal agencies increased powers to monitor digital communications in order to prevent terrorism. It also specified that pen/trap restrictions apply to routing information from electronic communications as well. This technically extends privacy protections, but also allows government agencies to compel ISPs to provide routing information instead of having to gather it themselves.

PROVISIONAL PRIVACY REGULATIONS
A number of US federal regulations protect certain types of consumer data. For instance, the Health Insurance Portability and Accountability Act (HIPAA) of 1996 regulates the collection and use of medical information. Organizations with access to someone's health care data may not disclose the data without permission from that person. HIPAA mainly applies to health care providers and pharmacies. Similarly, the Fair Credit Reporting Act of 1970 limited the use of individual personal and financial information by consumer credit reporting agencies. Other such laws include the Privacy Act (1974), the

Tax Reform Act (1976), and the Electronic Fund Transfer Act (1978). These laws were not necessarily designed to protect electronic data. Nevertheless, they form the basis of Internet privacy regulations. However, many Americans feel that more general privacy laws are necessary.

OWNERSHIP OF DATA AND STATE LAWS
One recent controversy in digital privacy concerns the ownership of digital data. Data transmitted through cell phones and ISPs become partially the property of the service provider. ISPs and social media websites have mined user data to market products to users and, in some cases, to share their information with third parties. As there are no specific federal laws against this, several state legislatures have restricted corporate access to digital data. California, Connecticut, and Delaware have all passed laws requiring commercial websites to clearly disclose corporate privacy policies and to comply with "Do Not Track" requests from users, especially when collecting personal information. In 2003, Minnesota prohibited ISPs from disclosing a user's Internet habits or history without their permission. These statutes were the first US state laws intended to protect individuals' Internet privacy. In October 2015, California adopted the

California Electronic Communications Privacy Act, hailed as the nation's most comprehensive digital privacy laws to date.

—*Micah L. Issitt*

BIBLIOGRAPHY

"Computer Crime Laws." *Frontline*. WGBH Educ. Foundation, 2014. Web. 28 Mar. 2016.

"Computer Fraud and Abuse Act (CFAA)." *Internet Law Treatise*. Electronic Frontier Foundation, 24 Apr. 2013. Web. 31 Mar. 2016.

Duncan, Geoff. "Can the Government Regulate Internet Privacy?" *Digital Trends*. Designtechnica, 21 Apr. 2014. Web. 28 Mar. 2016.

"Health Information Privacy." *HHS.gov*. Dept. of Health and Human Services, n.d. Web. 28 Mar. 2016. 4 Science Reference Center™ Privacy Regulations

"State Laws Related to Internet Privacy." *National Conference of State Legislatures*. NCSL, 5 Jan. 2016. Web. 28 Mar. 2016.

"USA Patriot Act." *Electronic Privacy Information Center*. EPIC, 31 May 2015. Web. 28 Mar. 2016.

Zetter, Kim. "California Now Has the Nation's Best Digital Privacy Law." *Wired*. Condé Nast, 8 Oct. 2015. Web. 28 Mar. 2016.

PROGRAMMING LANGUAGES

FIELDS OF STUDY

Computer Engineering; Software Engineering; Applications

ABSTRACT

A programming language is a code used to control the operation of a computer and to create computer programs. Computer programs are created by sets of instructions that tell a computer how to do small but specific tasks, such as performing calculations or processing data.

PRINICIPAL TERMS

- **Abstraction:** a technique used to reduce the structural complexity of programs, making them easier to create, understand, maintain, and use.
- **declarative language:** language that specifies the result desired but not the sequence of operations needed to achieve the desired result.
- **imperative language:** language that instructs a computer to perform a particular sequence of operations.
- **semantics:** rules that provide meaning to a language.
- **syntax:** rules that describe how to correctly structure the symbols that comprise a language.
- **Turing complete:** a programming language that can perform all possible computations.

WHAT ARE PROGRAMMING LANGUAGES?

Programming languages are constructed languages that are used to create computer programs. They relay sets of instructions that control the operation of a computer. A computer's central processing unit operates through machine code. Machine code is based on numerical instructions that are incredibly difficult to read, write, or edit. Therefore, higher-level programming languages were developed to simplify the creation of computer programs. Programming languages are used to write code that is then converted to machine code. The machine code is then executed by a computer, smartphone, or other machine.

There are many different types of programming languages. First-generation, or machine code, languages are processed by computers directly. Such languages are fast and efficient to execute. However, they are difficult for humans to read and require advanced knowledge of hardware to use. Second-generation languages, or assembly languages, are more easily read by humans. However, they must be converted into machine code before being executed by a computer. Second-generation languages are used more often than first-generation ones because they are easier for humans to use while still interacting quickly and efficiently with hardware.

Both first- and second-generation languages are low-level languages. Third-generation languages

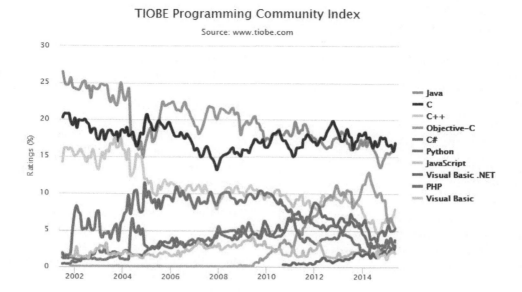

On the TIOBE index for various programming languages, C and Java are the highest-rated languages. By TIOBE Software B.V., CC BY-SA 4.0 (http://creativecommons.org/licenses/by-sa/4.0), via Wikimedia Commons.

are the most widely used programming languages. Third-generation, or high-level programming, languages are easier to use than low-level languages. However, they are not as fast or efficient. Early examples of such languages include Fortran, COBOL, and ALGOL. Some of the most widely used programming languages in the twenty-first century are third-generation. These include C++, C#, Java, JavaScript, and BASIC. Programming languages with higher levels of Abstraction are sometimes called fourth-generation languages. Higher levels of Abstraction increases platform independence. Examples include Ruby, Python, and Perl.

Programming languages can be based on different programming paradigms or styles. Imperative languages, such as COBOL, use statements to instruct the computer to perform a specific sequence of operations to achieve the desired outcome. Declarative languages specify the desired outcome but not the specific sequence of operations that will be used to achieve it. Structured Query Language (SQL) is one example.

Programming languages can also be classified by the number of different computations they can perform. Turing complete programming languages can perform all possible computations and algorithms.

Most programming languages are Turing complete. However, some programming languages, such as Charity and Epigram, can only perform a limited number of computations and are therefore not Turing complete.

HOW PROGRAMMING LANGUAGES ARE STRUCTURED

The basic structural rules of a programming language are defined in its syntax and semantics. A programming language's syntax is the grammar that defines the rules for how its symbols, such as words, numbers, and punctuation marks, are used. A programming language's semantics provide the rules used to interpret the meaning of statements constructed using its syntax. For example, the statement 1 + pizza might comply with a programming language's syntax. However, adding a number and a word together (as opposed to adding two numbers) might be semantically meaningless. Programs created with a programming language must comply with the structural rules established by the language's syntax and semantics if they are to execute correctly.

Abstraction reduces the structural complexity of programs. More Abstract languages are easier to

---

SAMPLE PROBLEM

The pseudocode below describes an algorithm designed to count the number of words in a text file and to delete the file if no words are found:

Open the file
For each word in file
    counter = counter + 1;
If counter = 0 Then
    Delete file
Close the file

Use the pseudocode to describe an algorithm that counts the number of words in a file and then prints the number of words if the number of words is greater than 300.

Answer:

Open the file
For each word in file
    wordcount = wordcount + 1;
If wordcount > 300 Then
    print wordcount
Close the file

The pseudocode above uses natural language for statements such as Open the file and programming language for statements such as wordcount = wordcount + 1. Pseudocode cannot be executed by a computer. However, it is helpful for showing the outline of a program or algorithm's operating principles.

---

one place, within the subroutine. Thus, it does not need to be duplicated.

USING PSEUDOCODE

The programming language statements that comprise a program are called "code." Code must comply with all of the programming language's syntax and semantic rules. Pseudocode uses a combination of a programming language and a natural language such as English to simply describe a program or algorithm. Pseudocode is easier to understand than code and is often used in textbooks and scientific publications.

THE FUTURE OF PROGRAMMING LANGUAGES

Programming languages are growing in importance with the continued development of the Internet and with the introduction of new programmable machines including household appliances, driverless cars, and remotely controlled drones. Such systems will increase the demand for new programming languages and paradigms designed for larger, more complex, and highly interconnected programs.

—*Maura Valentino, MSLIS*

BIBLIOGRAPHY

Belton, Padraig. "Coding the Future: What Will the Future of Computing Look Like?" *BBC News*. BBC, 15 May 2015. Web. 24 Feb. 2016.

Friedman, Daniel P., and Mitchell Wand. *Essentials of Programming Languages*. Cambridge: MIT P, 2006. Print.

Harper, Robert. *Practical Foundations for Programming Languages*. Cambridge: Cambridge UP, 2013. Print.

MacLennan, Bruce J. *Principles of Programming Languages: Design, Evaluation, and Implementation*. Oxford: Oxford UP, 1999. Print.

Scott, Michael L. *Programming Language Pragmatics*. Burlington: Kaufmann, 2009. Print.

Van Roy, Peter. *Concepts, Techniques, and Models of Computer Programming*. Cambridge: MIT P, 2004. Print.

Watt, David A. *Programming Language Design Concepts*. West Sussex: Wiley, 2004. Print.

Woods, Dan. "Why Adopting the Declarative Programming Practices Will Improve Your Return from Technology." *Forbes*. Forbes.com, 17 Apr. 2013. Web. 2 Mar. 2016.

understand and use. Abstraction is based on the principle that any piece of functionality that a program uses should be implemented only once and never duplicated. Abstraction focuses only on the essential requirements of a program. Abstraction can be implemented using subroutines. A subroutine is a sequence of statements that perform a specific task, such as checking to see if a customer's name exists in a text file. If Abstraction is not used, the sequence of statements needed to check if a customer's name exists in the file would need to be repeated every place in the program where such a check is needed. With subroutines, the sequence of statements exists in only

PROLOG

FIELDS OF STUDY

Programming Languages

ABSTRACT

The PROLOG language was released in 1971 and is a completely declarative language, although it is also capable of carrying out mathematical calculations. Numerous implementations of PROLOG have been produced, with two main variations. Programs are structured with only onr data type called the *term*, with four kinds of *terms*. Relations are structured on the pattern of "well-formed formulas" in logic, using *rules* and *facts*. A program is executed as a query run on a relation, and the result is evaluated as true or false by determining whether the query disagrees with any of the rules. PROLOG is used extensively in the field of artificial intelligence, and is more popular with European researchers, while the similar language LISP is more popular witrh North American researchers.

PRINICIPAL TERMS

- **artificial intelligence:** the intelligence exhibited by machines or computers, in contrast to human, organic, or animal intelligence.
- **character:** a unit of information that represents a single letter, number, punctuation mark, blank space, or other symbol used in written language.
- **declarative language:** language that specifies the result desired but not the sequence of operations needed to achieve the desired result.
- **programming languages:** sets of terms and rules of syntax used by computer programmers to create instructions for computers to follow. This code is then compiled into binary instructions for a computer to execute.
- **syntax:** rules that describe how to correctly structure the symbols that comprise a language.

HISTORY AND CHARACTERISTICS OF PROLOG

PROLOG was released in 1971 after being developed at l'Université de Marseilles, in France. The name is a contraction of the words *PROgrammation en LOGique* (in English, PROgramming in LOGic). It is one of the first logic programming languages, and although not as old as LISP, it is one of the oldest still in common usage. PROLOG is more commonly used in Europe, while LISP is more popular with North American researchers and developers. Both PROLOG and LISP are used extensively for research and development in the field of artificial intelligence. Numerous implementations of PROLOG have been produced , the two major variations being ISO Prolog and Edinburgh Prolog. PROLOG is geared to natural language processing, or the use of computers to process natural language. PROLOG can be implemented in any programming language or environment that is capable of calling a dynamic linked library, or .dll, file. This includes languages such as C, C++, C#, PHP, Java, Visual BASIC, Delphi and .NET, as well as other similar languages. PROLOG is particularly useful for database manipulation, symbolic math and language parsing applications.

PROGRAM FEATURES OF PROLOG

PROLOG is strictly a declarative language rather than imperative. The syntax of the language is based on logic notation. Program logic is expressed as relations in terms of *facts* and *rules*, and each process is initiated as a *query*. The execution of a PROLOG program uses an inverse logic approach to the resolution of a *query*. The program computes by searching for a resolution of the negated query, known as SLD resolution. The truth of a relation is not evaluated as to whether it is in agreement with the query. Rather, it is tested to determine if it agrees with any condition that counts as failure of the query. If accord with failure of the query is not found, then the relation is deemed to be true by default. PROLOG has just a single data type, called a *term*, with relations being defined as a *clause*. There are just four types of *term* in PROLOG, called *atom*, *number*, *variable*, or *compound_term*. The *atom* is just a general purpose term used as any variable name would be in another language. As in LISP, a *number* in PROLOG is either a float or an integer. The *variable* term always begins with an underscore or an upper case letter and is made up of a character string consisting of letters, numbers and underscores. The variable term functions like the variable term in a logic statement. The *compound_term* has a form similar to a function call in other languages. It consists of an *atom* with a list of arguments in parentheses, as for example

atom (arg1, arg2,...)

A list can also be used as an argument, as can a string of characters enclosed in quotation marks, as for example

atom (arg1, [arg2, arg3], "some text")

CLAUSE STRUCTURE IN PROLOG

Relations in PROLOG are stated as *clauses*, which are described as a *fact* or a *rule*. A *rule* essentially states a logical relationship of the type "this head statement is true if the following body conditions are true," and has the form

head statement :- body condition statement(s)

The symbol ":-" is read as "is true if." A simple logical statement of this form might be something like "(a = 2) is true if (a = b) and (b = 2)." In this simple example, the conditions described by "(a = b) and (b = 2)" would correspond to the body statements, and would be written as *facts* in the form

a (b) :- true.

b (2) :- true.

Given these *facts*, one could then pose the query

?- b(2).

equivalent to the question "is b equal to 2?," which would then return the value statement "yes" to indicate the "true" condition.

PRACTICE AND APPLICATON

Write a rule, facts, and a query statement to test the fact of whether a cat is green.

Answer

facts: cat (red).
         cat (blue).
         cat (brown).
         cat (tabby).
         cat (black)
rule: green :- (cat (green)).
query: ?- cat(green).

The *facts* specify that cats are red, blue, brown, tabby and black but no other colors are defined and therefore are not "true." The rule states that cats are green only if they are defined to be green. The query asks "are cats green?," and because comparisons with the facts all fail the program returns the value statement "no."

PROLOG AND LOGIC

Scientific method relies strictly on the application of pure logic to achieve results in any field. Strictly speaking, scientific method, in any area of application, seeks to eliminate as many unpredictable variables as possible in determining the answer to a specific question. In logic, this depends on the construction of "well-formed formulas," and such are the essential components of PROLOG programs. This is a central aspect in the development of artificial intelligence programs.

—*Richard M. Renneboog M.Sc.*

BIBLIOGRAPHY

Batchelor, Bruce. *Intelligent Image Processing in PROLOG*. London, UK: Springer-Verlag, 1991. Print.

Brauer, Max. *Logic Programming With Prolog*. New York, NY: Springer, 2005. Print.

Coelho, Helder, and José C. Cotta. *Prolog by Example: How to Learn, Teach and Use It*. New York, NY: Springer, 1996. Print.

Deransart, P., A. Ed-Dbali, and L. Cervoni. *PROLOG: The Standard Reference Manual*. New York, NY: Springer, 1996. Print.

Jones, M. Tim, *Artificial Intelligence: A Systems Approach*. Sudbury, MA: Jones and Bartlett, 2009. Print.

Mathews, Clive. *An Introduction to Natural Language Processing Through Prolog* New York, NY: Routledge, 2014. Print.

Scott, Michael L. *Programming Language Pragmatics*. Waltham, MA: Morgan Kaufmann, 2016. Print.

Sterling, Leon, ed. *The Practice of Prolog*. Boston, MA: MIT Press,1990. Print.

Q

QUANTUM COMPUTERS

FIELDS OF STUDY

Computer engineering

ABSTRACT

Quantum computers are computing devices that can theoretically have computing power that is many orders of magnitude greater than that of conventional computers. The basic unit of data in a quantum computer is the quantum bit, or qubit, that is the quantum state of electrons in an atom. Qubits can theoretically exist in several superposed states simultaneously, enabling them to carry far more information than id available using conventional two-state bits. The mathematical basis of the proportionality of qubit states is similar to that of the input weights of neural networks. There has been some successful development of quantum computer technology, but a great deal of research and development remains to be done before quantum computers become viable as a mainstream technology, and there are arguments as to why this eventuality can never be achieved.

PRINICIPAL TERMS

- **Constraints:** limitations on values in computer programming that collectively identify the solutions to be produced by a programming problem.
- **Control unit design:** describes the part of the cpu that tells the computer how to perform the instructions sent to it by a program.
- **Entanglement:** the phenomenon in which two or more particles' quantum states remain linked even if the particles are later separated and become part of distinct systems.
- **Logic implementation:** the way in which a cpu is designed to use the open or closed state of combinations of circuits to represent information.

- **Quantum bit (qubit):** a basic unit of quantum computation that can exist in multiple states at the same time, and can therefore have multiple values simultaneously.
- **Quantum logic gate:** a device that alters the behavior or state of a small number of qubits for the purpose of computation.
- **State:** a technical term for all of the stored information, and the configuration thereof, that a program or circuit can access at a given time; a complete description of a physical system at a specific point in time, including such factors as energy, momentum, position, and spin.
- **Superposition:** the principle that two or more waves, including waves describing quantum states, can be combined to give rise to a new wave state with unique properties. This allows a qubit to potentially be in two states at once.

A CONTROVERSIAL SUBJECT

The supposed difference in processing power between a semiconductor transistor-based computer and a quantum computer has been likened to reading all of the books in the library of congress one after the other versus reading them all at the same time. But quantum computers and their viability is probably the most controversial topic in computer science. Even while significant research is carried out to design and construct physical components such as a functioning quantum logic gate, there are those who offer valid conceptual reasons why quantum computers will never be either functional or viable. Indeed, the quantum computer currently exists only as a theoretically possible entity based solely on the mathematics of advanced quantum theory. However, history has amply demonstrated that theoretical possibility has become practical reality on many occasions. Accordingly, the 'jury is still out' on the potential future of quantum computers. In both concept and eventual practicality, the quantum computer is an alien device in comparison to current semiconductor transistor-based computer technology. Comprehending the functionality of a quantum computer requires a complete paradigm shift for computer science. The quantum bit (qubit) is entirely unlike the bit of current computer technology. The control unit design of a quantum computer can have little or no resemblance to current control unit design that is based on bit manipulation. The constraints of programming are entirely different for quantum computer program operation than for current bit-based computer programs. The logic implementation of quantum computers, based on quantum theory and mathematical principles, bears no resemblance to the boolean logic of transistor-based computers. Transistor logic as it is used in current computer technology uses just three kinds of transistor gate, called and, or, and not, that are effectively nothing more than electronic on-off switches for the flow of electrons through a semiconducting material. Accordingly, they have just one macroscopic state for each bit.

BITS AND QUBITS
Bits are controlled by the 'clock speed' of the particular computer, and is defined as being either the high state or the low state for the finite period of time determined by one clock cycle. A quantum bit (a qubit), however, is defined by the particular quantum state of the atom and can have several values at the same time, determined by the superposition or linear combination of several different states of the same atom. The qubit therefore exists as a quantum electronic state rather than as a macroscopic property. A transistor gate functions simply by switching the flow of electrons to change the state of a bit according to boolean relationships. A quantum gate, however, uses the quantum property of entanglement to change the state of a qubit. This is a very strange action that is not explained outside of quantum mathematics. Quantum entanglement occurs when two particles stay connected in a way such that any action on one particle equally affects the other particle, even when they are separated by great distances. Digital computer information is coded as strings of bits. In a quantum computer, the elements that carry the information are quantum states. This does not include

just the ground and excited states, but also linear combinations and superpositions of states. This allows making use of quantum parallelism techniques that would be far more powerful than even the massively parallel techniques of digital computing. A fair question at this point would be 'what are quantum states?'. In 1897, electrons and protons were positively identified as component particles of atoms, and though it was theorized at the same time, the existence of the neutron was not conclusively demonstrated until 1932. In the early 1900s, this model of atomic structure was refined using quantum mechanics to account for and describe the energies of electrons in atoms. Quantum mechanics describes the energy states that electrons in atoms are allowed to occupy. Each of these energy levels and the corresponding behavior and distribution of electrons is thus termed a 'quantum state'. An electron can move from one quantum state to another by absorbing or emitting a 'quantum' of the appropriate energy. A quantum is the minimum 'particle size' of energy required for a specific change of quantum state, and each atom has numerous possible quantum states. By using quantum states instead of the binary states of transistors, the amount of information that can be carried through the system increases exponentially with each additional state.

QUANTUM COMPUTERS AND NEURAL NETWORKS
A comparison can be made between the mathematics of qubits and the mathematics of neural networks. In particular, the functioning of an artificial neural network is determined in part by the 'weight' assigned to each input in relation to the 'threshold value' of the output. The logic of an artificial neural network triggers an output when a 'true' value (1) of the sum of the weighted inputs equal to or greater than the given threshold value is obtained. A false value (0) and no output is obtained otherwise. In a similar manner, the state of a quantum computer is described by a multidimensional vector in which each term has a complex coefficient. The state of a quantum computer is described when the sum of the squared value of the vector coefficients is equal to 1, analogous to the threshold state of an artificial neural network. This similarity has led to some speculation that the brain, a true neural network, may also function using

quantum computing principles, though no evidence of this has been found.

TYPES OF QUANTUM COMPUTERS AND THE PROGRESS SO FAR

Research into quantum-scale phenomena is time-consuming and expensive. Nevertheless, many agencies in government and industry are actively pursuing research into the development of quantum computers, for several reasons. The most important reason, of course, is the vastly superior speed of computation that quantum computers should exhibit, enabling them to solve computational problems that cannot be solved by conventional computers. Quantum computers would be far more effective at dealing with and analyzing large quantities of data, as well as for service in the field of cryptanalysis and encryption. As might be expected, the different theoretical avenues by which quantum states can be described have given rise to several different approaches to defining qubits and achieving quantum computing devices. These include

- Use of superconducting josephson junctions
- Use of the internal electronic state of trapped ions
- Use of the internal electronic state of neutral atoms in an optical lattice
- Use of quantum dots and either the spin state or position of trapped electrons
- Use of nuclear spin state in nuclear magnetic resonance
- Use of electron spin
- Use of nuclear spin of phosphorus donors in silicon
- Use of bose-einstein condensates
- Use of light modes through various linear elements
- Use of carbon nanoparticles and fullerenes

Each of these methods has had some success, in accord with the principles of scientific investigation that seek to test s single hypothesis by the strict control of variables. None, however, have yet indicated a general or universal method of generating qubits for the creation of a quantum computer. In recent years, there has been some successful commercialization of quantum computer technology by the canadian company d-wave. The device does function using quantum properties, but has yet to achieve greater performance than is available using classical computers. Nasa revealed the $15 million device publicly in december, 2015. More recently, in august, 2016, computer scientists at university of maryland constructed the first working quantum computer that is capable of being reprogrammed.

THE MAJOR PROBLEM FACING THE DEVELOPMENT OF QUANTUM COMPUTERS

That a quantum computer that functions at least partially as theorized has been built is perhaps verification of a paraphrased physical law that says 'for every theory there is an equal and opposite set of empirical results'. On one hand, the strong investment into the development of quantum computers supports the view that quantum computer technology will eventually provide computing power that is "hundreds of orders of magnitude greater than that of conventional computers." On the other hand, that such an increase has not been indicated by the performance of quantum technology so far supports the view that quantum computers will never be feasible. The principle argument for this latter view is based on the concept of 'noise'. In conventional computers, that function on a macroscopic scale, electronic interactions from both internal and external sources produce a constant background of low energy random signals occasionally punctuated by individual events of higher energy. This is electrical 'noise'. On the macroscopic scale, this interference can be readily filtered our or minimized such that the desired signals for data manipulation are essentially clean signals. On the quantum scale, however, there is no way to 'filter out' the quantum effects that arise from noise. The other major argument against the increased power of quantum computers is that even though the qubits of the process can be simultaneously in numerous states, their value can only be read as one of two states, making them no better than ordinary bits in terms of computing power. The reality is that, at the present time, neither argument for or against quantum computers has been demonstrated successfully, and a great deal of research is still required to either bring quantum computers out of the realm of theory and into the mainstream of technology or relegate them to the realm of failed scientific concepts.

—*richard m. Renneboog m.sc.*

BIBLIOGRAPHY

Miszczak, jarosław adam (2012) *high-level structures for quantum computing* williston, vt: morgan and claypool. Print.

Hirvensalo. Mike (2001) *quantum computing* new york, ny: springer. Print.

Rieffel, eleanor and polak, wolfgang (2011) *quantum computing. A gentle introduction* cambridge, ma: mit press. Print.

Berman, gennady p., doolen, gary d., mainieri, ronnie and tsifrinovich, vladimir i. (1998) *introduction to quantum computers* river edge, nj: world scientific publishing. Print.

Kaye, phillip, laflamme, raymond and mosea, michele (2007) *an introduction to quantum computing* new york, ny: oxford university press. Print.

Mermin, n. David (2007) *quantum computer science. An introduction* new york, ny: cambridge university press. Print.

QUANTUM COMPUTING

FIELDS OF STUDY

Computer Science; System-Level Programming; Computer Engineering

ABSTRACT

Quantum computing is an emerging field of computer engineering that uses charged particles rather than silicon electrical circuitry to process signals. Existing quantum computers are only in the experimental stage. Engineers believe that quantum computing has the potential to advance far beyond the limitations of traditional computing technology.

PRINICIPAL TERMS

- **entanglement:** the phenomenon in which two or more particles' quantum states remain linked even if the particles are later separated and become part of distinct systems.
- **quantum logic gate:** a device that alters the behavior or state of a small number of qubits for the purpose of computation.
- **quantum bit (qubit):** a basic unit of quantum computation that can exist in multiple states at the same time, and can therefore have multiple values simultaneously.
- **state:** a complete description of a physical system at a specific point in time, including such factors as energy, momentum, position, and spin.
- **superposition:** the principle that two or more waves, including waves describing quantum states, can be combined to give rise to a new wave state with unique properties. This allows a qubit to potentially be in two states at once.

SUBATOMIC COMPUTATION THEORIES

Quantum computing is an emerging field of computing that uses subatomic particles rather than silicon circuitry to transmit signals and perform calculations. Quantum physics studies particle behavior at the subatomic scale. At extremely small scales, subatomic particles such as photons (the basic unit of light) exhibit properties of both particles and waves simultaneously. This phenomenon, called wave-particle duality, gives subatomic particles unique properties. Traditional computer algorithms are constrained by the physical properties of digital electrical signals. Engineers working on quantum computing hope that quantum algorithms, based on the unique properties of quantum mechanics, will be able to complete computations faster and more efficiently.

THE BASICS OF QUANTUM DATA

Digital computing uses electrical signals to create binary data. Binary digits, called bits, have two possible values: 0 or 1. Digital computers also use logic gates. These are electronic circuits that process bits of data by amplifying or changing signals. Logic gates in digital computers accept one or more inputs and produce only one output. In quantum computing, digital bits are replaced by quantum bits (qubits). Qubits are created by manipulating subatomic particles.

The value of a qubit represents its current quantum state. A quantum state is simply all known data about a particle, including its momentum,

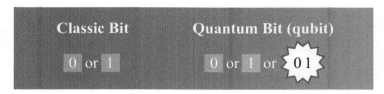

Classic Bit	Quantum Bit (qubit)
0 or 1	0 or 1 or 01

Quantum computing uses quantum bits (qubits). Classic bits can be in one of two states, 0 or 1, but qubits can be in state 0, state 1, or superstate 01. EBSCO illustration.

physical location, and energetic properties. To be used as a qubit, a particle should have two distinct states, representing the binary 0 and 1. For example, if the qubit is a photon, the two states would be horizontal polarization (0) and vertical polarization (1). However, a particle in a quantum system can exist in two or more states at the same time. This principle is called superposition. Thus, a qubit is not limited to binary values of either 0 or 1. Instead, it can have a value of 0, 1, or any superposition of both 0 and 1 at the same time.

Quantum particles also display a property known as entanglement. This is when two or more particles are linked in such a way that changing the state of one particle changes the state of the other(s), even after they are physically separated. Entanglement could potentially allow for the development of quantum computers that can instantly transmit information across great distances and physical barriers.

PRACTICAL DESIGN OF QUANTUM COMPUTERS
Current designs for quantum computers use energetic particles such as electrons or photons as qubits. The states of these particles are altered using quantum logic gates, much like digital logic gates alter electrical signals. A quantum gate may operate using energy from lasers, electromagnetic fields, or several other methods. These state changes can then be used to calculate data.

One avenue of research is the potential derivation of qubits from ion traps. Ions are atoms that have lost or gained one or more electrons. Ion traps use electric and magnetic fields to catch, keep, and arrange ions.

THE POTENTIAL OF QUANTUM COMPUTING
As of 2016, the practical value of quantum computing had only been demonstrated for a small set of potential applications. One such application is Shor's

algorithm, created by mathematician Peter Shor, which involves the mathematical process of factorization. Factorization is used to find two unknown prime numbers that, when multiplied together, give a third known number. Shor's algorithm uses the properties of quantum physics to speed up factorization. It can perform the calculation twice as fast as a standard algorithm. Researchers have also demonstrated that quantum algorithms might improve the speed and accuracy of search engines. However, research in this area is incomplete, and the potential benefits remain unclear.

There are significant challenges to overcome before quantum computing could become mainstream. Existing methods for controlling quantum states and manipulating particles require highly sensitive materials and equipment. Scientists working on quantum computers argue that they may make the biggest impact in technical sciences, where certain math and physics problems require calculations so extensive that solutions could not be found even with all of the computer resources on the planet. Special quantum properties, such as entanglement and superposition, mean that qubits may be able to perform parallel computing processes that would be impractical or improbable with traditional computer technology.

—Micah L. Issitt

BIBLIOGRAPHY
Ambainis, Andris. "What Can We Do with a Quantum Computer?" *Institute Letter* Spring 2014: 6–7. *Institute for Advanced Study.* Web. 24 Mar. 2016.

Bone, Simon, and Matias Castro. "A Brief History of Quantum Computing." *SURPRISE* May–June 1997: n. pag. *Department of Computing, Imperial College London.* Web. 24 Mar. 2016.

Crothers, Brooke. "Microsoft Explains Quantum Computing So Even You Can Understand." *CNET.* CBS Interactive, 25 July 2014. Web. 24 Mar. 2016.

Gaudin, Sharon. "Quantum Computing May Be Moving out of Science Fiction." *Computerworld.* Computerworld, 15 Dec. 2015. Web. 24 Mar. 2016.

"The Mind-Blowing Possibilities of Quantum Computing." *TechRadar.* Future, 17 Jan. 2010. Web. 26 Mar. 2016.

"A Quantum Leap in Computing." *NOVA.* WGBH/PBS Online, 21 July 2011. Web. 24 Mar. 2016.

R

RANDOM-ACCESS MEMORY

FIELDS OF STUDY

Computer Engineering; Information Technology

ABSTRACT

Random-access memory (RAM) is a form of memory that allows the computer to retain and quickly access program and operating system data. RAM hardware consists of an integrated circuit chip containing numerous transistors. Most RAM is dynamic, meaning it needs to be refreshed regularly, and volatile, meaning that data is not retained if the RAM loses power. However, some RAM is static or nonvolatile.

PRINICIPAL TERMS

- **direct-access storage:** a type of data storage in which the data has a dedicated address and location on the storage device, allowing it to be accessed directly rather than sequentially.
- **dynamic random-access memory (DRAM):** a form of RAM in which the device's memory must be refreshed on a regular basis, or else the data it contains will disappear.
- **nonvolatile random-access memory (NVRAM):** a form of RAM in which data is retained even when the device loses access to power.
- **read-only memory (ROM):** a type of nonvolatile data storage that can be read by the computer system but cannot be modified.
- **shadow RAM:** a form of RAM that copies code stored in read-only memory into RAM so that it can be accessed more quickly.
- **static random-access memory (SRAM):** a form of RAM in which the device's memory does not need to be regularly refreshed but data will still be lost if the device loses power.

HISTORY OF RAM

The speed and efficiency of computer processes are among the most areas of greatest concern for computer users. Computers that run slowly (lag) or stop working altogether (hang or freeze) when one or more programs are initiated are frustrating to use. Lagging or freezing is often due to insufficient computer memory, typically random-access memory (RAM). RAM is an essential computer component that takes the form of small chips. It enables computers to work faster by providing a temporary space in which to store and process data. Without RAM, this data would need to be retrieved from direct-access storage or read-only memory (ROM), which would take much longer.

Computer memory has taken different forms over the decades. Early memory technology was based on vacuum tubes and magnetic drums. Between the 1950s and the mid- 1970s, a form of memory called "magnetic-core memory" was most common. Although RAM chips were first developed during the same period, they were initially unable to replace core memory because they did not yet have enough memory capacity.

A major step forward in RAM technology came in 1968, when IBM engineer Robert Dennard patented the first dynamic random-access memory (DRAM) chip. Dennard's original chip featured a memory cell consisting of a paired transistor and capacitor. The capacitor stored a single bit of binary data as an electrical charge, and the transistor read and refreshed the charge thousands of times per second. Over the following years, semiconductor companies such as Fairchild and Intel produced DRAM chips of varying capacities, with increasing numbers of memory cells per chip. Intel also introduced DRAM with three transistors per cell, but over time the need for smaller and smaller computer components

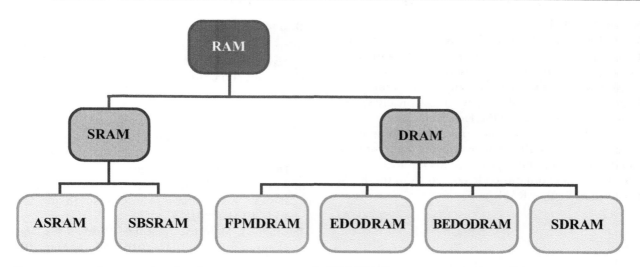

There are two major categories of random-access memory: static RAM (SRAM) and dynamic RAM (DRAM). Static RAM may be asynchronous SRAM (ASRAM) or synchronous SRAM with a burst feature (SBSRAM). Dynamic RAM may come in one of four types: fast page mode DRAM (FPMDRAM), extended data out DRAM (EDODRAM), extended data out DRAM with a burst feature (BEDODRAM), or synchronous DRAM (SDRAM). EBSCO illustration.

made this design less practical. In the 2010s, commonly used RAM chips incorporate billions of memory cells.

TYPES OF RAM

Although all RAM serves the same basic purpose, there are a number of different varieties. Each type has its own unique characteristics. The RAM most often used in personal computers is a direct descendant of the DRAM invented by Dennard and popularized by companies such as Intel. DRAM is dynamic, meaning that the electrical charge in the memory cells, and thus the stored data, will fade if it is not refreshed often. A common variant of DRAM is speed-focused double data rate synchronous DRAM (DDR SDRAM), the fourth generation of which entered the market in 2014.

RAM that is not dynamic is known as static random-access memory (SRAM). SRAM chips contain many more transistors than their DRAM counterparts. They use six transistors per cell: two to control access to the cell and four to store a single bit of data. As such, they are much more costly to produce. A small amount of SRAM is often used in a computer's central processing unit (CPU), while DRAM performs the typical RAM functions.

Just as the majority of RAM is dynamic, most RAM is also volatile. Thus, the data stored in the RAM will disappear if it is no longer being supplied with electricity—for instance, if the computer in which it is installed has been turned off. Some RAM, however, can retain data even after losing power. Such RAM is known as nonvolatile random-access memory (NVRAM).

USING RAM

RAM works with a computer's other memory and storage components to enable the computer to run more quickly and efficiently, without lagging or freezing. Computer memory should not be confused with storage. Memory is where application data is processed and stored. Storage houses files and programs. It takes a computer longer to access program data stored in ROM or in long-term storage than to access data stored in RAM. Thus, using RAM enables a computer to retrieve data and perform requested functions faster. To improve a computer's performance, particularly when running resource-intensive programs, a user may replace its RAM with a higher-capacity chip so the computer can store more data in its temporary memory.

Shadow RAM

While RAM typically is used to manage data related to the applications in use, at times it can be used to assist in performing functions that do not usually involve RAM. Certain code, such as a computer's basic input/output system (BIOS), is typically stored within the computer's ROM. However, accessing data saved in ROM can be time consuming. Some computers can address this issue by copying data from the ROM and storing the copy in the RAM for ease of access. RAM that contains code copied from the ROM is known as shadow RAM.

—*Joy Crelin*

Bibliography

Adee, Sally. "Thanks for the Memories." *IEEE Spectrum.* IEEE, 1 May 2009. Web. 10 Mar. 2016.

Hey, Tony, and Gyuri Pápay. *The Computing Universe: A Journey through a Revolution.* New York: Cambridge UP, 2015. Print.

ITL Education Solutions. *Introduction to Information Technology.* 2nd ed. Delhi: Pearson, 2012. Print.

"Shadow RAM Basics." *Microsoft Support.* Microsoft, 4 Dec. 2015. Web. 10 Mar. 2016.

Stokes, Jon. "RAM Guide Part I: DRAM and SDRAM Basics." *Ars Technica.* Condé Nast, 18 July 2000. Web. 10 Mar. 2016.

"Storage vs. Memory." *Computer Desktop Encyclopedia.* Computer Lang., 1981–2016. Web. 10 Mar. 2016.

REMOVABLE MEMORY

FIELDS OF STUDY

Information Technology; Computer Engineering; Digital Media

ABSTRACT

Removable memory systems allow the user to store digital data and transfer them between computer systems. Using magnetic, optical, and circuit-based systems, removable memory has evolved throughout the digital age and continues to be widely used, though cloud storage may eventually take its place.

PRINICIPAL TERMS

- **magnetic storage:** a device that stores information by magnetizing certain types of material.
- **main memory:** the primary memory system of a computer, often called "random access memory" (RAM), accessed by the computer's central processing unit (CPU).
- **optical storage:** storage of data by creating marks in a medium that can be read with the aid of a laser or other light source.
- **read-only memory (ROM):** type of computer data storage that is typically either impossible to erase and alter or requires special equipment to do so; it is usually used to store basic operating information needed by a computer.
- **solid-state storage:** computer memory that stores information in the form of electronic circuits, without the use of disks or other read/write equipment.
- **volatile memory:** memory that is created and maintained only while a computer is active and is generally erased when the computer is powered down.

TEMPORARY AND PERMANENT MEMORY

Computer memory, sometimes called "storage," refers to the hardware used to store both computer instructions and digital data for processing. There are two basic types of computer memory. Volatile memory, including a computer's random access memory (RAM), is stored as electrical signals and so requires power. RAM, sometimes called main memory, is created each time a computer is powered on. It provides rapid access to instructions and data. These are already stored as electrical signals and so do not have to be "read" and converted from another storage medium. Nonvolatile memory is stored physically, such that powering down the computer does not erase the data. Removable memory refers to certain

Removable memory comes in many forms, partly because of changes in technology and partly because they solve different challenges for the user. Most external data storage devices have some balance of portability and durability. By avaragado from Cambridge, CC BY 2.0 (http://creativecommons.org/licenses/by/2.0), via Wikimedia Commons

subsets of both of these types of memory. Nonvolatile removable memory involves methods of storage that can be removed or transferred from one computer to another. Volatile removable memory is expansion memory that can be installed on a computer to increase temporary memory capacity. One of the most familiar examples of removable memory is the SIM cards used in some mobile phones. Until recently, most mobile devices used removable memory cards to store user data. The card could be removed and installed on another mobile device, allowing users to transfer personal data between mobile devices.

MAGNETIC AND OPTICAL STORAGE MEDIA

The earliest forms of computer memory consisted of punch cards that represented digital data physically. The data could then be read and interpreted by a computer. In the 1950s computer engineers began using magnetic storage. Magnetic storage uses electronic heads to write and read data on a tape or other material coated with a magnetically sensitive substance. The floppy disks that were popular in the 1970s and

1980s for storing data and applications were a form of magnetic storage. In the 1980s computer scientists developed the first optical storage systems. In these systems, data was stored by making physical marks on a medium coated with a light-sensitive material. A laser could imprint the storage medium with digital data. It could then read the data and translate it into electrical signals that could be interpreted by a computer. Commonly used optical storage media included CD-ROMs and later DVD technology. Blu-ray discs are also examples of optical storage technology. Meanwhile, advances in magnetic storage led to the debut of shingled magnetic recording (SMR) devices. SMR devices can create layers of magnetic data that partially overlap like roof shingles. This gives them vastly increased storage capacity.

SOLID-STATE STORAGE

Reading and writing on magnetic or optical media requires complex moving parts. For instance, a magnetic or optical disk must be turned and a moving laser or electronic head must be used to read the encoded data. These delicate moving parts are prone to malfunction. To avoid this problem, engineers in the 1980s pioneered the use of self-contained solid-state storage systems in which data could be encoded in the form of electrical signals and circuits.

Among the earliest types of solid-state storage was read-only memory (ROM). This typically refers to a storage system that is built into a computer and cannot be erased or altered by a user. Some modern ROM systems, however, can be altered by the user with specialized tools. A computer's ROM usually contains the basic instructions that the computer needs to function. CD-ROMs are so-called because the information they hold also cannot be changed or erased, although they are not solid-state technology.

In the 1990s a variety of solid-state removable memory systems were introduced to the consumer market. These included flash memory chips and drives that store data as electrical signals. The popular thumb drives, USB drives, and other memory cards and drives are solid-state storage media. Engineers have also created multi-level cells (MLC), which can store more digital data in each electrical cell and so provide vast increases in storage capacity.

CLOUDS AND VIRTUAL COMPUTING

Removable media, while still used for many applications, is rapidly changing due to growth of virtual data storage. Cloud networks let users save data to the Web and later retrieve the data using any Internet-enabled device. They are rapidly replacing the use of removable storage and external hard drives for maintaining personal and organizational data. In addition, engineers have begun using virtual networks to increase processing capability by distributing tasks across several different computers. As cloud storage and virtual networks handle more of the storage and processing, removable storage may be used less and less by consumers while data centers have greater need of physical storage.

—*Micah L. Issitt*

BIBLIOGRAPHY

Alcorn, Paul. "SMR (Shingled Magnetic Recording) 101." *Tom's IT Pro*. Purch, 10 July 2015. Web. 7 Mar. 2016.

Blanco, Xiomara. "Top Tablets with Expandable Storage." *CNET*. CBS Interactive, 25 Jan. 2016. Web. 7 Mar. 2016.

Edwards, Benj. "From Paper Tape to Data Sticks: The Evolution of Removable Storage." *PCWorld*. IDG Consumer & SMB, 7 Feb. 2010. Web. 7 Mar. 2016.

Goodwins, Rupert. "The Future of Storage: 2015 and Beyond." *ZDNet*. CBS Interactive, 1 Jan. 2015. Web. 7 Mar. 2016.

Mercer, Christina. "5 Best Storage Devices for Startups: Removable, SSD or Cloud, What Type of Storage Should Your Business Use?" *Techworld*. IDG UK, 22 Dec. 2015. Web. 7 Mar. 2016.

Taylor, Ben. "Cloud Storage vs. External Hard Drives: Which Really Offers the Best Bang for your Buck?" *PCWorld*. IDG Consumer and SMB, 10 July 2014. Web. 7 Mar. 2016.

S

SCALING SYSTEMS

FIELDS OF STUDY

Information Systems; Information Technology; System-Level Programming

ABSTRACT

System scalability is the ability of a computer system or product to continue to function correctly when it is altered in size or volume due to user need. To be considered scalable, a system must be able not only to function properly while scaled but also to adapt to and take advantage of its new environment and increased capacity.

PRINICIPAL TERMS

- **caching:** the storage of data, such as a previously accessed web page, in order to load it faster upon future access.
- **heterogeneous scalability:** the ability of a multiprocessor system to scale up using parts from different vendors.
- **horizontal scaling:** the addition of nodes (e.g., computers or servers) to a distributed system.
- **proxy:** a dedicated computer server that functions as an intermediary between a user and another server.
- **vertical scaling:** the addition of resources to a single node (e.g., a computer or server) in a system.

SCALABILITY

As computer technology rapidly advances, computer systems must handle increasingly large and complex demands. Rather than replacing an existing system each time these demands exceed its capacity, one can simply scale up or out.

Scalability is the ability of a computer system to adapt to and accommodate an increased workload. Horizontal scaling works by distributing the workload across multiple servers or systems. Vertical scaling involves adding resources, such as hardware upgrades, to an existing server or system. Scalability allows data to be processed at a greater rate, decreasing load times and increasing productivity.

HORIZONTAL VERSUS VERTICAL SCALING

Both horizontal and vertical scaling play key roles in how networks and computers operate. How fast and how well a program or web application (app) functions depends on the resources at its disposal. For example, a new web app may have only a small number of users per day, who can easily be accommodated by a single server. However, as the number of users increases, that server will at some point become unable to support them all. Users may find that their connection is much slower, or they may be unable to connect to the server at all. At this point, the server must be scaled either vertically, such as by adding more memory, or horizontally, by adding more servers.

Vertical scaling is limited by the physical size of the existing system and its components. Fortunately, most modern computers employ open architecture, which supports the addition of hardware from different vendors. This allows for greater flexibility when upgrading components or even rebuilding part of the system to accommodate more components. One example of vertical scaling might be adding another processing unit to a single-processor system so that it can carry out parallel operations. An open-architecture system could use processors from different vendors or even different types of processors that perform different functions, such as a central processing unit (CPU) and a graphic processing unit (GPU). A system that can be scaled using such heterogeneous components is said to have heterogeneous scalability.

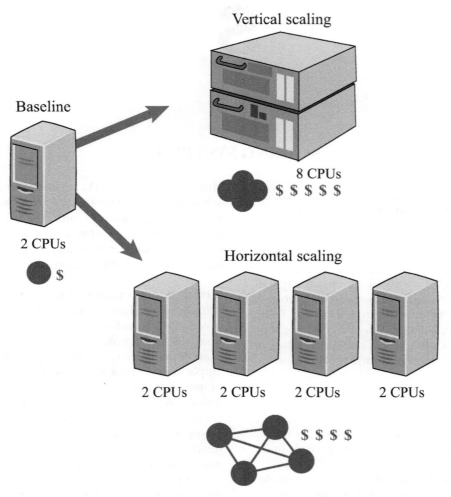

Vertical and horizontal scaling have their pros and cons. Vertical scaling increases computing power with a more powerful (and often more costly) computer. Horizontal scaling increases computing power with more computers of equal power (and equal price), but a more complex network and programming is required. EBSCO illustration.

Ultimately, however, vertical scaling can only go so far. For larger-scale operations such as cloud computing, horizontal scaling is more common. Cloud computing connects multiple hardware or software components, such as servers or networks, so they can work as a single unit. The downside of horizontal scaling is that it is not immediate. It requires careful planning and preparation to make the components work logically together. While adding more servers is a good way to deal with increased traffic, for instance, if the system cannot properly balance the load across the available servers, a user may see little improvement.

OTHER SCALING TECHNIQUES
Scalability refers to more than just processing power. It also includes various techniques to improve system efficiency and provide the best performance even when usage levels are high. One way to reduce server load during periods of high usage is to implement caching. A cache is a block of memory in which recently accessed files are stored briefly for easy

retrieval. If a user requests a file from the server that was accessed recently and is still in the cache, the system can provide the cached version instead of the original. This provides a faster response time and reduces the load on the server.

Another way is to use a proxy server. This is particularly useful for a system with multiple servers. A proxy functions as an intermediary between the user and the main servers. It receives requests from users and directs them to the appropriate servers. A proxy can also combine requests to speed up processing. If multiple users request the same data, a system without a proxy has to perform multiple retrievals of that data. However, if the requests are filtered through a proxy server, the proxy can perform a single retrieval and then forward the data to each user.

SCALING FOR THE CLOUD

Unlike other computing systems, many cloud-computing services automatically scale up or down to be more efficient. Extra computing resources, such as additional servers, are provided when usage is high and then removed when they are no longer needed. This is known as "auto-scaling." It was first introduced by Amazon Web Services and has since been adopted by other cloud-computing services, such as Microsoft Azure and Google Cloud Platform.

For large Internet companies such as these, which support massive distributed storage and cloud-computing systems, simply auto-scaling is not enough. They must be able to scale up from a handful of servers to thousands rapidly, efficiently, and resiliently, without server outages that might impact user experience. This type of architecture is described as "hyper-scale." Hyper-scale data-center architecture is mainly software-based, replacing much of the hardware of a traditional data center with virtual machines. In addition to supporting rapid, efficient auto-scaling, this greatly reduces infrastructure costs for both established and new companies. Large companies that support hundreds of millions of daily users do not have to pay to operate the massive data centers that traditional architecture would require. Meanwhile, new start-ups can launch with just a few servers and then easily scale up to a few thousand when necessary.

Receive-side scaling (RSS) is another way in which companies can scale up. This directs incoming network traffic to various CPUs for processing, thereby speeding up the network. RSS has become a feature in some cloud-computing services, such as Microsoft Azure.

—*Daniel Horowitz*

BIBLIOGRAPHY

De George, Andy. "How to Autoscale an Application." *Microsoft Azure*. Microsoft, 7 Dec. 2015. Web. 17 Feb. 2016.

El-Rewini, Hesham, and Mostafa Abd-El-Barr. *Advanced Computer Architecture and Parallel Processing*. Hoboken: Wiley, 2005. Print.

Evans, Kirk. "Autoscaling Azure—Virtual Machines." *.NET from a Markup Perspective*. Microsoft, 20 Feb. 2015. Web. 17 Feb. 2016.

Hill, Mark D. "What Is Scalability?" *Scalable Shared Memory Multiprocessors*. Ed. Michel Dubois and Shreekant Thakkar. New York: Springer, 1992. 89–96. Print.

Matsudaira, Kate. "Scalable Web Architecture and Distributed Systems." *The Architecture of Open Source Applications*. Ed. Amy Brown and Greg Wilson. Vol. 2. N.p.: Lulu, 2012. PDF file.

Miniman, Stuart. "Hyperscale Invades the Enterprise." *Network Computing*. UBM, 13 Jan. 2014. Web. 8 Mar. 2016.

Peterson, Larry L., and Bruce S. Davie. *Computer Networks: A Systems Approach*. 5th ed. Burlington: Morgan, 2012. Print.

SIGNAL PROCESSING

FIELDS OF STUDY

Algorithms; Information Technology; Digital Media

ABSTRACT

"Signal processing" refers to the various technologies by which analog or digital signals are received, modified, and interpreted. A signal, broadly defined, is data transmitted over time. Signals permeate everyday life, and many modern technologies operate by acquiring and processing these signals.

PRINICIPAL TERMS

- **analog signal:** a continuous signal whose values or quantities vary over time.
- **filter:** in signal processing, a device or procedure that takes in a signal, removes certain unwanted elements, and outputs a processed signal.
- **fixed point:** a type of digital signal processing in which numerical data is stored and represented with a fixed number of digits after (and sometimes before) the decimal point, using a minimum of sixteen bits.
- **floating point:** a type of digital signal processing in which numerical data is stored and represented in the form of a number (called the mantissa, or significand) multiplied by a base number (such as base-2) raised to an exponent, using a minimum of thirty-two bits.
- **Fourier transform:** a mathematical operator that decomposes a single function into the sum of multiple sinusoidal (wave) functions.
- **linear predictive coding:** a popular tool for digital speech processing that uses both past speech samples and a mathematical approximation of a human vocal tract to predict and then eliminate certain vocal frequencies that do not contribute to meaning. This allows speech to be processed at a faster bit rate without significant loss in comprehensibility.

DIGITAL AND ANALOG SIGNALS

Signals can be analog or digital. Analog signals are continuous, meaning they can be divided infinitely into different values, much like mathematical functions. Digital signals, by contrast, are discrete. They consist of a set of values, and there is no information "between" two adjacent values. A digital signal is more like a list of numbers.

Typically, computers represent signals digitally, while devices such as microphones represent signals in an analog fashion. It is possible to convert a signal from analog to digital by "sampling" the signal—that is, recording the value of the signal at regular intervals. A digital signal can be converted to analog by generating an electrical signal that approximates the digital signal.

One example of analog-to-digital signal conversion is the setup used to record a music performance to CD. The sound waves produced by the musicians' voices and instruments are picked up by microphones plugged into a computer. The microphones convert the sound waves to analog electrical signals, which the computer then converts to digital ones. Each signal can then be processed separately. For example, the signal from the drum can be made quieter, or that from the guitar can be processed to add distortion or reverb. Once the individual signals are processed, they are combined into a single signal, which is then converted to the CD format.

PROCESSING SIGNALS

The core concept in signal processing is the Fourier transform. A transform is a mathematical operator that changes how data are expressed without changing the value of the data themselves. In signal processing, a transform changes how a signal is represented. The Fourier transform allows signals to be broken down into their constituent components. It turns a single signal into superimposed waves of different frequencies, just as a music chord consists of a set of superimposed musical notes. The signals can then be quickly and efficiently analyzed to suppress unwanted components ("noise"), extract features contained in an image, or filter out certain frequencies.

In signal processing, a filter is a device or process that takes in a signal, performs a predetermined operation on it (typically removing certain unwanted elements), and outputs a processed signal. For example, a low-pass filter removes high frequencies from a signal, such as the sound produced by a

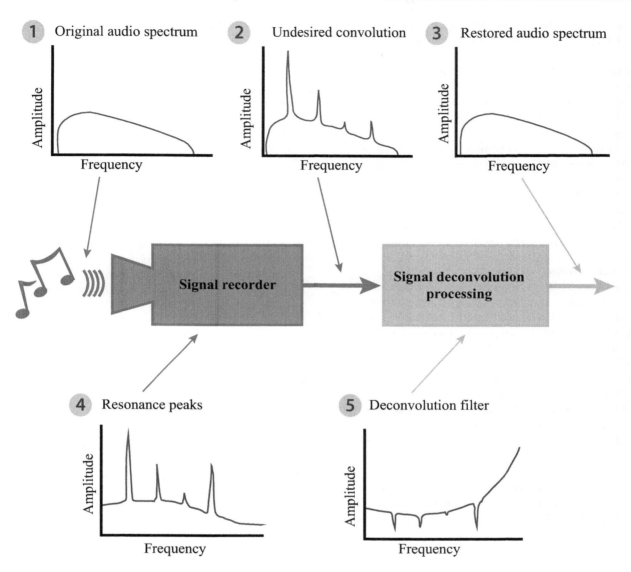

Signal processing involves sensors that receive a signal input (1). Programming to record the input may include deconvolution filters (5) to remove any undesired convolutions (2), such as harmonic resonance (4), so that the signal output (3) is restored. EBSCO illustration.

piccolo, and leaves only the low frequencies intact, such as those from a bass guitar. A high-pass filter does the opposite. It removes low frequencies from signals, keeping only the high frequencies.

When processing digital signals, one important consideration is how the data should be stored. The discrete values that make up a digital signal may be stored as either fixed point or floating point. Both formats have advantages and disadvantages. A digital signal processor (DSP) is optimized for one or the other. Fixed-point DSPs are typically cheaper and require less processing power but support a much smaller range of values. Floating-point DSPs offer greater precision and a wider value range, but they are costlier and consume more memory and power. Floating-point DSPs can also handle fixed-point values, but because the system is optimized for floating point, the advantages of fixed point are lost.

EXTRACTING INFORMATION FROM SIGNALS
Sometimes a signal contains encoded information that must be extracted. For example, a bar-code reader extracts the bar code from an acquired image. In this case, the image is the signal. It varies in space rather than in time, but similar techniques can be applied. Filters can be used to remove high spatial frequencies (such as sudden changes in pixel value) or low spatial frequencies (such as very gradual changes in pixel value). This allows the bar-code reader to find and process the bar code and determine the numerical value that it represents.

APPLICATIONS
Signal processing is everywhere. Modems take incoming analog signals from wires and turn them into digital signals that one computer can use to communicate with another. Cell phone towers work similarly to let cell phones communicate. Computer vision is used to automate manufacturing or tracking. Signals are stored in CDs, DVDs, and solid-state drives (SSDs) to represent audio, video, and text.

One particular application of signal processing is in speech synthesis and analysis. Speech signal processing relies heavily on linear predictive coding, which is based on the source-filter model of speech production. This model posits that humans produce speech through a combination of a sound source (vocal cords) and a linear acoustic filter (throat and mouth). The filter, in the rough shape of a tube,

modifies the signal produced by the sound source. This modification produces resonant frequencies called "formants," which make speech sound natural but carry no inherent meaning. Linear predictive coding uses a mathematical model of a human vocal tract to predict these formants. They can then be removed for faster analysis of human speech or produced to make synthesized speech sound more natural.

—*Andrew Hoelscher, MEng, John Vines, and Daniel Horowitz*

BIBLIOGRAPHY
Boashash, Boualem, ed. *Time-Frequency Signal Analysis and Processing: A Comprehensive Reference.* 2nd ed. San Diego: Academic, 2016. Print.
Lathi, B. P. *Linear Systems and Signals.* 2nd rev. ed. New York: Oxford UP, 2010. Print.
Owen, Mark. *Practical Signal Processing.* New York: Cambridge UP, 2012. Print.
Prandoni, Paolo, and Martin Vetterli. *Signal Processing for Communications.* Boca Raton: CRC, 2008. Print.
Proakis, John G., and Dimitris G. Manolakis. *Digital Signal Processing: Principles, Algorithms, and Applications.* 4th ed. Upper Saddle River: Prentice, 2007. Print.
Shenoi, Belle A. *Introduction to Digital Signal Processing and Filter Design.* Hoboken: Wiley, 2006. Print.
Vetterli, Martin, Jelena Kovacevic, and Vivek K. Goyal. *Foundations of Signal Processing.* Cambridge: Cambridge UP, 2014. Print.

SMART HOMES

FIELDS OF STUDY

Network Design; Computer Engineering; Embedded Systems

ABSTRACT

Smart-home technology encompasses a wide range of everyday household devices that can connect to one another and to the Web. This connectivity allows owners to program simple daily tasks and, in some cases, to control device operation from a distance. Designed for convenience, smart homes also hold the promise of improved independent living for elderly people and those with disabilities.

PRINICIPAL TERMS

- **dual-mesh network:** a type of mesh network in which all nodes are connected both through wiring and wirelessly, thus increasing the reliability of the system. A node is any communication intersection or endpoint in the network (e.g., computer or terminal).
- **electronic interference:** the disturbance generated by a source of electrical signal that affects an electrical circuit, causing the circuit to degrade or malfunction. Examples include noise, electrostatic discharge, and near-field and far-field interference.
- **Internet of things:** a wireless network connecting devices, buildings, vehicles, and other items with network connectivity.
- **mesh network:** a type of network in which each node relays signal through the network. A node is any communication intersection or endpoint in the network (e.g., computer or terminal).
- **ubiquitous computing:** a trend of embedding microprocessors and transmitters in everyday objects, enabling the objects to communicate with other computing devices.

THE EMERGING FIELD OF HOME AUTOMATION

Smart homes, or automated homes, are houses in which household electronics, environmental controls, and other appliances are connected in a network. Home automation is a growing trend in the 2010s that allows owners to monitor their homes from afar, automate basic home functions, and save money in the long run on utility payments and other costs. Creating a smart home generally involves buying a hub and then various smart devices, appliances, plugs, and sensors that can be linked through that hub to the home network. Setting up a smart home is costly and requires strong network infrastructure to work well.

CREATING THE HOME NETWORK

Not all smart devices are designed to connect in the same way. Wi-Fi enabled devices connect to Wi-Fi networks, which use super-high frequency (SHF) radio signals to link devices together and to the Internet via routers, modems, and range extenders. Other smart devices have Bluetooth connectivity. Bluetooth also uses SHF radio waves but operates over shorter ranges. Assorted Bluetooth devices may be connected to a smart-home network for remote in-home control but cannot be operated outside the home. Essentially, a Wi-Fi network is necessary for Internet-based control.

More advanced options for smart-home networking involve mesh networks. In these networks, each node can receive, repeat, and transmit signals to all others in the network. Mesh networks may be wired, with cables connecting the nodes. More often they are wireless, using Bluetooth or Wi-Fi to transmit radio signals between nodes. Smart-home network company Insteon has created products linked into dual-mesh networks. Such networks use a dual-band connection, with both wired and wireless connections between the nodes. Dual-mesh networks are designed to avoid electronic interference from outside signals coming from sources such as microwave ovens and televisions. In a dual-mesh network, interruption to the power lines entering the home or to the home's wireless network will not prevent signal from reaching a networked device. Mesh networks also work with devices from different manufacturers, while other types of networks may not be compatible with all of an owner's devices.

CONNECTING SMART DEVICES AND UTILITIES

Different types of devices can be connected to smart-home networks. Smart light switches and dimmers can

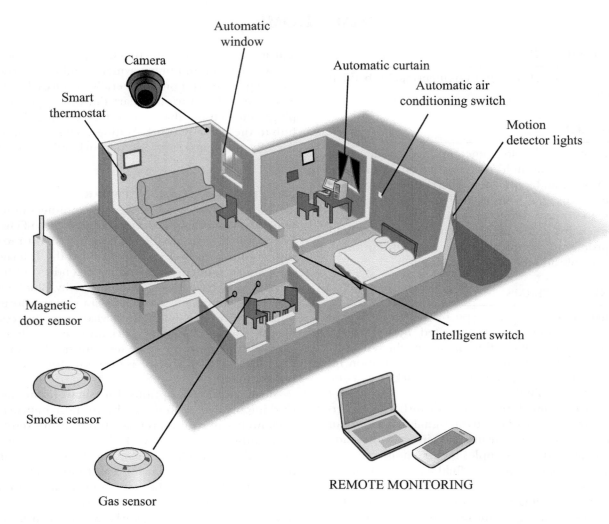

There are a number of devices and technologies available to improve a home's efficiency and sustainability, allowing individuals to customize their environment at any time, anywhere. Some products, such as programmable thermostats, can be programmed to run only at the optimal time needed to reduce resource use. Others such as motion-sensored lights are designed to respond instantly to demand. There are even devices that sync to a smartphone and allow the homeowner to control security, lighting, and many other devices remotely. EBSCO illustration.

be used to control existing light fixtures, turning lights on or off or connecting them to a scheduler for timed activation. A smart-home owner might program their lights to coincide with the sunrise/sunset cycle, for example. The use of smart plugs also allows for the control of existing, ordinary devices. Smart kitchen appliances and entertainment units are becoming available as well.

Other devices can be used to monitor home utilities. For instance, smart water sensors can detect moisture or leaks. Smart locking devices placed on doors and windows can alert the owner when a door or window is unsecured and allow them to lock and unlock doors and windows remotely. Sensors and controls can measure heat within each room and adjust thermostat settings. Smart-home owners can use programmable thermostats to lower temperatures during sleep periods and to raise heat levels before they wake. While many thermostats have timers, automated systems also allow owners to adjust heat and energy consumption from their smartphone or computer, even when away from the home. Some even feature machine learning, in which the device adjusts its behavior over time in response to repeated usage patterns. By shutting off air conditioning, heat, or electricity when not needed, automated monitoring and control can save owners large amounts of money in utility costs each year.

GOALS AND CHALLENGES

Smart-home technology is part of a technological movement called the Internet of things (IoT). IoT is based on the idea that devices, vehicles, and buildings will all someday be linked into a wireless network and collect and communicate data about the environment. This concept relies on ubiquitous computing, the practice of adding microprocessors and Internet connectivity to basic devices and appliances. Ubiquitous computing aims to make data processing a constant background activity embedded in daily life.

While still in its infancy, smart-home technology is intended to enhance security and convenience for homeowners. A home could be configured, for instance, so that motion sensors detecting activity in the morning open window shades, turn on a coffee maker, and raise the heat. Although still largely a hobbyist industry, home automation is also beginning to help elderly people and those with disabilities. Features such as voice activation, mobile apps, and touch screens make it easier for those owners to control their environment and live independently.

Consumers have been slow to adopt smart-home technology for several reasons. One is fear that manufacturers will use such devices to collect and sell data about their personal habits. Another is fear that the security of such systems could be hacked and 4 Science Reference Center™ Smart Homes lead to safety issues such as burglary. The multiple users' preferences, power dynamics, and rapid shifts in schedule that present in family life pose further challenges for smart-home systems. With the debut of wearable digital devices like smart watches, some companies are experimenting with linking wearable and smart-home devices. Wearable devices that can record biometric data could be used to verify a homeowner's identity for increased security and personalization, for example.

—Micah L. Issitt

BIBLIOGRAPHY

Clark, Don. "Smart-Home Gadgets Still a Hard Sell." *Wall Street Journal*. Dow Jones, 5 Jan. 2016. Web. 12 Mar. 2016.

Glink, Ilyce. "10 Smart Home Features Buyers Actually Want." *CBS News*. CBS Interactive, 11 Apr. 2015. Web. 12 Mar. 2016.

Gupta, Shalene. "For the Disabled, Smart Homes Are Home Sweet Home." *Fortune*. Fortune, 1 Feb. 2015. Web. 15 Mar. 2016.

Higginbotham, Stacey. "5 Reasons Why the 'Smart Home' Is Still Stupid." *Fortune*. Fortune, 19 Aug. 2015. Web. 12 Mar. 2016.

Taylor, Harriet. "How Your Home Will Know What You Need Before You Do." *CNBC*. CNBC, Jan 6 2016. Web. 11 Mar. 2016.

Vella, Matt. "Nest CEO Tony Fadell on the Future of the Smart Home." *Time*. Time, 26 June 2014. Web. 12 Mar. 2016.

SOFTWARE ARCHITECTURE

FIELDS OF STUDY

Applications; Software Engineering; Operating Systems

ABSTRACT

Software architecture refers to the specific set of decisions that software engineers make to organize the complex structure of a computer system under development. Sound software architecture helps minimize the risk of the system faltering as well as optimizing its performance, durability, and reliability.

PRINICIPAL TERMS

- **agile software development:** a method of software development that addresses changing requirements as they arise throughout the process.
- **component-based development:** an approach to software design that uses standardized software components to create new applications and new software.
- **functional requirement:** a specific function of a computer system, such as calculating or data processing.
- **nonfunctional requirement:** also called extrafunctional requirement; an attribute of a computer system, such as reliability, scalability, testability, security, or usability, that reflects the operational performance of the system.
- **plug-in:** an application that is easily installed (plugged in) to add a function to a computer system.
- **separation of concerns:** a principle of software engineering in which the designer separates the computer program's functions into discrete elements.

CHALLENGES OF SOFTWARE ARCHITECTURE

Building architects deal with static structures that maintain their structural integrity over the long term. In contrast, the architecture of a computer program must be able to change, grow, and be modified. "Software architecture" refers to the internal operations of a system under development and how its elements will ultimately function together. It is a blueprint that helps software engineers avoid and troubleshoot potential problems. These problems are far easier to address while the system is in development rather than after it is operational.

Software architects examine how a system's functional requirements and nonfunctional requirements relate to each other. Functional requirements control what processes a system is able to perform. Nonfunctional requirements control the overall operation of a program rather than specific behaviors. They include performance metrics such as manageability, security, reliability, maintainability, usability, adaptability, and resilience. Nonfunctional requirements place constraints on the system's functional requirements.

The main challenge for the software architect is to determine which requirements should be optimized. If a client looking for a new software program is asked what requirements are most critical—usability, performance, or security, for instance—they are most likely going to say all of them. Because this cannot be done without exorbitant costs, the software architect prioritizes the list of requirements, knowing that there must be trade-offs in the design of any application. If, for instance, the project is a long-term home-loan application program or a website for purchasing airplane tickets, the architect will most likely focus on security and modification. If the project is short term, such as a seasonal marketing campaign or a website for a political campaign, the architect will focus instead on usability and performance.

METHODOLOGY

Software architecture design is an early-stage Abstract process that allows for the testing of an assortment of scenarios in order to maximize the functionality of a system's most critical elements. Well-planned software architecture helps developers ascertain how the system will operate and how best to minimize potential risks or system failures. In most cases, software architecture factors in modifiability, allowing the system to grow over time to prevent obsolescence and anticipate future user needs.

Software architecture projects can involve weeks, months, or even years of development. Software architects often develop a basic skeletal system early in the development process. While this early system lacks the depth and reach of the desired program,

SOFTWARE ARCHITECTURE DESIGN PRINCIPLES

| Build for adaptivity rather than longevity | Use computer modeling to analyze risk and dependencies | Use visualizations for communication and collaboration | Minimize complexity through key engineering decisions |

Software architecture is the structural design of the software. When designing new software, it is important to build in flexibility, use models to analyze the design and reduce risks, use visualizations to communicate and collaborate with stakeholders, and maintain a focus on reducing complexity. EBSCO illustration.

it provides critical insights into the system's developing functionality. Incremental modifications are then made and tested to ensure that the emerging system is performing properly. Multiple iterations of the system are made until full functionality is achieved. This approach is known as agile software development.

Software architects work to understand how the system will ultimately operate to meet its performance requirements. Programs should be designed to be flexible enough to grow into more sophisticated and more specialized operations. Plug-ins can be used to add features to existing systems. By designing software functions as discrete elements, software architects can make the system easier to adapt. This design principle is known as separation of concerns. It allows one element of the program to be upgraded without dismantling the entire superstructure, for example. This has the potential to save considerable time and money. Component-based development is a related idea that borrows from the industrial assembly-line model. It involves using and reusing standardized software components across different programs.

IMPLICATIONS

Software architecture aims to balance the end user's needs with the system's behavioral infrastructure and the expectations of the company that will support the software. Software architects start the development process by looking at the broadest possible implications of a proposed system's elements and their relationships to one another. This distinguishes them from code developers, whose vision is often relatively narrow and specified. Software architects evaluate the critical needs of the software, consider the needs of both the client and the end user, and draft system blueprints that maximize those requirements while managing the practical concerns of time and cost.

—*Joseph Dewey*

BIBLIOGRAPHY

Bass, Len, Paul Clements, and Rick Kazman. *Software Architecture in Practice*. 3rd ed. Upper Saddle River: Addison, 2012. Print.

Cervantes, Humberto, and Rick Kazman. *Designing Software Architectures: A Practical Approach*. Upper Saddle River: Addison, 2016. Print.

Clements, Paul, et al. *Documenting Software Architectures: Views and Beyond*. 2nd ed. Upper Saddle River: Addison, 2011. Print.

Langer, Arthur M. *Guide to Software Development: Designing and Managing the Life Cycle*. New York: Springer, 2012. Print.

Mitra, Tilak. *Practical Software Architecture: Moving from System Context to Deployment*. Indianapolis: IBM, 2015. Print.

Rozanski, Nick, and Eoin Woods. *Software Systems Architecture: Working with Stakeholders Using Viewpoints and Perspectives.* 2nd ed. Upper Saddle River: Addison, 2012. Print.

Sonmez, John Z. *Soft Skills: The Software Developer's Life Manual.* Shelter Island: Manning, 2015. Print.

Taylor, Richard N., Nenad Medvidovi , and Eric M. Dashofy. *Software Architecture: Foundations, Theory, and Practice.* Hoboken: Wiley, 2010. Print

SOFTWARE REGULATIONS

FIELDS OF STUDY

Software Engineering

ABSTRACT

Software developers must keep numerous legal and regulatory considerations in mind when creating software. Organizations that use the software should also be aware of these regulations in order to remain compliant, both in their record keeping and in their use of internal-use, proprietary, and open-source software.

PRINICIPAL TERMS

- **compliance:** adherence to standards or specifications established by an official body to govern a particular industry, product, or activity.
- **Health Insurance Portability and Accountability Act (HIPAA):** a 1996 law that established national standards for protecting individuals' medical records and other personal health information.
- **internal-use software:** software developed by a company for its own use to support general and administrative functions, such as payroll, accounting, or personnel management and maintenance.
- **open-source software:** software that makes its source code available to the public for free use, study, modification, and distribution.
- **proprietary software:** software owned by an individual or company that places certain restrictions on its use, study, modification, or distribution and typically withholds its source code.
- **Sarbanes-Oxley Act (SOX):** a 2002 law that requires all business records, including electronic records and electronic messages, to be retained for a certain period of time.

SOFTWARE REGULATIONS AND LEGAL STANDARDS

Until 1983, computer programs could not copyrighted in the United States. A software developer could copyright source code, but not the binary program produced when this code is compiled. This is because the compiled program was viewed as a "utilitarian good" generated from the code rather than a creative work. In order to assert a copyright, the developer had to make the source code available with the program. While publishing the source code gave a developer greater control, it also made it easier for others to copy and modify the program.

Copyright rules began to change with the US Court of Appeals' decision in *Apple Computer, Inc. v. Franklin Computer Corp.* (1983). It was the first appellate court ruling to state that machine-readable code is subject to copyright. Prior to this, developers had no reason to withhold source code. Computers were not standardized enough to make large-scale development profitable, and software often had to be modified to run on different computers. Introducing software copyright allowed for greater profit potential and provided new incentives for standardization. It also gave developers a reason to keep source code private: now they had a copyright to protect.

Since then, several laws have been passed to regulate both software development and its end use. Because computer technology is constantly evolving, new regulations are often needed. One such law is the Health Insurance Portability and Accountability Act (HIPAA). HIPAA includes provisions designed to protect patients' health information, particularly when using software developed for health-care providers. Any software or application (app) that collects or stores personally identifiable health information or shares it with certain covered medical entities,

Act	Maintaining privacy, confidentiality, and integrity of . . .
BASEL II	Personal financial information and transactions transmitted and stored by financial institutions
GLBA	Personal financial information stored by financial institutions
HIPAA	Health care information
PCI	Credit card information stored by merchants
Sarbanes-Oxley	Financial data in publicly traded corporations
SB 1386	Customers' personal information stored by any organization that does business in the state of California

Without a single overarching regulatory organization for online personal information privacy and confidentiality, a number of acts have been adopted to protect consumers. Each act focuses on a different aspect of online privacy protection. EBSCO illustration.

such as doctors and hospitals, must be in compliance with HIPAA.

Another law, the Sarbanes-Oxley Act (SOX), covers information retention. It states that all organizations, regardless of size, must retain certain business records for at least five years. E-mails and electronic records are included in this category.

TYPES OF SOFTWARE

There are several types of software, distinguished by which license governs their use. A software license is a legal instrument that states how copyrighted software can be used. Open-source software makes its source code available, with no restrictions on how it may be used. Its license gives users the right to modify the program, make copies, and distribute it to others. Open-source software is usually, but not always, free of charge.

Proprietary software is software on which the copyright holder has placed certain restrictions. It typically comes with a license agreement. This is an implied contract between the copyright holder and the end user. The license agreement spells out what the user can and cannot to do with the software. It may also include a disclaimer of responsibility should the

software damage the user's computer in some way. As a legal contract, license agreements can, in theory, be enforced in court. In practice, enforceability may depend on the terms of the agreement, how and when the user consented to it, and even which court has jurisdiction.

Other types of software include freeware, shareware, and internal-use software. Freeware can be freely used, copied, and distributed but does not permit modification of source code. Shareware is a type of proprietary software that is initially provided for no cost and can be freely copied and distributed, but continued use under certain conditions requires the purchase of a license. Internal-use software, or private software, is developed for a company's own internal use but not made publicly available.

MOBILE AND SMARTPHONE SOFTWARE

In 2008, Google released the first version of Android, a smartphone operating system (OS) based on Linux. Android is open-source software, with source code available through the Android Open Source Project. As a result, a large community has formed in which developers modify and distribute their own versions of Android. These modified versions often provide updates and bug fixes ahead of official releases. Others are designed to support older devices or to run on devices designed for other OSs.

The Food and Drug Administration (FDA) has said that it does not intend to regulate mobile medical apps and consumer devices to the same extent as other medical software. Official guidelines state that unless an app or device makes disease-specific claims, it will receive no or low-level oversight, depending on how much risk it poses to patients. Any app that shares health information with covered medical entities must be HIPAA compliant.

THE VALUE OF SOFTWARE REGULATION

Software regulations and standards provide numerous benefits, including limiting flaws in software and lessening users' exposure to viruses. They are also geared toward protecting users' privacy. Regulation is about ensuring the confidentiality, accessibility, availability, and integrity of information. It is a form of accountability that will allow both proprietary and open-source software to improve as technology moves forward.

—*Daniel Horowitz*

BIBLIOGRAPHY

Aziz, Scott. "With Regulation Looming, It's Time for Industry to Raise the Bar for Software Quality." *Wired.* Condé Nast, 28 Aug. 2014. Web. 31 Mar. 2016.

Balovich, David. "Sarbanes-Oxley Document Retention and Best Practices." *Creditworthy News.* 3JM Company, 5 Sept. 2007. Web. 1 Apr. 2016.

"Categories of Free and Nonfree Software." *GNU Operating System.* Free Software Foundation, 1 Jan. 2016. Web. 31 Mar. 2016.

Gaffney, Alexander. "FDA Confirms It Won't Regulate Apps or Devices Which Store Patient Data." *Regulatory Affairs Professionals Society.* Regulatory Affairs Professional Soc., 6 Feb. 2015. Web. 31 Mar. 2016.

Rouse, Margaret. "Sarbanes-Oxley Act (SOX)." *TechTarget.* TechTarget, June 2014. Web. 31 Mar. 2014.

Wang, Jason. "HIPAA Compliance: What Every Developer Should Know." *Information Week.* UBM, 11 July 2014. Web. 26 Feb. 2016.

SOFTWARE TESTING

FIELDS OF STUDY

Software Engineering; Applications; Computer Engineering

ABSTRACT

Software testing ensures that software and applications perform well and meet users' needs. It is intended to expose problems such as defects (often referred to as bugs) or inefficient design. Software testing allows project managers to make informed decisions about moving forward or making changes on a product before releasing it.

PRINICIPAL TERMS

- **agile software development:** an approach to software development that addresses changing requirements as they arise throughout the process, with programming, implementation, and testing occurring simultaneously.
- **black-box testing:** a testing technique in which function is analyzed based on output only, without knowledge of structure or coding.
- **dynamic testing:** a testing technique in which software input is tested and output is analyzed by executing the code.
- **phased software development:** an approach to software development in which most testing occurs after the system requirements have been implemented.
- **static testing:** a testing technique in which software is tested by analyzing the program and associated documentation, without executing its code.
- **white-box testing:** a testing technique that includes complete knowledge of an application's coding and structure.

THE IMPORTANCE OF SOFTWARE TESTING

Software testing contributes to the development of programs and applications (apps) by detecting gaps between the software's stated requirements and its actual performance. This testing informs software engineers about any poor functionality, missing requirements, errors, and defects. Originally, programmers tested software at the end of the development process. However, the popularity of the graphical user interface (GUI) and other more user-friendly software highlighted the need for more thorough testing. Consumers expect programs to work properly from the start. Therefore, programmers started taking a more careful and systematic approach. Soon, testing throughout the development process became the industry standard.

Manufacturers value software testing because it prevents flawed products from entering the market. Delivering software or launching an application only to have customers discover bugs can have both short- and long-term negative effects. Poorly tested software can lead to declining sales, damage to the company's brand, and possibly even harm to a customer's business operations. In addition, testing can save time and money. It is much easier to correct errors during the development process than after a release. Without testing, manufacturers risk lost opportunities, as users who have a bad experience often switch to a competing product.

SOFTWARE TESTING METHODS

"Software testing" is broad term describing a variety of specific methods. Programmers test different aspects of applications, including design, response to inputs, usability, stability, and results. The type of test they choose is based on factors such as the programming language and what they need to learn.

Agile software development is a development process that emphasizes teamwork, customer involvement, and user testing of portions of the system. Agile practitioners follow the views of the "Agile Manifesto" (2001), written by a group of developers who wanted to change the software development process. According to the agile methodology, testing is not a phase. Instead, tests are performed continuously as the product is developed and requirements change. In contrast, phased software development follows the original requirements, includes the features that matter most, and tests when development is completed. New requirements can be established after the original software is in use.

White-box testing is also referred to as glass-box or clear-box testing. It produces test data by examining the program's code. White-box testing is useful for

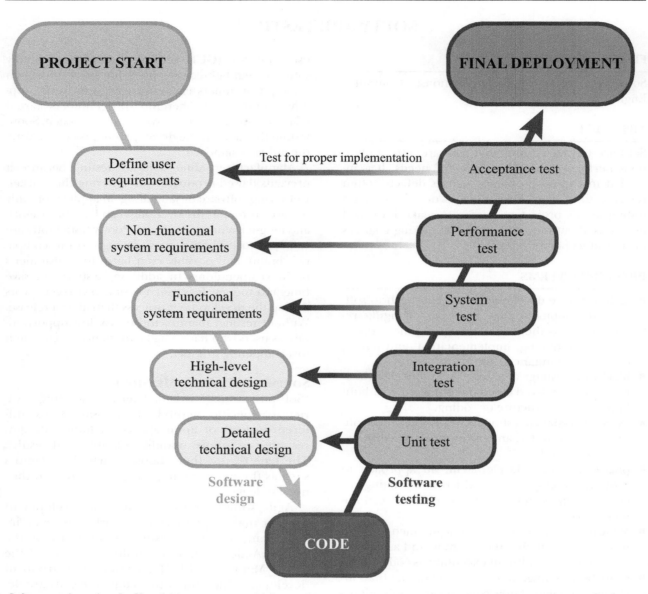

Software testing using the V-model incorporates multiple tests for each stage of software development to ensure all levels of software design, from user requirements to low-level technical design requirements, are implemented properly. Image adapted from Infinite Computing Systems, Inc EBSCO illustration.

finding errors in hidden code. However, it can be expensive and requires in-depth knowledge of the program or app's programming language. The opposite of white-box testing is black-box testing, which tests functionality using only the specifications, not the code. Testers know what the app is supposed to do but not how it operates.

There are dozens of additional approaches to software testing. In static testing, code reviews and analyses are used to discover inconsistencies and variables in the code. In dynamic testing, the code is actually executed and its behavior is analyzed. Static testing is for verification, while dynamic testing is for validation.

AUTOMATING THE PROCESS

Because the testing process is critical and detail oriented as well as repetitive, tools have been developed to assist in the process. These include:

- test management tools
- static analysis tools
- test data preparation tools
- test execution tools
- performance testing tools
- coverage measurement tools
- incident management tools

Software testing tools can manage and schedule the testing process. These tools also log defects, track changes, monitor performance, and conduct static testing, analysis, and design. Specialized software performs repetitive tasks and tests that are difficult or time consuming to perform manually.

WHEN TESTING IS INSUFFICIENT

Releasing an app that does not function as expected can cause serious repercussions, even for a respected company. When Apple first introduced the Apple Maps app, users soon discovered that certain cities were mislabeled or had disappeared, familiar landmarks had moved, and satellite images were obscured by clouds. As a result, the company faced ridicule, and Apple's CEO was compelled to issue an apology.

To ensure proper testing and avoid such problems, programmers follow a series of steps, starting with reviewing software specifications and developing a test plan. They also 4 Science Reference Center™ Software Testing write test cases. Test cases are sets of actions that verify program functions. Writing effective test cases can save time and money throughout the testing process. As testing proceeds, bugs are discovered, such as when an expected result is not the actual result. These bugs are logged and assigned to a developer for fixing.

Programmers approach testing from many directions. Some attempt to make an app perform functions it should not do. Others imagine various scenarios, such as what would happen if someone tried to use the software maliciously or with no knowledge of the application. By putting themselves in the place of various users, testers can uncover defects and improve the user experience.

MOVING SOFTWARE TESTING TO THE CLOUD

Information technology (IT) firms spend hundreds of billions every year on software testing because consumers and businesses expect the software and apps they purchase to work as they expect. These firms invest in servers, databases, storage, operating systems, and testing tools, all of which need to be upgraded and maintained. The cost is generally accepted because the risk of delivering a flawed product greatly outweighs the cost of thorough testing.

As IT companies face greater competition and pressures to earn profits, testing will require new solutions. Cloud-based systems provide speed and cost savings because service providers can maintain systems, tools, storage, and databases in a more cost-effective manner.

—*Teresa E. Schmidt*

BIBLIOGRAPHY

Holland, Bill. "Software Testing: A History." *SitePoint*. SitePoint, 15 Feb. 2012. Web. 9 Feb. 2016.

Jorgensen, Paul C. *Software Testing: A Craftsman's Approach*. 4th ed. Boca Raton: CRC, 2014. Print.

Mili, Ali, and Fairouz Tchier. *Software Testing: Concepts and Operations*. Hoboken: Wiley, 2015. Print.

Mitchell, Jamie L., and Rex Black. *Advanced Software Testing*. 2nd ed. 3 vols. Santa Barbara: Rocky Nook, 2015. Print.

Murphy, Chris. "How to Save $150 Billion: Move All App Dev and Testing to the Cloud." *Forbes*. Forbes.com, 3 Feb. 2016. Web. 10 Feb. 2016.

Singh, Amandeep. "Top 13 Tips for Writing Effective Test Cases for Any Application." *Quick Software Testing*. QuickSoftwareTesting, 23 Jan. 2014. Web. 11 Feb. 2016.

Smith, Catharine. "Tim Cook Issues Apology for Apple Maps." *Huffpost Tech*. TheHuffingtonPost.com, 28 Sept. 2012. Web. 11 Feb. 2016.

"Why Do You Need to Test Software?" *BBC Academy*. BBC, 23 Feb. 2015. Web. 7 Feb. 2016.

SOFTWARE-DEFINED RADIO

FIELDS OF STUDY

Digital Media; Software Engineering

ABSTRACT

By modifying the hardware of a standard radio set with layers of operational software, a software-defined radio (SDR) system can greatly expand its versatility and power efficiently and cost-effectively. Digital and SDR technology provide radio reception a stronger, more durable signal. SDR may also allow for the centralization of common applications that rely on radio communication.

PRINICIPAL TERMS

- **amplifier:** a device that strengthens the power, voltage, or current of a signal.
- **converter:** a device that expands a system's range of reception by bringing in and adapting signals that the system was not originally designed to process.
- **mixer:** a component that converts random input radio frequency (RF) signal into a known, fixed frequency to make processing the signal easier.
- **modulator:** a device used to regulate or adjust some aspect of an electromagnetic wave.
- **software-defined antennas:** reconfigurable antennas that transmit or receive radio signals according to software instructions.
- **sound card:** a peripheral circuit board that enables a computer to accept audio input, convert signals between analog and digital formats, and produce sound.

HOW RADIO WORKS

Radio remains the most useful and most widely used communication technology for sharing both entertainment and information. The basic technology of the conventional radio is simple: radio waves are enriched with information—music, talk, weather, news—and sent wirelessly from a transmitter to a receiver. The receiver might be down the street, or halfway around the earth. The source emits a steady radio signal whose frequency or amplitude is modulated to add the message information. Then, signal is beamed from large antennas to smaller ones that grab the signal and relay it to a receiver. The receiver uses a mixer to convert the carrier radio frequency (RF) to an intermediate one for easier processing. A filter removes distortion, and the signal is then released into the system's amplifiers, thus recreating the original radio programming. The varieties of radio wave technology—the familiar AM/FM radio bands, walkie-talkies, ham radio systems, emergency notification systems, transportation communication systems—attest to its vital role in moving information efficiently and effectively.

Although such radio systems revolutionized communication, this analog system did present major problems. The signal (that is, the energy wave itself) was tied to the information being sent. If the wave is disrupted or compromised, the information itself is compromised. This has led to frustrations with reception, the clarity of a signal. Signals are vulnerable to distortions within the atmosphere, most notably from rain, fog, or solar activity. They can also be disrupted or even lost due to surfaces coming between transmitter and receiver. Nearly any solid object, from trees to buildings, can destabilize a radio signal. And the strength of the original broadcast signal and the range of receiver also impact the reach (and quality) of the signal. Because the information is relayed by waves, the analog radio set relies on an array of hardware components, each with a radio chip designed to perform one crucial function. The physical layout of each component—antenna, mixer, amplifier, modulator, and sometimes, converters—can be considerable. Analog radio, although useful and remarkably versatile, is thus cumbersome, costly, and not entirely reliable.

Software-defined radio (SDR), like analog radio, transmits acoustic information from a source to a destination and uses similar equipment. However, SDR addresses several longstanding issues, from sound quality to cost.

DIGITAL AND SOFTWARE-DEFINED RADIO

In digital radio, information is encoded as binary numbers. These are then sent as on-off pulses along radio waves. The information is thus generated (or defined) by a computer and in turn, the transmission is decoded by another computing device. More

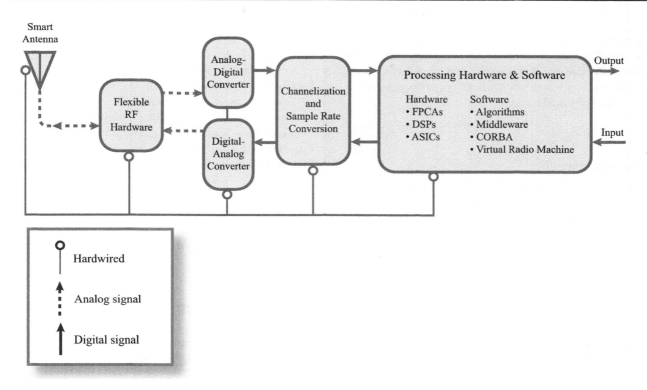

Software defined radio replaces much of the hardware needed for standard analog radio processing with computer hardware and software. The software requires conversion of the signal from analog to digital before processing. Adapted from McSush derivation of Tarkvaralise raadio plokkskeem.

importantly, the transmission is sent again and again in a kind of barrage of information. This guarantees a stronger, more consistent signal reception because the receiver can assemble the reproduced signal from the many information fragments recovered from the signal field. That process takes fractions of seconds longer but greatly improves the sound quality and signal reliability.

The SDR process begins with the microphone where the initial audio data are created. As in analog radio, an amplifier is fed the signal and strengthens it for transmission. That signal is sent through an analog-to-digital converter that recasts the information as ASCII data. ASCII is an encoding system that converts English language content into numbers. This makes the transfer of the signal to another computer possible. A modulator then presses that ASCII data onto the designated RF carrier. Amplifiers strengthen the signal to the degree necessary to broadcast it from a software-defined antenna. A software-defined

receiving antenna retrieves the signal, which is converted to a stable frequency. Then, a demodulator separates the ASCII data from the RF carrier. A digital-to-analog converter in a sound card reproduces the information. An amplifier then projects it into a speaker system, headphone, or earpiece. Software is used in as much of the signal processing as possible in SDR systems.

IMPLICATIONS

SDR is perhaps best appreciated for its implications. A single outfit might serve as a radio, certainly. But just by adding various software applications, it might also provide cell phone reception, fax and ham radio capabilities, weather and traffic information, global positioning, and web access. In addition, SDR could greatly enhance business operations via videoconferencing. All of those functions use different frequencies in different bands of the electromagnetic spectrum. SDR expands the radio's ability to receive and process those various frequencies. SDR is already used for military operations and cellular networks. Its versatility and flexibility may someday make SDR a cost-effective way for anyone to access virtually all communication systems without greatly altering a radio set's physical parts.

—*Joseph Dewey*

BIBLIOGRAPHY

Dillinger, Markus, Kambiz Madani, and Nancy Alonistioti. *Software Defined Radio: Architectures, Systems, and Functions.* Hoboken: Wiley, 2003. Print.

Ewing, Martin. *The ABCs of Software Defined Radio.* Hartford: Amer. Radio Relay League, 2012. Print.

Grayver, Eugene. *Implementing Software Defined Radio.* New York: Springer, 2012. Print.

Johnson, C. Richard, and William A. Sethares. *Telecommunications Breakdown: Concepts of Communication Transmitted via Software-Defined Radio.* New York: Prentice, 2003. Print.

Pu, Di, and Alexander M. Wyglinski. *Digital Communication Systems Engineering with Software-Defined Radio.* London: Artech, 2013. Print.

Reed, Jeffrey H. *Software Radio: A Modern Approach to Radio Engineering.* New York: Prentice, 2002. Print.

SPEECH-RECOGNITION SOFTWARE

FIELDS OF STUDY

Software Engineering; Applications; Algorithms

ABSTRACT

Speech-recognition software records, analyzes, and responds to human speech. The earliest such systems were used for speech-to-text programs. Speech recognition became commonplace in the 2010s through automated assistants. Speech recognition depends on complex algorithms that analyze speech patterns and predict the most likely word from various possibilities.

PRINICIPAL TERMS

- **deep learning:** an emerging field of artificial intelligence research that uses neural network algorithms to improve machine learning.
- **hidden Markov model:** a type of model used to represent a dynamic system in which each state can only be partially seen.
- **neural network:** in computing, a model of information processing based on the structure and function of biological neural networks such as the human brain.
- **phoneme:** a sound in a specified language or dialect that is used to compose words or to distinguish between words.
- **speaker independent:** describes speech software that can be used by any speaker of a given language.

THE BASICS OF SPEECH RECOGNITION

Speech-recognition software consists of computer programs that can recognize and respond to human speech. Applications include speech-to-text software that translates speech into digital text for text messaging and document dictation. This software is also used by automated personal assistants such as Apple's Siri and Microsoft's Cortana, which can respond to spoken commands. Speech-recognition software development draws on the fields of linguistics, machine learning, and software engineering. Researchers first began investigating the possibility of speech-recognition software in the 1950s. However, the first such programs only became available to the public in the 1990s.

Speech-recognition software works by recognizing the phonemes that make up words. Algorithms are used to identify the most likely word implied by the sequence of phonemes detected. The English language has forty-four phonemes, which can be combined to create tens of thousands of different words. A particularly difficult aspect of speech recognition is distinguishing between homonyms (or homophones). These are words that consist of the same phonemes

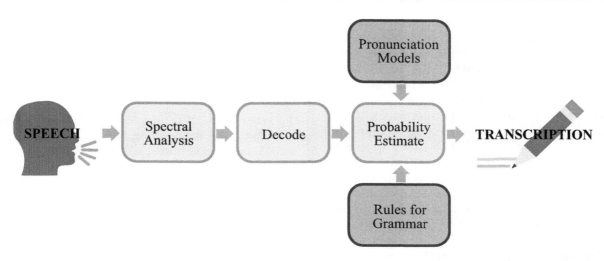

Speech recognition software must perform a number of processes to convert the spoken word to the written word. A spoken sentence must first go through spectral analysis to identify the unique soundwave, then the soundwaves are decoded into potential words, and finally the words are run through probability algorithms that estimate their likelihood based on the rules of grammar and pronunciation models. The end result is the most likely transcription of what was spoken. EBSCO illustration.

but are typically spelled differently. Examples include "addition" versus "edition" and "scent" versus "cent." Distinguishing between homonyms requires an understanding of context. Speech-recognition software must be able to evaluate surrounding words to discern the most likely homonym intended by the speaker.

Some speech-recognition software uses training, in which the speaker first reads text or a list of vocabulary words to help the program learn particularities of their voice. Training increases accuracy and decreases the error rate. Software that requires training is described as speaker dependent. Speaker-independent software does not require training, but it may be less accurate. Speaker-adaptive systems can alter some operations in response to new users.

SPEECH-RECOGNITION ALGORITHMS

Research into speech-recognition software began in the 1950s. The first functional speech-recognition programs were developed in the 1960s and 1970s. The first innovation in speech-recognition technology was the development of dynamic time warping (DTW). DTW is an algorithm that can analyze and compare two auditory sequences that occur at different rates.

Speech recognition advanced rapidly with the invention of the hidden Markov model (HMM). The

HMM is an algorithm that evaluates a series of potential outcomes to a problem and estimates the probability of each one. It is used to determine the "most likely explanation" of a sequence of phonemes, and thus the most likely word, given options taken from a speaker's phonemes. Together, HMMs and DTW are used to predict the most likely word or words intended by an utterance.

Speech recognition is based on predictive analysis. An important part of developing a predictive algorithm is feature engineering. This is the process of teaching a computer to recognize features, or relevant characteristics needed to solve a problem. Raw speech features are shown as waveforms. Speech waveforms are the 2-D representations of sonic signals produced when various phonemes are said.

An emerging feature in speech recognition is the use of neutral networks. These computing systems are designed to mimic the way that brains handle computations. Though only beginning to affect speech recognition, neural networks are being combined with deep learning algorithms, which make use of raw features, to analyze data.

APPLICATIONS AND FUTURE DIRECTIONS

Deep neural network algorithms and other advancements have made speech recognition software more accurate and efficient. The most familiar applications for speech-recognition technology include the voice-to-text and voice-to-type features on many computers and smartphones. Such features automatically translate the user's voice into text for sending text messages or composing documents or e-mails. Most speech-recognition programs rely on cloud computing, which is the collective data storage and processing capability of remote computer networks. The user's speech is uploaded to the cloud, where computers equipped with complex algorithms analyze the speech before returning the data to the user. Automated assistant programs such as Siri and Cortana can use an array of data collected from a user's device to aid comprehension. For instance, if a user tells a speech-recognition program "bank," the program uses the Internet and global positioning systems to return data on nearby banks or banks that the user has visited in the past.

Experts predict that speech-recognition apps and devices will likely become ubiquitous. Fast, accented, or impeded speech and slang words pose much less of a challenge than they once did. Speech-recognition software has become a basic feature in many new versions of the Mac and Windows operating systems. These programs also help make digital technology more accessible for people with disabilities. In future, as voice recognition improves and becomes commonplace, a wider range of users will be able to use advanced computing features.

—*Micah L. Issitt*

BIBLIOGRAPHY

Gallagher, Sean. "Cortana for All: Microsoft's Plan to Put Voice Recognition behind Anything." *Ars Technica*. Condé Nast, 15 May 2015. Web. 21 Mar. 2016.

Information Resources Management Association, ed. *Assistive Technologies: Concepts, Methodologies, Tools, and Applications*. Vol. 1. Hershey: Information Science Reference, 2014. Print.

"How Speech-Recognition Software Got So Good." *Economist*. Economist Newspaper, 22 Apr 2014. Web. 21 Mar. 2016.

Kay, Roger. "Behind Apple's Siri Lies Nuance's Speech Recognition." *Forbes*. Forbes. com, 24 Mar. 2014. Web. 21 Mar. 2016.

Manjoo, Farhad. "Now You're Talking!" *Slate*. Slate Group, 6 Apr. 2011. Web. 21 Mar. 2016.

McMillan, Robert. "Siri Will Soon Understand You a Whole Lot Better." *Wired*. Condé Nast, 30 June 2014. Web. 21 Mar. 2016.

Pinola, Melanie. "Speech Recognition through the Decades: How We Ended Up with Siri." *PCWorld*. IDG Consumer & SMB, 2 Nov. 2011. Web. 21 Mar. 2016.

T

TURING MACHINE

FIELDS OF STUDY

Algorithms; computer science

ABSTRACT

A Turing machine is a mathematical tool that is the equivalent of a digital computer. It is the most widely used model of computation in computability and complexity theory in the computer world today. In its simplest form, it consists of an input/output relationship that the machine computes—the input is in binary format on a tape that the machine uses to compute the function, and the output is the contents of the tape when the machine stops. Studying this machine led to the study of classes of language and ultimately led to the development of computer programming languages.

PRINICIPAL TERMS

- **algorithm:** a set of step-by-step instructions for performing computations.
- **character:** a unit of information that represents a single letter, number, punctuation mark, blank space, or other symbol used in written language.
- **coding theory:** the study of codes and their use in certain situations for various applications.
- **function:** instructions read by a computer's processor to execute specific events or operations.
- **process:** the execution of instructions in a computer program.
- **programming languages:** sets of terms and rules of syntax used by computer programmers to create instructions for computers to follow. This code is then compiled into binary instructions for a computer to execute.
- **Turing complete:** a programming language that can perform all possible computations.

BIRTH OF THE TURING MACHINE

In the 1930s, before digital computers were even thought of, several mathematicians began to think about what it means to compute a function. Alan Turing was one of these mathematicians, and his Turing machine became part of the definition of a computable function: "A function is computable if it can be computed by a Turing machine." He is called the father of modern computing because the Turing machine is the precursor to all modern computers.

He developed these ideas at King's College, Cambridge, between the years 1932 to 1935, when many free-thinking mathematicians congregated there. His mentors include Christopher Morcom, Philip Hall, M.H.A. Newman, and Bertrand Russell. In 1936-37, he published a groundbreaking paper, *On computable numbers, with an application to the Entscheidungsproblem*, describing the theory of Turing machines and defining computability. One sentence reads "It is possible to invent a single machine which can be used to compute any computable sequence." He called this machine a "universal computing machine," and defined it as a Turing machine that could read the description of any other Turing machine and to carry out what that Turing machine would have done. One might think that complex tasks need complex machines, but that is not true; a simple machine can perform extremely complex functions when given enough time. Computers are now programmed to perform extremely complicated operations very quickly, but none of them can outdo a basic Turing machine in regards computing a function. Now that we all have computers with us constantly, we can see what Turing only imagined: He came up with the idea of a Universal Turing Machine 10 years before it could even be implemented!

A model of a Turing Machine as seen at the Go Ask Alice exhibit at the Harvard Collection of Historical Scientific Instruments. By GabrielF (Own work) [CC BY-SA 3.0 (http://creativecommons.org/licenses/by-sa/3.0)], via Wikimedia Commons

WHAT EXACTLY IS A TURING MACHINE?

This simple machine is made up of a tape (a ribbon of paper) containing an algorithm and a programmable read-write head (an instrument that can read the symbols on the tape, write a new symbol and move, for example, left/right or up/down). These terms may seem a bit archaic, but they give a sense of what kinds of machines existed when Turing invented this machine. The input on the tape must consist of a finite number of symbols; however, the tape, ideally, is infinitely long. Turing imagined this type of input to show that there are tasks that these machines cannot perform, even with unlimited time and working memory.

The Turing machine starts out in a specific state, then a program is written, on the tape, consisting of a list of transitions, telling the head what the next state should be and where it should move. The transition tells the head to do three things: (1) print something on the tape, (2) move to the right or left by one cell, and (3) change to a new state. The machine can also stop if there is no unique transition, for example, there is nothing on the tape.

Even Turing himself did not think that this machine would be the model one would use to actually build a practical computing machine. He called his idea a machine of the mind, a thought experiment, something that was created to explore problems in logic and discover the limits of the human mind as it applied to machines. In fact, he himself did not promote his ideas about Turing machines very much in the fields of computer science or mathematical logic. Others, such as Martin Davis and Marvin Minksy, saw the potential for this idea in mainstream logic and computer sciences. This idea gained further ground in the 1970s with complexity theory and in the 1980s with the development of quantum computing.

HOW DOES A TURING MACHINE WORK?

There are really only six operations that a basic Turing machine can perform:

- Read the symbol that is under the head
- Write a symbol on the tape under the head
- Move the tape one square left
- Move the tape one square right
- Change its state
- Stop

How a Turing machine acts is completely dependent on (1) the current state of the machine, (2) the symbol on the tape that is currently being read by the head, and (3) a table of transition rules, or the program, for the machine.

We could write this as $State_{current}$, Symbol, $State_{next}$, Action. If the machine is in a Current State and the tape contains the recognized Symbol, the machine will then move into the Next State and take Action.

The tape serves as the memory of the machine while the head is the mechanism by which data is accessed and updates the action of the machine.

Here is an example of a program for a Turing machine which starts with a blank, endless tape. This function tells the machine to perform the process of printing alternating characters, 0s and 1s, on the tape, leaving a blank space in between each numeral. The machine has four possible states, A, B, C, or D.

State	Print	Move	Next State
A	0	R	B
B		R	C
C	1	R	D
D		R	A

With this simple programming, the Turing machine will print an endless tape full of alternating 1s and 0s, leaving a blank space in between each numeral.

WHY IS A TURING MACHINE SO IMPORTANT?

The advent of the Turing machine got people started thinking about coding theory for computers, what kinds of problems computers are able to solve, and what kinds of programming languages we can use to communicate with them, such as Turing complete. Any computer is, at its base, a Turing machine. It is a model of computation that captures the idea of what computability is very simply, without needing to think about all the parts your computer currently contains. According to the Church-Turing thesis, no computing device is more powerful than a Turing machine. Some scientists even find examples of Turing machines in nature; for example, our cells have ribosomes that translate RNA into proteins, using much the same process as a Turing machine.

Alan Turing was a large part of the Enigma project during World War II. This project was the subject of several movies such as *The Imitation Game* and *Enigma*.

—*Marianne Moss Madsen, MS*

BIBLIOGRAPHY

Bernhardt, Chris. *Turing's Vision: The Birth of Computer Science*. MIT Press, 2016.

Boyle, David. *Alan Turing: Unlocking the Enigma*. CreateSpace Independent Publishing Platform, 2014.

Dyson, George. *Turing's Cathedral: The Origins of the Digital Universe*. Vintage, 2012.

Hodges, Andrew. *Alan Turing: The Enigma: The Book That Inspired the Film "The Imitation Game."* Princeton University Press, 2014.

McKay, Sinclair. *The Secret Lives of Codebreakers: The Men and Women Who Cracked the Enigma Code at Bletchley Park*. Plume, 2012.

Nicolelis, Miguel A. and Ronald M. Cicurel. *The Relativistic Brain: How it works and why it cannot be simulated by a Turing machine*. CreateSpace Independent Publishing Platform, 2015.

Petzold, Charles. *The Annotated Turing: A Guided Tour Through Alan Turing's Historic Paper on Computability and the Turing Machine*. Wiley, 2008.

Soare, Robert I. *Turing Computability: Theory and Applications (Theory and Applications of Computability)*. Springer, 2016.

TURING TEST

FIELDS OF STUDY

Computer Science; Robotics

ABSTRACT

The Turing test is a game proposed by computer scientist and mathematician Alan Turing in 1950 in order to determine whether machines can think. Despite criticism, the test has shaped the development and study of artificial intelligence ever since.

PRINICIPAL TERMS

- **artificial intelligence:** the intelligence exhibited by machines or computers, in contrast to human, organic, or animal intelligence.
- **automaton:** a machine that mimics a human but is generally considered to be unthinking.
- **chatterbot:** a computer program that mimics human conversation responses in order to interact with people through text; also called "talkbot," "chatbot," or simply "bot."
- **imitation game:** Alan Turing's name for his proposed test, in which a machine would attempt to respond to questions in such a way as to fool a human judge into thinking it was human.

CAN MACHINES THINK?

In 1950, British mathematician Alan Turing (1912–54) wrote a paper titled "Computing Machinery and Intelligence" in which he asked, "Can machines think?" The question was too difficult to answer directly, so instead he thought up a simple game to compare a machine's ability to respond in conversation to that of a human. If the machine could fool a human participant into believing it was human, then it could be considered functionally indistinguishable from a thinking entity (i.e., a person). This game later came to be known as the Turing test.

In Turing's time, digital computers and automata already existed. However, the notion of developing machines with programming sophisticated enough to engender something like consciousness was still very new. What Turing envisioned would, within the decade, become the field of artificial intelligence (AI): the quest for humanlike, or at least human-equivalent, intelligence in a machine.

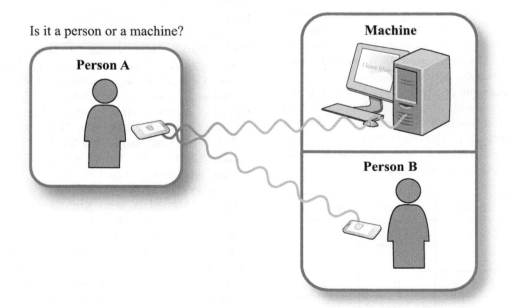

The Turing test of artificial intelligence should allow person A to determine whether they are corresponding via technology with a human or a machine entity. EBSCO illustration.

THE IMITATION GAME

To settle the question of whether or not machines can think, Turing proposed what he called an imitation game. He based it on a party game of the time, played by a man, a woman, and a third party (the judge). In the game, the judge stays in a separate room from the contestants and asks them questions, which they answer via writing or an impartial intermediary. Based on their answers, the judge must determine which player is which. The man tries to fool the judge, while the woman attempts to help him or her.

In Turing's game, instead of a man and a woman, the players would be a human and a machine. Computer terminals would be used to facilitate anonymous text-based communication. If the machine could fool a human judge a sufficient percentage of the time, Turing argued, it would be functionally equivalent to a human in terms of conversational ability. Therefore, it must be considered intelligent.

Later versions of the game eliminated the need for a human to compete against the machine. Instead, the judge is presented with a text-based interface and told to ask it questions to determine if he or she is talking to a person or a machine. This version has been the basis of the Loebner Prize chatterbot competition since 1995.

NATURAL LANGUAGE PROCESSING

Turing's challenge spurred the development of a subfield of AI known as "natural language processing" (NLP), devoted to creating programs that can understand and produce humanlike speech. The 1966 program ELIZA, written by German American computer scientist Joseph Weizenbaum (1923–2008), was very convincing in its ability to imitate a Rogerian psychotherapist ("How does that make you feel?"). Programs like ELIZA are called "chatterbots," or simply "bots." These programs became a feature of early text-based Internet communication. With the advent of smartphones, both Apple and Google have developed advanced NLP to underpin the voice-activated functions of their devices, most famously Apple's AI assistant Siri.

CRITICISM OF THE TURING TEST

The Turing test has been a point of contention among AI researchers and computer scientists for decades. Objections fall into two main camps. Some object to the idea that the Turing test is a reasonable test of the presence of intelligence, while others object to the idea that machines are capable of intelligence at all. With regard to the former, some critics have argued that a machine without intelligence could pass the test if it had sufficient processing power and memory to automatically respond to questions with prewritten answers. Others feel that the test is too difficult and too narrow a goal to help guide AI research.

LEGACY OF THE TURING TEST

Despite objections, the Turing test has endured as a concept in AI research and robotics. Many similar, though not identical, tests are also informally called Turing tests. In 2014, sixty years after Turing's death, a program named Eugene Goostman made headlines for passing a version of the Turing test.

However, the Turing test's lasting contribution to AI is not its utility as a test, but the context in which Turing proposed it. With his proposal, Turing became the first major advocate for the idea that humanlike machine intelligence was indeed possible. Though he died before AI emerged as a proper field of study, his ideas shaped its development for decades thereafter.

—*Kenrick Vezina, MS*

BIBLIOGRAPHY

Ball, Phillip. "The Truth about the Turing Test." *Future*. BBC, 24 July 2015. Web. 18 Dec. 2015.

Myers, Courtney Boyd, ed. *The AI Report. Forbes*. Forbes.com, 22 June 2009. Web. 18 Dec. 2015.

Olley, Allan. "Can Machines Think Yet? A Brief History of the Turing Test." *Bubble Chamber*. U of Toronto's Science Policy Working Group, 23 June 2014. Web. 18 Dec. 2015.

Oppy, Graham, and David Dowe. "The Turing Test." *Stanford Encyclopedia of Philosophy (Spring 2011 Edition)*. Ed. Edward N. Zalta. Stanford U, 26 Jan. 2011. Web. 18 Dec. 2015.

Turing, Alan M. "Computing Machinery and Intelligence." *Mind* 59.236 (1950): 433–60. Web. 23 Dec. 2015.

"Turing Test." *Encyclopædia Britannica*. Encyclopædia Britannica, 23 Sept. 2013. Web. 18 Dec. 2015.

U

UNICODE

FIELDS OF STUDY

Computer Science; Software Engineering; Information Technology

ABSTRACT

Unicode is a character-encoding system used by computer systems worldwide. It contains numeric codes for more than 120,000 characters from 129 languages. Unicode is designed for backward compatibility with older character-encoding standards, such as the American Standard Code for Information Interchange (ASCII). It is supported by most major web browsers, operating systems, and other software.

PRINICIPAL TERMS

- **glyph:** a specific representation of a grapheme, such as the letter *A* rendered in a particular typeface.
- **grapheme:** the *smallest* unit used by a writing system, such as alphabetic letters, punctuation marks, or Chinese characters.
- **hexadecimal:** a base-16 number system that uses the digits 0 through 9 and the letters *A, B, C, D, E,* and *F* as symbols to represent numbers.
- **normalization:** a process that ensures that different code points representing equivalent characters will be recognized as equal when processing text.
- **rendering:** the process of selecting and displaying glyphs.
- **script:** a group of written signs, such as Latin or Chinese characters, used to represent textual information in a writing system.
- **special character:** a character such as a symbol, emoji, or control character.

CHARACTER-ENCODING SYSTEMS

In order for computer systems to process text, the characters and other graphic symbols used in written languages must be converted to numbers that the computer can read. The process of converting these characters and symbols to numbers is called "character encoding." As the use of computer systems increased during the 1940s and 1950s, many different character encodings were developed.

To improve the ability of computer systems to interoperate, a standard encoding system was developed. Released in 1963 and revised in 1967, the American Standard Code for Information Interchange (ASCII) encoded ninety-five English language characters and thirty-three control characters into values ranging from 0 to 127. However, ASCII only provided support for the English language. Thus, there remained a need for a system that could encompass all of the world's languages.

Unicode was developed to provide a character encoding system that could encompass all of the scripts used by current and historic written languages. By 2016, Unicode provided character encoding for 129 scripts and more than 120,000 characters. These include special characters, such as control characters, symbols, and emoji.

UNDERSTANDING THE UNICODE STANDARD

The Unicode standard encodes graphemes and not glyphs. A grapheme is the smallest unit used by a writing system, such as an alphabetic letter or Chinese character. A glyph is specific representation of a grapheme, such as the letter *A* rendered in a particular typeface and font size. The Unicode standard provides a code point, or number, to represent each grapheme. However, Unicode leaves the rendering of the glyph that matches the grapheme to software programs. For example, the Unicode value

Graphic character symbol	Hexadecimal character value										
0020	0 0030	@ 0040	P 0050	` 0060	p 0070	00A0	° 00B0	À 00C0	Ð 00D0	à 00E0	ð 00F0
! 0021	1 0031	A 0041	Q 0051	a 0061	q 0071	¡ 00A1	± 00B1	Á 00C1	Ñ 00D1	á 00E1	ñ 00F1
" 0022	2 0032	B 0042	R 0052	b 0062	r 0072	¢ 00A2	² 00B2	Â 00C2	Ò 00D2	â 00E2	ò 00F2
# 0023	3 0033	C 0043	S 0053	c 0063	s 0073	£ 00A3	³ 00B3	Ã 00C3	Ó 00D3	ã 00E3	ó 00F3
$ 0024	4 0034	D 0044	T 0054	d 0064	t 0074	¤ 00A4	´ 00B4	Ä 00C4	Ô 00D4	ä 00E4	ô 00F4
% 0025	5 0035	E 0045	U 0055	e 0065	u 0075	¥ 00A5	µ 00B5	Å 00C5	Õ 00D5	å 00E5	õ 00F5
& 0026	6 0036	F 0046	V 0056	f 0066	v 0076	¦ 00A6	¶ 00B6	Æ 00C6	Ö 00D6	æ 00E6	ö 00F6
' 0027	7 0037	G 0047	W 0057	g 0067	w 0077	§ 00A7	· 00B7	Ç 00C7	× 00D7	ç 00E7	÷ 00F7
(0028	8 0038	H 0048	X 0058	h 0068	x 0078	¨ 00A8	¸ 00B8	È 00C8	Ø 00D8	è 00E8	ø 00F8
) 0029	9 0039	I 0049	Y 0059	i 0069	y 0079	© 00A9	¹ 00B9	É 00C9	Ù 00D9	é 00E9	ù 00F9
* 002A	: 003A	J 004A	Z 005A	j 006A	z 007A	ª 00AA	º 00BA	Ê 00CA	Ú 00DA	ê 00EA	ú 00FA
+ 002B	; 003B	K 004B	[005B	k 006B	{ 007B	« 00AB	» 00BB	Ë 00CB	Û 00DB	ë 00EB	û 00FB
, 002C	< 003C	L 004C	\ 005C	l 006C	| 007C	¬ 00AC	¼ 00BC	Ì 00CC	Ü 00DC	ì 00EC	ü 00FC
- 002D	= 003D	M 004D	] 005D	m 006D	} 007D	- 00AD	½ 00BD	Í 00CD	Ý 00DD	í 00ED	ý 00FD
. 002E	> 003E	N 004E	^ 005E	n 006E	~ 007E	® 00AE	¾ 00BE	Î 00CE	Þ 00DE	î 00EE	þ 00FE
/ 002F	? 003F	O 004F	_ 005F	o 006F	007F	¯ 00AF	¿ 00BF	Ï 00CF	ß 00DF	ï 00EF	ÿ 00FF

The Unicode Standard is a universally recognized coding system for more than 120,000 characters, using either 8-bit (UTF-8) or 16-bit (UTF-16) encoding. This chart shows the character symbol and the corresponding hexadecimal UTF-8 code. For the first 127 characters, UTF-8 and ASCII are identical. EBSCO illustration.

of U+0041 (which represents the grapheme for the letter *A*) might be provided to a web browser. The browser might then render the glyph of the letter *A* using the Times New Roman font.

Unicode defines 1,114,112 code points. Each code point is assigned a hexadecimal number ranging from 0 to 10FFFF. When written, these values are typically preceded by U+. For example, the letter *J* is assigned the hexadecimal number 004A and is written U+004A. The Unicode Consortium provides charts listing all defined graphemes and their associated code points. In order to allow organizations to define their own private characters without conflicting with assigned Unicode characters, ranges of code points are left undefined. One of these ranges includes all of the code points between U+E000 and U+F8FF. Organizations may assign undefined code points to their own private graphemes.

One inherent problem with Unicode is that certain graphemes have been assigned to multiple code points. In an ideal system, each grapheme would be assigned to a single code point to simplify text processing. However, in order to encourage the adoption of the Unicode standard, character encodings such as ASCII were supported in Unicode. This resulted in certain graphemes being assigned to more than one code point in the Unicode standard.

Unicode also provides support for normalization. Normalization ensures that different code points that represent equivalent characters will be recognized as equal when processing text. For example, normalization ensures that the character *é* (U+00E9) and the combination of characters *e* (U+0065) and (U+0301) are treated as equivalent when processing text.

SAMPLE PROBLEM

Using a hexadecimal character chart as a reference, translate the following characters into their Unicode code point values: <, 9, ?, E, and @.

Then select an undefined code point to store a private grapheme.

Answer:

Unicode uses the hexadecimal character code preceded by a U+ to indicate that the hexadecimal value refers to a Unicode character. Using the chart, <, 9, ?, E, and @ are associated with the following hexadecimal values: 003C, 0039, 003F, 0045, and 0040. Their Unicode code point values are therefore: U+003C, U+0039, U+003F, U+0045, and U+0040.

A private grapheme may be assigned any code point value within the ranges U+E000 to U+F8FF, U+F0000 to U+FFFFF, and U+100000 to U+10FFFD. These code points are left undefined by the Unicode standards.

USING UNICODE TO CONNECT SYSTEMS WORLDWIDE

Since its introduction in 1991, Unicode has been widely adopted. Unicode is supported by major operating systems and software companies including Microsoft and Apple. Unicode is also implemented on UNIX systems as well. Unicode has become an important encoding system for use on the Internet. It is widely supported by web browsers and other Internet-related technologies. While older systems such as ASCII are still used, Unicode's support for multiple languages makes it the most important character-encoding system in use. New languages, pictographs, and symbols are added regularly. Thus, Unicode remains poised for significant growth in the decades to come.

—*Maura Valentino, MSLIS*

BIBLIOGRAPHY

Berry, John D. *Language Culture Type: International Type Design in the Age of Unicode.* New York: Graphis, 2002. Print.

Gillam, Richard. *Unicode Demystified: A Practical Programmer's Guide to the Encoding Standard.* Boston: Addison-Wesley, 2002. Print.

Graham, Tony. *Unicode: A Primer.* Foster City: M&T, 2000. Print.

Korpela, Jukka K. *Unicode Explained.* Sebastopol: O'Reilly Media, 2006. Print.

"The Unicode Standard: A Technical Introduction." *Unicode.org.* Unicode, 25 June 2015. Web. 3 Mar. 2016.

"Unicode 8.0.0." *Unicode.org.* Unicode, 17 June 2015. Web. 3 Mar. 2016.

"What Is Unicode?" *Unicode.org.* Unicode, 1 Dec. 2015. Web. 3 Mar. 2016.

UNIX

FIELDS OF STUDY

Operating Systems; Computer Science; Information Technology

ABSTRACT

UNIX is a computer operating system originally developed by researchers at Bell Laboratories in 1969. The term is also used to refer to later operating systems based in part on its source code. Several of the original UNIX operating system's features, such as its hierarchical file system and multiuser support, became standard in later systems.

PRINICIPAL TERMS

- **command-line interpreter:** an interface that interprets and carries out commands entered by the user.
- **hierarchical file system:** a directory with a treelike structure in which files are organized.
- **kernel:** the central component of an operating system.
- **multitasking:** capable of carrying out multiple tasks at once.
- **multiuser:** capable of being used by multiple users at once.
- **operating system (OS):** a specialized program that manages a computer's functions.

ORIGIN OF UNIX

As computer technology rapidly developed in the mid-twentieth century, programmers sought to create means of interfacing with computers and making use of their functions in a more straightforward, intuitive way. Chief among the goals of many programmers was the creation of an operating system (OS). OSs are specialized programs that manage all of computers' processes and functions. Although many different OSs were created over the decades, UNIX proved to be one of the most influential. UNIX inspired numerous later OSs and continues to be used in various forms into the twenty-first century.

Development of the original UNIX OS began in 1969 at Bell Laboratories, a research facility then owned by AT&T. Researchers at Bell had been working on the Multiplexed Information and Computing Service (Multics) project in collaboration with the Massachusetts Institute of Technology and General Electric. The group had focused on creating systems that allowed multiple users to access a computer at once. After AT&T left the project, Bell programmers Ken Thompson and Dennis Ritchie began work on an OS. Thompson coded the bulk of the system, which consisted of 4,200 lines of code, in the summer of 1969, running it on an outdated PDP-7 computer.

The OS was initially named the Unmultiplexed Information and Computing Service, a play on the name of the Multics project. That name was later shortened to UNIX. Thompson, Ritchie, and their colleagues continued to work on UNIX over the next several years. In 1973 they rewrote the OS in the new programming language C, created by Ritchie. UNIX gained popularity outside of Bell Laboratories in the mid-1970s. It subsequently inspired the creation of many UNIX variants and UNIX-like OSs.

UNDERSTANDING UNIX

While different editions of UNIX vary, the majority of UNIX OSs have some common essential characteristics. UNIX's kernel is the core of the OS. The kernel is responsible for executing programs, allocating memory, and otherwise running the system. The user interacts with what is known as the "shell." The shell is an interface that transmits the user's commands to the kernel. The original UNIX shell was a command-line interpreter, a text-based interface into which the user types commands. Over time programmers developed a variety of different shells for UNIX OSs. Some of these shells were graphical user interfaces that enabled the user to operate the computer by interacting with icons and windows. Files saved to a computer running UNIX are stored in a hierarchical file system. This file system used a treelike structure that allowed folders to be saved within folders.

USING UNIX

In keeping with its origins as a project carried out by former Multics researchers, UNIX was designed to have multiuser capabilities. This enabled multiple people to use a single computer running UNIX at the same time. This was an especially important feature in the late 1960s and early 1970s. At this time,

The creators of UNIX were Dennis Ritchie (standing) and Ken Thompson (sitting). By Peter Hamer, CC BY-SA 2.0 (http://creativecommons.org/licenses/by-sa/2.0), via Wikimedia Commons.

computers were not personal computers but large, expensive mainframes that took up a significant amount of space and power. Multiuser capabilities made it possible for the organization using the UNIX OS to maximize the functions of their computers.

UNIX is also a multitasking system, meaning it can carry out multiple operations at once. One of the first programs designed for UNIX was a text-editing program needed by the employees of Bell Laboratories. Over time, programmers wrote numerous programs compatible with the OS and its later variants, including games, web browsers, and design software.

UNIX VARIANTS AND UNIX-LIKE OPERATING SYSTEMS

At the time that UNIX was first developed, AT&T was prohibited from selling products in fields other than telecommunications. The company was therefore unable to sell its researchers' creation. Instead, the company licensed UNIX's source code to various institutions. Programmers at those many institutions rewrote portions of the OS's code, creating UNIX variants that suited their needs. Perhaps the most influential new form of UNIX was the Berkeley Software Distribution variant, developed at the University of California, Berkeley, in the late 1970s.

In the early twenty-first century, the multitude of UNIX variants and UNIX-derived OSs are divided into two main categories. Those that conform to standards established by an organization known as the Open Group, which holds the trademark to the UNIX name, may be referred to as "certified UNIX operating systems." Systems that are similar to UNIX but do not adhere to the Open Group's standards are typically known as "UNIX-like operating systems." The latter category includes Apple's OS X and the free, open-source system Linux. The mobile OSs Android and Apple iOS also fall under this category and account for hundreds of millions of users, arguably making UNIX-derived OSs the most widely used systems ever.

—*Joy Crelin*

BIBLIOGRAPHY

Gancarz, Mike. *The UNIX Philosophy*. Woburn: Butterworth, 1995. Print.

"History and Timeline." *Open Group*. Open Group, n.d. Web. 28 Feb. 2016.

Raymond, Eric S. "Origins and History of Unix, 1969–1995." *The Art of UNIX Programming*. Boston: Pearson Education, 2004. Print.

Stonebank, M. "UNIX Introduction." *University of Surrey*. U of Surrey, 2000. Web. 28 Feb. 2016.

Toomey, Warren. "The Strange Birth and Long Life of Unix." *IEEE Spectrum*. IEEE, 28 Nov. 2011. Web. 28. Feb. 2016.

"What Is Unix?" *Knowledge Base*. Indiana U, 2015. Web. 28 Feb. 2016.

Worstall, Tim. "Is Unix Now the Most Successful Operating System of All Time?" *Forbes*. Forbes.com, 7 May 2013. Web. 7 Mar. 2016.

WEB DESIGN PROGRAMMING TOOLS

FIELDS OF STUDY

Information Technology; Graphic Design

ABSTRACT

Web design refers to the visual design, layout, and coding of a website. It is a subset of the larger discipline of web development. Web designers are responsible for building effective websites that effectively communicate ideas and provide users with the information they need. Web design programming tools help designers create attractive, well-organized websites.

PRINICIPAL TERMS

- **HTML editor:** a computer program for editing web pages encoded in hypertext markup language (HTML).
- **JavaScript:** a flexible programming language that is commonly used in website design.
- **jQuery:** a free, open-source JavaScript library.
- **screen-reading program:** a computer program that converts text and graphics into a format accessible to visually impaired, blind, learning disabled, or illiterate users.

From Early HTML to Responsive Design

The Internet has come a long way from the time when black computer screens displayed plain text and a blinking cursor. Websites did not exist before 1989, when English physicist Tim Berners-Lee began developing what would soon become the World Wide Web. In the process, he invented hypertext markup language (HTML), which is the code that describes the structure of a website. HTML was extremely simple at first, with fewer than twenty defined tags. However, it soon expanded as users' needs grew.

While HTML is still the backbone of web design, other web markup languages have since been developed to supplement it. JavaScript, introduced in 1995, resolved some of HTML's limitations while increasing interactivity between the site and the user. It became useful for game development, desktop applications, and animated and interactive web functions. The development of JavaScript libraries such as jQuery eliminated the need for developers to create their own libraries, making coding much easier and faster.

Adobe Flash, previously known as Macromedia Flash from 1996 to 2005, also allowed designers to animate web graphics and improve user engagement. However, Flash effects require a great deal of processing power and can take a long time to load, making the technology unsuitable for mobile devices and smartphones.

Cascading Style Sheets (CSS) was introduced in 1998. CSS structures the design separately from the content, which is managed through HTML editors. This removed more limitations for designers, allowing them have greater control over the appearance of websites.

With the popularity of mobile devices came more challenges for web designers, from screen size to data load speed. Responsive web design solved these issues. Responsive design uses the same content across devices, but with different layouts for each one. Usually, the width of the web browser window determines which layout is used. Other determining factors include whether or not JavaScript or certain HTML or CSS features are supported. JavaScript frameworks such as jQuery can be used to test for these features. Designers have also addressed the need for different layouts for each type of device by adopting flat design, a minimalist approach that simplifies visual

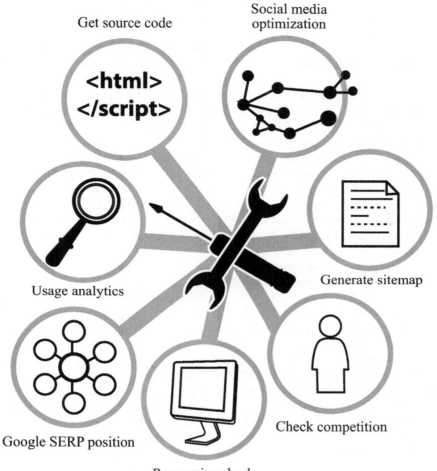

Get source code

Social media
optimization

<html>
</script>

Usage analytics

Generate sitemap

Google SERP position

Responsive check

Check competition

Web design tools available through outside experts may provide the solutions needed to design a superior website. Such tools include usage analytics, source-code generators, search engine optimization, sitemap generators, competitive analysis, layout responsiveness, and social media optimization. EBSCO illustration.

elements and emphasizes the message itself. The lack of complex elements makes alternate layouts easier to design and allows them to load faster.

WHY WEB DESIGN MATTERS

While it is important for websites to look appealing to users, web design includes more than just aesthetics. The designer's role is to make a website as user-friendly as possible. The elements of good web design include:

- Navigation: how website information is located. Good navigation makes it easy for users to move through the site and find what they need.
- Organization: how information is presented. Ideally, information should be presented in order from most important to least important.
- Appearance. A site that looks appealing, with proper use of color, space, type, and images, builds trust and engagement.

Good web design creates an environment in which users are comfortable, can find what they need, and feel that their time was well spent.

POWERFUL TOOLS MAKE WEBSITE DESIGN EASIER

The Internet is where business is conducted, learning takes place, and connections are made. Businesses and organizations need to have a web presence so their customers can learn about their products and services. The development of web design programming tools has made it possible for professional designers to create attractive and user-friendly websites in much less time, and thus at a lower cost.

Tools such as website builders also allow amateurs with no knowledge of coding to successfully create and launch their own websites without having to learn HTML or CSS. As of 2016, popular cloud-based website builders include Squarespace, Webflow, Weebly, Jimdo, WordPress, and Wix. Each of these tools uses HTML5, the fifth version of the HTML standard. Many feature a variety of predesigned, customizable templates. Templates simplify the entire process by allowing users to plug their content into a preexisting layout. Web designers have already completed both the visible design and the invisible coding that makes everything work. Using a website template may have some disadvantages, such as a limit on customization. However, the ease and cost savings of these tools and templates appeal to many consumers.

MAKING THE WEB MORE ACCESSIBLE

Because the Internet is a primarily visual experience, it is important that web designers make websites more accessible for people with impaired vision. Screen-reading programs translate on-screen text and output it to a text-to-speech system or a refreshable braille display. Text-to-speech programs can also improve accessibility for those who are illiterate or learning disabled.

In addition to textual content, screen-reading programs can convey visual elements as well. To facilitate this, web designers should label images and other decorative elements with "alt text"—alternative text that describes these elements for those who cannot see them. This text is also displayed on the screen if the element it describes fails to load.

THE FUTURE OF WEB DESIGN

As web technology advances, web design tools will be developed to take advantage of new possibilities, as seen with wearable technology, cloud computing, and smartphone applications. Programming tools will use artificial intelligence to determine the purpose of content and automatically alter the design for optimal results or for specific users' needs. The way people interact with computers will continue to drastically change, just as it has done from the beginning.

—*Teresa E. Schmidt*

BIBLIOGRAPHY

Brownlee, John. "The History of Web Design Explained in 9 GIFs." *Co.Design*. Fast Co., 5 Dec. 2014. Web. 22 Feb. 2016.

"Choose a Website Builder: 14 Top Tools." *Creative Bloq*. Future, 8 Feb. 2016. Web. 22 Feb. 2016.

"Designing for Screen Reader Compatibility." *WebAIM*. Center for Persons with Disabilities, Utah State U, 19 Nov. 2014. Web. 22 Feb. 2016.

"HTML Tutorial." *Tutorials Point*. Tutorials Point, 2016. Web. 22 Feb. 2016.

Laszlo, Arp. "Why Is Website Design So Important?" *Sunrise Pro Websites & SEO*. Sunrise Pro Websites, 22 Jan. 2016. Web. 22 Feb. 2016.

Luenendonk, Martin. "Top Programming Languages Used in Web Development." *Cleverism*. Cleverism, 21 June 2015. Web. 22 Feb. 2016.

"Pros & Cons of Website Templates." *Entheos*. Entheos, n.d. Web. 22 Feb. 2016

Weller, Nathan. "A Look into the Future of Web Design: Where Will We Be in 20 Years?" *Elegant Themes Blog*. Elegant Themes, 9 May 2015. Web. 22 Feb. 2016.

WEB GRAPHIC DESIGN

FIELDS OF STUDY

Graphic Design; Digital Media; Programming Language

ABSTRACT

Web graphic design is the use of graphic design techniques in designing websites. Web graphic designers must balance the marketing aspects of a website with aesthetic design criteria. They also attempt to increase the likelihood that the website will be found in search results and therefore be an effective advertising tool.

PRINICIPAL TERMS

- **logotype:** a company or brand name rendered in a unique, distinct style and font; also called a "wordmark."
- **search engine optimization (SEO):** techniques used to increase the likelihood that a website will appear among the top results in certain search engines.
- **tableless web design:** the use of style sheets rather than HTML tables to control the layout of a web page.
- **typography:** the art and technique of arranging type to make language readable and appealing.
- **wireframe:** a schematic or blueprint that represents the visual layout of a web page, without any interactive elements.

DESIGNING FOR THE WEB

Web graphic design is a subfield of graphic design that focuses on designing for the web. It typically involves a blend of graphic design techniques and computer programming. Many websites are used as marketing materials for businesses and organizations. Consequently, web designers often incorporate business logos and other promotional materials. They also use search engine optimization (SEO) techniques to increase the likelihood that the web page will be found by search engines.

WEB DESIGN FUNCTIONALITY

Most websites, whether business, advocacy, news, or personal sites, serve as both marketing and informational tools. A website that represents a specific brand will incorporate iconic logos, logotypes, or combination marks to aid brand recognition. An iconic logo is a symbol or emblem that represents a person, business, or organization. A logotype is a company or brand name rendered in a unique or proprietary font and style. Combination marks combine icons with logotypes.

In addition to being informative, a website should be visually appealing and easy to understand. Skilled use of typography helps web designers achieve these goals. For example, while logotypes are designed to catch the eye, important information should be presented in a font that is aesthetically pleasing yet unobtrusive. A font that draws attention to itself will detract from the message of the text. Typography techniques such as this help web designers make websites easy to read and navigate.

In addition, web designers must be familiar with SEO, which relies on elements such as keywords and links to ensure that users can find a website using a search engine. SEO techniques are most effective when incorporated into the overall design of a website. As such, many web graphic designers also help users optimize their websites for better search results.

By the 2010s, web design had begun to focus on designing for the mobile web, creating websites and e-commerce sites that could be viewed and accessed using mobile devices. Many do-it-yourself (DIY) website builders started offering mobile web design templates and conversions.

DEVELOPMENT OF WEB DESIGN TECHNIQUES

Modern web design began in the 1990s, with the creation and adoption of hypertext markup language (HTML). This markup language specifies the location and appearance of objects displayed on a web page. In the mid-1990s, web designers began using HTML tables. These consist of static cells that can be arranged on a page to specify the location of text or objects. By placing tables within tables, designers could create a richer experience for websites.

The programming language JavaScript was developed in 1995 to code website behavior. This made individual web pages interactive for the first time. With JavaScript, web designers could embed

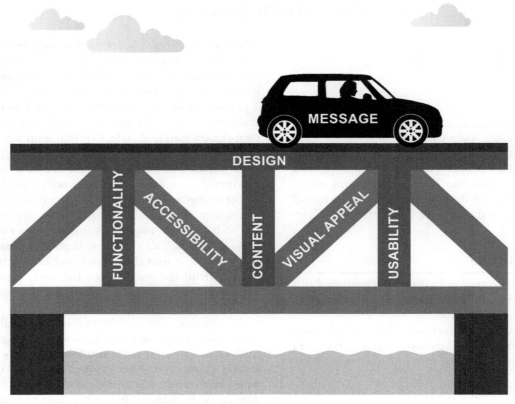

A quality website design ensures that the message gets across. Good web design incorporates accessible content, provided in a visually appealing way, through strong website back-end functionality and front-end usability. EBSCO illustration.

image galleries, add functions such as drop-down menus, and make websites respond visually when a user clicked on text or images. Flash, developed in 1996, allowed designers to add animated elements to websites. However, not all browsers supported Flash, so it was not as universally useful as JavaScript.

The next major step in web graphic design was the introduction of tableless web design. This approach to web design does not rely on HTML tables. Instead, it uses style sheets to format pages. The Cascading Style Sheets (CSS) language, first introduced in 1996, allows designers to separate the visual elements of the design from the content. CSS can determine the appearance and spacing, while individual elements are still described in HTML. Personal Home Page (PHP), another language used in web design, was developed in 1994 but gained in

popularity during the late 1990s. It provides further options for making web pages interactive. When used with HTML, PHP allows websites to create new content based on user information and to collect information from visitors.

In the 2010s, most web design is still based on a combination of CSS and HTML. Programming in PHP and the database server software MySQL provides options for greater flexibility and interactivity. New versions of web design languages, such as CSS4 and HTML5, offer more ways of incorporating multimedia, better support for multilingual websites, and a wider array of aesthetic options for designers.

USER-GENERATED WEB DESIGN
The popular website WordPress is actually a content management system (CMS) built on PHP and

MySQL. It debuted in 2003 as a blogging website, offering users templates and tools to design blogs. The site soon became popular as a basic web design system, allowing users with limited knowledge of HTML or other coding to build basic and functional, if aesthetically rudimentary, websites. Soon web designers began tailoring WordPress sites for customers. This has essentially served as a shortcut for the web design industry. As of 2015, WordPress remained among the most popular web design systems in the world, accounting for about half of all CMS-based websites on the Internet.

The debut of WordPress launched a new era in DIY web graphic design. Other websites such as Squarespace, Wix, and Weebly soon began offering users the ability to design websites quickly and easily, without the need to understand programming languages. Most such sites provide users templates in the form of wireframes. Wireframes are structural layouts that specify the location of images, text, and interactive elements on a web page but are not themselves interactive. Users can then insert their own images and text and, depending on the underlying program, also rearrange the layout of the wireframe. Like most elements of online commerce and marketing, web graphic design is moving toward user-generated and user-influenced design in order to open up to broader audiences.

—*Micah L. Issitt*

BIBLIOGRAPHY

Allanwood, Gavin, and Peter Beare. *User Experience Design: Creating Designs Users Really Love.* New York: Fairchild, 2014. Print.

Cezzar, Juliette. "What Is Graphic Design?" *AIGA.* Amer. Inst. of Graphic Arts, 2016. Web. 16 Mar. 2016.

Hagen, Rebecca, and Kim Golombisky. *White Space Is Not Your Enemy: A Beginner's Guide to Communicating Visually through Graphic, Web & Multimedia Design.* 2nd ed. Burlington: Focal, 2013. Print.

Malvik, Callie. "Graphic Design vs. Web Design: Which Career Is Right for You?" *Rasmussen College.* Rasmussen Coll., 25 July 2013. Web. 16 Mar. 2016.

Schmitt, Christopher. *Designing Web & Mobile Graphics: Fundamental Concepts for Web and Interactive Projects.* Berkeley: New Riders, 2013. Print.

Williams, Brad, David Damstra, and Hal Stern. *Professional WordPress: Design and Development.* 3rd ed. Indianapolis: Wiley, 2015. Print.

WINDOWS OPERATING SYSTEM

FIELDS OF STUDY

Operating Systems; Information Technology; Computer Science

ABSTRACT

Windows is an operating system (OS) developed by Microsoft. It has been one of the dominant OSs for personal computing since the 1990s. Windows is designed to operate across a variety of platforms, including personal computers, smartphones, and the Xbox One video-gaming console.

PRINICIPAL TERMS

- **graphical user interface (GUI):** a type of computer interface that uses graphic icons and a pointer to access system functions and files.
- **language interface packs:** programs that translate interface elements such as menus and dialog boxes into different languages.
- **multitasking:** in computing, the ability to perform or plan multiple operations concurrently.
- **multiuser:** capable of being accessed or used by more than one user at once.
- **operating system shell:** a user interface used to access and control the functions of an operating system.
- **platform:** the underlying computing system on which applications can run, which may be the computer's hardware or its operating system.

HISTORY OF THE WINDOWS OPERATING SYSTEM

Windows is an operating system (OS) developed by Microsoft Corporation for personal computers (PCs). Windows features a graphical user interface (GUI). A GUI allows users to interact with the OS using a pointer, dropdown menus, graphic icons, and movable windows representing folders and drives. Microsoft debuted Windows in the early 1980s. Since then, Windows has dominated the PC market. By some estimates, 90 percent of computers worldwide run some version of Windows.

The first PCs sold on the consumer market used a text-based disk operating system (DOS). Microsoft's version of DOS, called MS-DOS, used a command-line interface in which users typed text commands to activate functions. Microsoft debuted the first

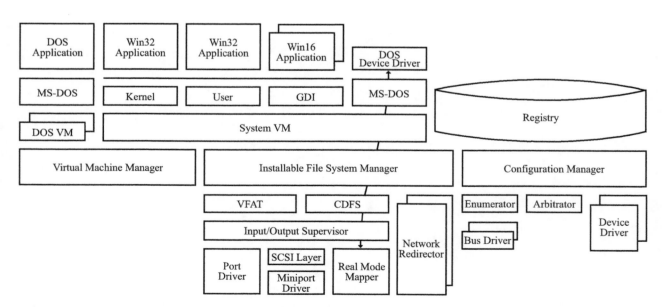

Architecture of Windows 95 as depicted by Ruud Koot. The development of Windows 95 improved upon MS-DOS by incorporating a 32-bit multitasking architecture. By Ruud Koot, CC BY-SA 3.0 (http://creativecommons.org/licenses/by-sa/3.0), via Wikimedia Commons

version of Windows in 1985. Windows 1.0 was an operating system shell that could be installed over the MS-DOS system. Windows featured one of the first GUIs. Windows was based on a model called WIMP (windows, icons, menus, pointer objects). GUIs allow users to navigate a virtual desktop with a mouse that controls a pointer icon. GUIs also enable users to click on text and graphic icons to activate programs. For instance, in Windows 1.0, clicking on a folder icon activated the underlying DOS command "dir" to display the contents of a directory.

A new version, Windows 95, introduced the company's Internet Explorer browser. Windows 95 also introduced the Start button and taskbar. The Windows 95 kernel (basic underlying programming) remained the standard for additional versions of Windows until 2001, with the release of Windows XP. Windows 7, released in 2009, was one of the company's most popular versions of the interface. Windows 7 featured a new, more intuitive layout. Windows 8 introduced a completely redesigned interface with tiles that are easier for users with touch-screen tablets. In 2015, Microsoft debuted Windows 10, the first version of their OS to be offered as a free upgrade. Windows 10 also introduced a new service model in which the OS would continually receive updates to various features and functions. This model is similar to that employed by Microsoft's competitor Apple. Windows 10 was also the first version of Windows that had a universal application architecture. This means apps on Windows 10 can be used on smartphones, tablets, and the Xbox One gaming system as well.

FEATURES OF WINDOWS

In computing, a platform can be defined as the computer system that a program uses. A platform can be the hardware architecture or an OS. The earliest versions of the Windows OS were cross-platform and could be used with different types of architecture. In 2016, the Windows OS is one of the primary platforms for software. The Windows OS is designed to be compatible with the X86-64 computer architecture, a 64-bit processor system created by AMD in 2000. Windows 10 can also be implemented on older x86 (32-bit) computers.

Windows OS is designed to be a multiuser environment in which more than one users can use the same OS. The Remote Desktop Connection system allows multiple users to use Windows at once. The Windows

OS also enables users to adjust OS settings for multiple accounts, essentially creating multiple versions of the OS on the same computer. It is also a multitasking system in which more than one processing job can run at a time. The Windows 10 system introduced a new automated assistant called Cortana in an effort to compete with the popularity of Apple's Siri.

DOMINANCE AND REINVENTION

More than 90 percent of PCs worldwide ran some version of Windows in 2015. The majority (52 percent) run Windows 7, which, as of 2016, was the company's most popular product. The release of Windows 10 in 2015 had little impact, accounting for 12 percent of the market in February 2016. Windows is sold around the world and comes formatted for a variety of languages. In addition, language interface packs are available for free download and offer support for languages not found in full versions. Each pack requires a base language that can be activated within the OS after installation.

Given the rising popularity of handheld computing devices, such as tablets and smartphones, the global computer market is no longer based solely on the PC market. While Microsoft continues to dominate the PC market, the company controls only 14 percent of the global market across all computing devices. Looking at smartphone OSs specifically, Windows controls less than 3 percent of that market, falling far behind Android and Apple iOS. As a result, with the release of Windows 10, Microsoft has made changes to its basic strategy. The company has reduced its focus on proprietary software. Instead it is increasingly embracing the potential of open-source software. Future versions of Windows will likely show an increased focus on cloud-based computing, intuitive touch-screen technology, and alternative interface controls such as voice activation and multitouch gestures.

—*Micah L. Issitt*

BIBLIOGRAPHY

Bishop, Todd. "Microsoft Exec Admits New Reality: Market Share No Longer 90%— It's 14%." *Geekwire*. GeekWire, 14 July 2014. Web. 30 Jan. 2015.

Gibbs, Samuel. "From Windows 1 to Windows 10: 29 Years of Windows Evolution." *Guardian.* Guardian News and Media, 2 Oct. 2014. Web. 2 Jan. 2016.

"A History of Windows." *Microsoft.* Microsoft, 2016. Web. 2 Jan. 2016.

McLellan, Charles. "The History of Windows: A Timeline." *ZDNet.* CBS Interactive, 14 Apr. 2014. Web. 15 Feb. 2016.

Protalinski, Emil. "Windows 10 Ends 2015 under 10% Market Share." *VentureBeat.* VentureBeat, 1 Jan. 2016. Web. 26 Feb. 2016.

Warren, Tom. "Windows Turns 30: A Visual History." *Verge.* Vox Media, 19 Nov. 2015. Web. 26 Feb. 2016.

WIREFRAMES

FIELDS OF STUDY

Computer Science; Graphic Design; Digital Media

ABSTRACT

A wireframe visually represents the basic elements of a proposed website before a web designer actually builds it. Creating a wireframe ensures that both the designer and the client can agree on the basic layout and functionality of the site. It also allows them to map out how the end user will ultimately interact with the site.

PRINICIPAL TERMS

- **high-fidelity wireframe:** an image or series of images that represents the visual elements of the website in great detail, as close to the final product as possible, but still does not permit user interaction.
- **information hierarchy:** the relative importance of the information presented on a web page.
- **interaction design:** the practice of designing a user interface, such as a website, with a focus on how to facilitate the user's experience.
- **portlet:** an independently developed software component that can be plugged into a website to generate and link to external content.
- **prototype:** a simplified, but visually very similar, version of the final website that simulates how the user will interact with the site.
- **typography:** the style, arrangement, and overall appearance of text on a website.
- **widget:** an independently developed segment of code that can be plugged into a website to display information retrieved from other sites.

WHAT IS A WIREFRAME?

A wireframe is a visual, noninteractive representation of a proposed website. A web designer may use a wireframe to determine how best to lay out the elements of the site. If the site is being designed for a client, the designer may show the client a wireframe to get approval for the final product.

A website is essentially a sophisticated information storage bank. Ultimately, it is designed for the end users, who will need easy, efficient access to the information it contains. To that end, a web designer will block out the basic elements of the proposed website first, most often drawing them on paper. This drawing is the wireframe. Though it may be no more than a set of squares, circles, columns, and lines that represent the individual elements of the pages, it allows the designer and the client to review the site concept together before it is built. Any changes in the website format can be easily made at this stage.

A wireframe also allows the designer to sort the content of the proposed website into an information hierarchy. The designer can then determine how best to arrange it on the page. Part of web design is knowing what information users expect to find and where they are most likely to look for it. Eye-tracking studies have determined that upon accessing a website, a user will first look at the top-left corner of the page, scan across the top, then look down the left side and scan partway across the screen. This essentially forms an F-shape. Thus, crucial information should be concentrated mainly in the upper left portion of the page. Such insights into how users interact with and experience websites are an important part of the design field known as interaction design.

Bootstrap Thumb Default

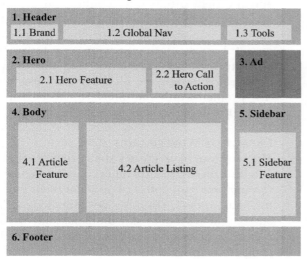

Bootstrap Thumb Portrait

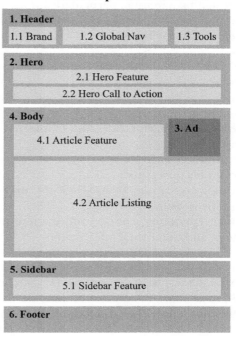

Bootstrap Thumb Smartphone

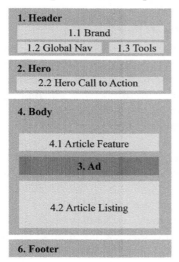

Wireframes of a website in three different screen layouts show how the same content can be presented in different ways to best fit the device. These high-fidelity wireframes indicate what type of content is presented in which locations as a first step in designing for responsiveness. EBSCO illustration.

WIREFRAMING IN WEB DESIGN

When designing a website, particularly for a business or other organization, certain elements are standard. These elements may include:

The client's logo, to identify the business or organization in a recognizable way

A navigation bar, to provide access to the different pages of the site

A front-page image or slide show of images

A search box

A registration or login box, if the site allows (or requires) users to register

A header containing an index of the major elements of the website

A footer containing contact information, privacy guidelines, and other legal information

Many websites also include external content such as news headlines, weather reports, or social media feeds. In most cases, the creator of this external content generates a portlet or widget that designers can simply plug into the code of the site. While portlets and widgets work in similar ways, they differ mainly in their execution. A portlet is executed on the server, so that the code it generates is not visible to the end user. A widget is executed in the browser.

A wireframe should incorporate any and all of elements that will be included in the final website. In most cases, the wireframe serves as a blueprint, with no distracting aesthetic features such as font, color, or images. The emphasis is on functionality—that is, how the proposed site will accommodate the end user. This type of wireframe is also called a "low-fidelity wireframe."

Once the client is onboard with the concept, the designer may produce a more sophisticated, high-fidelity wireframe. This type of wireframe is usually created using dedicated software. It represents the final website in much more detail. It incorporates such design elements as images, color schemes, and typography. Sequences of images are used to represent how users will interact with the site. However, the wireframe itself is still noninteractive.

Sometimes, between the high-fidelity wireframe and the final product, a designer will produce an interactive prototype. While a prototype will usually look very similar to the final product, its appearance matters less than its functionality. A prototype should simulate the user experience with the completed website. Prototypes are ideal for early user testing. Any problems with the interface can be addressed before the designer begins coding for the final site.

IMPACT OF WIREFRAMES

Much as an architect creates a blueprint for a skyscraper or a director storyboards a film, a web designer uses a wireframe to spot potential problems before building the final product. This saves both the designer's time and the client's money and, ultimately, shields the end user from frustration—all critical factors in the highly competitive world of Internet marketing.

—*Joseph Dewey*

BIBLIOGRAPHY

Greenberg, Saul, et al. *Sketching User Experiences: The Workbook.* Waltham: Morgan, 2012. Print. 4 Science Reference Center™ Wireframes

Hamm, Matthew J. *Wireframing Essentials: An Introduction to User Experience Design.* Birmingham: Packt, 2014. Print.

Klimczak, Erik. *Design for Software: A Playbook for Developers.* Hoboken: Wiley, 2013. Print.

Krug, Steve. *Don't Make Me Think, Revisited: A Common Sense Approach to Web Usability.* 3rd ed. Berkeley: New Riders, 2014. Print.

Marsh, Joel. *UX for Beginners: A Crash Course in 100 Short Lessons.* Sebastopol: O'Reilly, 2016. Print.

Nielsen, Jakob. "F-Shaped Pattern for Reading Web Content." *Nielsen Norman Group.* Nielsen Norman Group, 17 Apr. 2006. Web. 4 Mar. 2016.

Norman, Don. *The Design of Everyday Things.* Rev. and expanded ed. New York: Basic, 2013. Print.

Treder, Marcin. "Wireframes vs. Prototypes: What's the Difference?" *Six Revisions.* Jacob Gube, 11 Apr. 2014. Web. 4 Mar. 2016.

WIRELESS NETWORKS

FIELDS OF STUDY

Information Technology; Network Design

ABSTRACT

Wireless networks allow computers and other Internet-enabled devices to connect to the Internet without being wired directly to a modem or router. They transmit signals across radio waves instead. Wireless networks make communication services available almost anywhere, without the need for wired connections.

PRINICIPAL TERMS

- **electromagnetic spectrum:** the complete range of electromagnetic radiation, from the longest wavelength and lowest frequency (radio waves) to the shortest wavelength and highest frequency (gamma rays).
- **local area network (LAN):** a network that connects electronic devices within a limited physical area.
- **microwaves:** electromagnetic radiation with a frequency higher than that of radio wave but lower than that of visible light.
- **node:** a point at which a communication network connects to another network, is redistributed, or terminates at a user interface.
- **personal area network (PAN):** a network generated for personal use that allows several devices to connect to one another and, in some cases, to other networks or to the Internet.
- **radio waves:** low-frequency electromagnetic radiation, commonly used for communication and navigation.

HOW WIRELESS NETWORKS WORK

Computer networks allow multiple computers to access information stored in other locations and to use common devices, such as printers and file servers. Setting up a traditional computer network requires cables, distributors, routers, and internal and external network cards.

As computers became smaller and therefore more mobile, cables and wires presented challenges. The development of wireless connectivity made it more convenient to use laptops, tablets, and other mobile devices in more places. Wireless networks provide the same connectivity and access as traditional networks without the need for physical cables.

Most wireless networks transmit signals using radio waves, the lowest-frequency waves in the electromagnetic spectrum. A computer sends data across a wireless network through nodes, which are all the hubs, switches, and devices connected to the network. The data are translated into a radio signal and transmitted to the wireless router. The router then sends the information to the Internet through an Ethernet cable. Multiple devices can access the same wireless router to connect to the Internet.

TYPES OF WIRELESS NETWORKS

Broadly speaking, the different types of wireless networks can be divided into four main categories, based on range and how they transmit information.

A personal area network (PAN) covers the shortest range. PANs are typically used by one person to transmit information between two or more devices. The network is generated by one device, such as a cell phone. Any device within range of that device can then communicate with it, or with any other device within its range. The range of a wireless PAN (WPAN) is generally no more than a few meters.

The next-largest type of network is a local area network (LAN). The term was initially used to refer to devices within a limited area, such as a single building, that are connected via cables to a central access point, such as a modem. A wireless LAN (WLAN) works much the same way. Instead of being connected to the individual devices by cables, however, the access point connects to a router that generates a wireless network. This network is omnidirectional, meaning that it is equally strong in all directions. As a result, the signal is weaker, and thus has a shorter range, than if it were focused in one direction.

A wireless wide area network (WWAN) can cover a range of several miles. To accomplish this, WWANs use directional antennas and typically transmit signals via microwaves. Microwaves have a higher frequency than radio waves, although there is some overlap. While microwaves can travel farther, they cannot penetrate obstacles as well as radio waves. Thus, to use the network, a device must have a direct line of sight

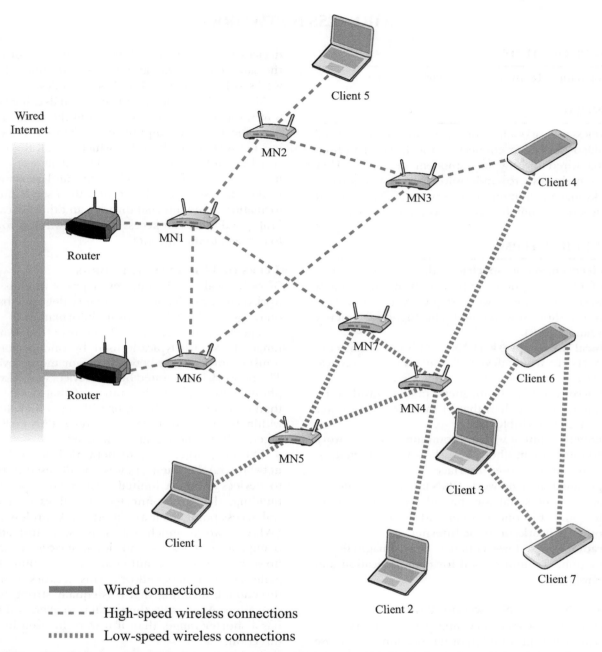

Wired connections
High-speed wireless connections
Low-speed wireless connections

Wireless networks allow multiple clients or terminals to connect with other terminals via wireless pathways. Depending on the type of router, the connection may be high speed or low speed. EBSCO illustration.

to the antenna or be connected to another node that does. A WWAN may be either point-to-point (P2P) or point-to-multipoint (P2MP). P2P, the simplest type, connects one node to another. P2MP connects multiple nodes to one central access point.

Finally, a mesh network is a wireless network in which all devices connect to each other and relay signals from device to device, without a wired infrastructure. In a mesh network, all nodes—that is, all devices connected to the network—transmit signals.

PROS AND CONS OF WIRELESS
WLANs gained popularity over regular LANs because of their convenience and cost savings. In fact, wireless networks became so prolific that Apple phased out Ethernet connection ports on its MacBook laptops. By investing in wireless networking hardware, businesses can avoid the trouble and expense of installing cables throughout a building. They can quickly expand the network when needed. Consumers can easily set up home networks for use with laptops, tablets, and smartphones. Users can be more productive by accessing the Internet wherever they are.

However, wireless networks have drawbacks as well. Their smaller range makes additional equipment, such as repeaters, necessary for larger buildings. In addition, radio waves are prone to interference, making reliability an issue. Security is also a concern. Since data are transmitted through the air, wireless networks involve greater risk.

A BIG IMPACT FROM INVISIBLE RADIO WAVES
Wireless networks have allowed more people around the world to access the Internet from more locations than ever. In business, wireless networks enable companies to operate with more speed, flexibility, and connection. In health care, they allow physicians to consult with faraway specialists or to check in with patients in rural areas. For individuals, wireless networks bring friends and families closer, enabling them to keep in more frequent contact and allow users to stream entertainment or download books while on the go. News is delivered as it happens. People immediately share their reactions, while families and friends can check on those affected.

Without wireless technology, there would be no smartphones and no texting, and there would be less access to social media. Social media enables people all over the world to express themselves, whether for personal or political reasons. From young voters in the United States to the political uprisings throughout the Middle East during the Arab Spring of 2011, social media has become an important force for political awareness and connection over shared interests.

THE EVOLUTION OF WIRELESS NETWORKS
Businesses are increasingly moving to cloud-based networks, which eliminate on-site hardware and move all infrastructure, administration, and processes to remote servers. Communities are creating mesh networks with their own infrastructure to share resources and connect through their mobile devices. Such mesh networks provide more reliable communication and connection during emergencies or natural disasters. At home, consumers are taking advantage of new technology to make wireless connection to the Internet easier and more convenient. It is clear that the proliferation of wireless networks has changed the way the world learns, communicates, and conducts business. As the cost of wireless technology decreases, the benefits will spread even further.

—*Teresa E. Schmidt*

BIBLIOGRAPHY
"Computer—Networking." *Tutorials Point.* Tutorials Point, 2016. Web. 22 Feb. 2016.
"DCN—Computer Network Types." *Tutorials Point.* Tutorials Point, 2016. Web. 22 Feb. 2016.
De Filippi, Primavera. "It's Time to Take Mesh Networks Seriously (and Not Just for the Reasons You Think)." *Wired.* Condé Nast, 2 Jan. 2014. Web. 22 Feb. 2016.
Lander, Steve. "Disadvantages or Problems for Implementing Wi-Fi Technology." *Small Business—Chron.com.* Hearst Newspapers, n.d. Web. 22 Feb. 2016.
Smith, Matt. "Wi-Fi vs. Ethernet: Has Wireless Killed Wired?" *Digital Trends.* Designtechnica, 18 Jan. 2013. Web. 22 Feb. 2016.
"Types of Wireless Networks." *Commotion.* Open Technology Inst., n.d. Web. 3 Mar. 2016.
"Wireless History Timeline." *Wireless History Foundation.* Wireless Hist. Foundation, 2016. Web. 22 Feb. 2016.

WORKPLACE MONITORING

FIELDS OF STUDY

Information Technology

ABSTRACT

Modern technology makes it possible for firms to monitor many aspects of the workplace, particularly employee activities and performance. Most monitoring systems used are electronic, including computer terminals, e-mail, Internet, telephone and smartphone systems, GPS, drones, and others. Challenges to workplace monitoring appear in the arena of employee rights and privacy legislation.

PRINICIPAL TERMS

- **packet sniffers:** a program that can intercept data or information as it moves through a network.
- **real-time monitoring:** a process that grants administrators access to metrics and usage data about a software program or database in real time.
- **remote monitoring:** a platform that reviews the activities on software or systems that are located off-site.
- **silent monitoring:** listening to the exchanges between an incoming caller and an employee, such as a customer service agent.
- **transparent monitoring:** a system that enables employees to see everything that the managers monitoring them can see.

MONITORING TECHNOLOGIES

The practice of monitoring employees at work is known as "workplace monitoring." According to the US Office of Technology Assessment, workplace monitoring is the collection, storage, analysis, and reporting of information about workers' activities.

The introduction of computers and other telecommunication technologies in the workplace brought great changes to monitoring practices. Workplace monitoring includes a wide variety of technologies, such as hidden or overt video cameras, global positioning systems (GPS), landline telephones, cell phones and smartphones, computer software and systems, the Internet, and drones. Many techniques have been developed to improve workplace monitoring. For example, packet sniffers are computer programs that can analyze communication flow across networks and intercept malicious files. Many businesses monitor all network traffic passing through their servers. Automatic programs can alert managers if a networked computer connects to a malicious domain, block particular websites, and scan e-mails for spam.

Many organizations monitor their workers in order to protect their property and information, protect employees, and measure the quality of their work and productivity. However, employers must also be mindful of legal and ethical considerations of workplace monitoring. Employers must work to maintain a sense of trust in their employees when engaging in monitoring. Monitoring practices can generate negative feeling among employees and create legal problems for the employer if they are not implemented carefully.

Further complicating the issue is the increasingly fluid nature of workplace boundaries. For instance, work is often spread across geographical spaces. Many employers have a large number of employees who work away from headquarters, either because they travel or work remotely. This has led to remote monitoring by way of varied technologies and devices, including GPS and drones.

Moreover, a growing amount of work is computerized, through intranet and Internet networks. As a result, employers face real risks of employee misuse of these systems. However, these same systems have led to improvements in workplace monitoring technology.

BENEFITS OF WORKPLACE MONITORING

Because of the growing instances of hacking and information theft, employers have legitimate concerns about the security of workplace information. They may also want to conduct quality controls and monitor employee productivity. For example, it is a common practice for call centers to record customer service calls, in a system commonly known as silent monitoring. Another type of system, real-time monitoring, allows managers to analyze usage on computer systems as it occurs. Real-time monitoring

WHY USE COMPUTER MONITORING SOFTWARE?	WHAT CAN PROGRAMS DO?
• **Prevent Data Leakage**	• **Track Document Usage**
• **Track Source of Leakages**	• **Track Device Usage**
• **Avoid Cyber Threats**	• **Record Website Browsing**
• **Block Web Distractions**	• **Record Chat and Instant Message Usage**
• **Determine Employee Productivity**	• **Monitor Application Activity**
• **Stays Effective When Offline**	• **Block Incoming and Outgoing Email**
	• **Record Screenshots**
	• **Monitor Network and Bandwidth Usage**
	• **Record File Printing Data**

Workplace monitoring software is designed to monitor employees' computer use in order to protect against external threats, limit distractions, prevent and track information leaks, and support productivity online and offline. These objectives can be accomplished through a number of possible software configurations. EBSCO illustration.

helps track not only employee activity but also other firm processes, such as sales data and other trends. It is often used in conjunction with remote monitoring. Remote monitoring involves tracking the activity of employees who work off-site.

Hundreds of software programs and other technologies are available to monitor the workplace. Some of these programs are free. Companies can establish their own monitoring systems. Others prefer to hire firms that provide the technology and analyze the data. The data retrieved by monitoring technologies may be used to identify information leaks or theft, evaluate employee performance, detect malware, or analyze business trends. Among the latest monitoring technologies are drones. Drones are mainly used in real estate, construction, and agriculture.

There are many ways in which electronic workplace monitoring is used. According to the American Management Association, 78 percent of major companies in the United States reported monitoring employee use of e-mail, Internet, or phone in 2015. Some employers also use monitoring technology in order to measure employee performance by measuring time spent at the computer or keystroke speed.

LEGAL AND ETHICAL CONSIDERATIONS
The increasing breadth of workplace monitoring has raised ethical and legal concerns. On the balance are the privacy rights of workers. Electronic workplace monitoring has become so common that labor rights advocates have raised concerns about the abuse or inappropriate use of monitoring practices. On the other hand, employers are concerned about liability costs and legal consequences.

In the opinion of some experts, electronic workplace monitoring is subject to insufficient government regulation. This may be due to its constant innovations and its relatively new status. Technically, employers are legally allowed to listen to, watch over, record, and read all work-related forms of communication. However, federal law stipulates that personal calls cannot be monitored.

Experts recommend companies ensure their monitoring practices are fair and consistent. It is crucial to implement measures to prevent managers from engaging in abuse of power and other illegal activities using monitoring technologies. Staff in charge of workplace monitoring must be adequately trained. Some organizations implement transparent monitoring in order to make employees feel more at ease with the process. Transparent monitoring welcomes the participation of employees in the process. Finally, when employers suspect employees of engaging in criminal action, experts recommend they alert the authorities rather than try to set up a sting operation on their own.

Debate continues about what employers should be allowed to monitor and to what extent employees have the right to know they are being monitored. Nevertheless, employees should always assume that they are being monitored and keep all private or personal communication in separate accounts and devices.

—*Trudy Mercadal, PhD*

BIBLIOGRAPHY

Alton, Larry. "Email Security in 2016: What You Need to Know." *Inc.* Mansueto Ventures, 18 Feb. 2016. 21 Feb. 2016.

Plumb, Charles. "Drones in the Workplace." *EmployerLINC.* McAfee and Taft, 14 Dec. 2015. Web. 21 Jan. 2016.

Smith, Eric N. *Workplace Security Essentials: A Guide for Helping Organizations Create Safe Work Environments.* Oxford: Butterworth-Heinemann, 2014. Print.

Stanton, Jeffrey, and Kathryn R. Slam. *The Visible Employee: Using Workplace Monitoring and Surveillance to Protect Information Assets without Compromising Employee Privacy or Trust.* Medford: Information Today, 2006. Print.

Tabak, Filiz, and William Smith. "Privacy and Electronic Monitoring in the Workplace: A Model of Managerial Cognition and Relational Trust Development." *Employee Responsibilities and Rights Journal* 17.3 (2005): 173–89. Print.

Yakowitz, Will. "When Monitoring Your Employees Goes Horribly Wrong." *Inc.* Mansueto Ventures, 14 Feb. 2016. Web. 21 Feb. 2016.

Yerby, Jonathan. "Legal and Ethical Issues of Employee Monitoring." *Online Journal of Applied Knowledge Management* 1.2 (2013): 44–54. Web. 21 Jan. 2016.

HISTORY OF EVENTS LEADING UP TO THE DEVELOPMENT OF MODERN COMPUTERS

circa 2400 BCE
The abacus – the first known calculator, was probably invented by the Babylonians as an aid to simple arithmetic around this time period.

circa 1115 BCE
The south-pointing chariot was invented in ancient China. It was the first known geared mechanism to use a differential gear.

circa 500 BCE
First known use of 0 (by mathematicians in ancient India around this date.

circa 500 BCE
Indian grammarian Pāṇini formulated the grammar of Sanskrit (in 3959 rules) known as the Ashtadhyayi. His grammar had the computing power equivalent to a Turing machine. The Panini-Backus form used to describe most modern programming languages is also significantly similar to Pāṇini's grammar rules.

circa 300 BCE
Indian mathematician/scholar/musician Pingala first described the binary number system which is now used in the design of essentially all modern computing equipment.

circa 200 BCE
The Chinese invented the suanpan (Chinese abacus) which was widely used until the invention of the modern calculator, and continues to be used in some cultures today.

circa 125 BCE
The Antikythera mechanism: A clockwork, analog computer believed to have been designed and built in the Corinthian colony of Syracuse. The mechanism contained a differential gear and was capable of tracking the relative positions of all then-known heavenly bodies.

circa 100 BCE
Chinese mathematicians first used negative numbers.

circa 60 BCE
Heron of Alexandria made numerous inventions, including "Sequence Control." This was, essentially, the first computer program. He also made numerous innovations in the field of automata, which are important steps in the development of robotics.

circa 200
Jaina mathematicians invented logarithms.

circa 600
Indian mathematician Brahmagupta was the first to describe the modern place-value numeral system.

724
Chinese inventor Liang Lingzan built the world's first fully mechanical clock; the earliest true computers, made a thousand years later, used technology based on that of clocks.

820
Persian mathematician, Muḥammad ibn Mūsā al-Khwārizmī, described the rudiments of modern algebra. The word "algorithm" is derived from al-Khwarizmi's Latinized name "Algoritmi".

circa 850
Arab mathematician, Al-Kindi (Alkindus gave the first known recorded explanation of cryptanalysis in "A Manuscript on Deciphering Cryptographic Messages". In particular, he is credited with developing the frequency analysis method whereby variations in the frequency of the occurrence of letters could be analyzed and exploited to break encryption ciphers.

850
The Banū Mūsā brothers, in their "Book of Ingenious Devices", invented an organ which played interchangeable cylinders automatically. They also invented an automatic flute player which appears to have been the first programmable machine.

996

Persian astronomer, Abū Rayhān al-Bīrūnī, invented the first geared mechanical astrolabe, featuring eight gear-wheels.

circa 1000

Abū Rayhān al-Bīrūnī invented the Planisphere, an analog computer as well as the first mechanical lunisolar calendar—an early example of a fixed-wired knowledge processing machine.

circa 1015

Arab astronomer, Abū Ishāq Ibrāhīm al-Zarqālī (Arzachel) of al-Andalus, invented the Equatorium, a mechanical analog computer device used for finding the longitudes and positions of the Moon, Sun and planets without calculation.

1020

The mechanical geared astrolabe earlier developed by Abū Rayhān al-Bīrūnī perfected by Ibn Samh. This can be considered an ancestor of the mechanical clock.

circa 1100

Arab astronomer, Jabir ibn Aflah (Geber), invented the Torquetum, an observational instrument and mechanical analog computer device used to transform between spherical coordinate systems. It was designed to take and convert measurements made in three sets of coordinates: horizon, equatorial, and ecliptic.

1206

Arab engineer, Al-Jazari, designed a humanoid-shaped mannequin which may have been the first scientific plan for a robot. His "castle clock" considered one of the earliest programmable analog computers, displayed the zodiac and solar and lunar orbits. The length of day and night could be re-programmed every day in order to account for the changing lengths of day and night throughout the year.

1235

Persian astronomer Abi Bakr of Isfahan invented a brass astrolabe with a geared calendar movement based on the design of Abū Rayhān al-Bīrūnī's mechanical calendar analog computer.

1300

Ramon Llull invented the Lullian Circle: a notional machine for calculating answers to philosophical questions via logical combinatorics. This idea was taken up by Leibniz centuries later, and is thus one of the founding elements in computing and information science.

circa 1400

Kerala school of astronomy and mathematics in South India invented the floating point number system.

circa 1400

Jamshīd al-Kāshī invented the "Plate of Conjunctions", an analog computer instrument used to determine the time of day at which planetary conjunctions will occur, and for performing linear interpolation.

circa 1400

Ahmad al-Qalqashandi gives a list of ciphers in his "Subh al-a'sha" which include both substitution and transposition, and for the first time, a cipher with multiple substitutions for each plaintext letter. He also gives an exposition on and worked example of cryptanalysis, including the use of tables of letter frequencies and sets of letters which cannot occur together in one word.

1492

Leonardo da Vinci produced drawings of a device consisting of interlocking cog wheels which can be interpreted as a mechanical calculator capable of addition and subtraction. Da Vinci also made plans for a mechanical man: an early design for a robot.

1588

Joost Buerghi discovered natural logarithms.

1614

Scotsman John Napier reinvented a form of logarithms and an ingenious system of movable rods (referred to as Napier's Rods or Napier's bones). These rods were based on lattice or gelosia multiplication algorithm and allowed the operator to multiply, divide and calculate square and cube roots by moving the rods around and placing them in specially constructed boards.

1622

William Oughtred developed slide rules based on natural logarithms as developed by John Napier.

1623

German polymath Wilhelm Schickard drew a device that he called a "Calculating Clock" on two letters that he sent to Johannes Kepler; one in 1623 and the other in 1624. A fire later destroyed the machine as it was being built in 1624 and he decided to abandon his project. This machine became known to the world only in 1957 when the two letters were discovered. Some replicas were built in 1961. This machine had no impact on the development of mechanical calculators.

1642

France

French polymath Blaise Pascal invented the mechanical calculator. Called "machine arithmétique", Pascal's calculator and eventually Pascaline, its public introduction in 1645 started the development of mechanical calculators first in Europe and then in the rest of the world. It was the first machine to have a controlled carry mechanism. Pascal built 50 prototypes before releasing his first machine (eventually twenty machines were built). The Pascaline inspired the works of Gottfried Leibniz (1671), Thomas de Colmar (1820) and Dorr E. Felt (1887).

1668

United Kingdom

Sir Samuel Morland (1625–1695), of England, produced a non-decimal adding machine, suitable for use with English money. Instead of a carry mechanism, it registered carries on auxiliary dials, from which the user re-entered them as addends.

1671

Germany

German mathematician, Gottfried Leibniz designed a machine which multiplied, the 'Stepped Reckoner'. It could multiply numbers of up to 5 and 12 digits to give a 16 digit result. Two machines were built, one in 1694 (it was discovered in an attic in 1879), and one in 1706.

1685

Germany

In an article titled ""Machina arithmetica in qua non additio tantum et subtractio sed et multiplicatio nullo, diviso vero paene nullo animi labore peragantur"", Gottfried Leibniz described a machine that used wheels with movable teeth which, when coupled to a Pascaline, could perform all four mathematical operations. There is no evidence that Leibniz ever constructed this pinwheel machine.

1709

Italy

Giovanni Poleni was the first to build a calculator that used a pinwheel design. It was made of wood and was built in the shape of a "calculating clock".

1726

United Kingdom

Jonathan Swift described (satirically) a machine ("engine") in his "Gulliver's Travels". The "engine" consisted of a wooden frame with wooden blocks containing parts of speech. When the engine's 40 levers are simultaneously turned, the machine displayed grammatical sentence fragments.

1774

Germany

Philipp Matthäus Hahn, in what is now Germany, made a successful portable calculator able to perform all four mathematical operations.

1775

United Kingdom

Charles Stanhope, 3rd Earl Stanhope, of England, designed and constructed a successful multiplying calculator similar to Leibniz's.

1786

Germany

J. H. Müller, an engineer in the Hessian army, first conceived of the idea of a difference engine.

1801
France
Joseph-Marie Jacquard developed an automatic loom controlled by punched cards.

1820
France
Charles Xavier Thomas de Colmar invented the 'Arithmometer' which after thirty more years of development became, in 1851, the first mass-produced mechanical calculator. An operator could perform long multiplications and divisions quickly and effectively by using a movable accumulator for the result. This machine was based on the earlier works of Pascal and Leibniz.

1822
United Kingdom
Charles Babbage designed his first mechanical computer, the first prototype of the decimal difference engine for tabulating polynomials.

1832
Russia
Semen Korsakov proposed the usage of punched cards for information storage and search. He designed several machines to demonstrate his ideas, including the so-called "linear homeoscope".

1832
United Kingdom
Babbage and Joseph Clement produced a prototype segment of his difference engine. The complete engine was planned to operate both on sixth-order differences with numbers of about 20 digits, and on third-order differences with numbers of 30 digits. Each addition would have been done in two phases, the second one taking care of any carries generated in the first. The output digits were to be punched into a soft metal plate, from which a printing plate might have been made. No more than this prototype piece was ever finished.

1834
United Kingdom
Babbage conceived, and began to design, his decimal 'Analytical Engine'. A program for it was to be stored on read-only memory, in the form of punched cards.

The machine envisioned would have been capable of an addition in 3 seconds and a multiplication or division in 2–4 minutes. It was to be powered by a steam engine. In the end, no more than a few parts were actually built.

1835
United States
Joseph Henry invented the electromechanical relay.

1842
France
Timoleon Maurel patented the Arithmaurel, a mechanical calculator with a very intuitive user interface, especially for multiplying and dividing numbers because the result was displayed as soon as the operands were entered. It received a gold medal at the French national show in Paris in 1849. Unfortunately its complexity and the fragility of its design prevented it from being manufactured.

1842
United Kingdom
Construction of Babbage's difference engine was cancelled as an official project. The cost overruns had been considerable (£17,470 was spent, which, in 2004 money, would be about £1,000,000).

1843
Sweden
Per Georg Scheutz and his son Edvard produced a third-order difference engine with printer; the Swedish government agrees to fund their next development.

1847
United Kingdom
Babbage designed an improved, simpler difference engine (the Difference Engine No.2), a project which took 2 years. The machine would have operated on 7th-order differences and 31-digit numbers, but nobody was found to pay to have it built. In 1989-1991 a team at London's Science Museum did build one from the surviving plans. They built components using modern methods, but with tolerances no better than Clement could have provided... and, after a bit of tinkering and detail-debugging, they found that the machine works properly. In 2000, the printer was also completed.

1848
United Kingdom
British Mathematician George Boole developed binary algebra (Boolean algebra) which has been widely used in binary computer design and operation, beginning about a century later.

1851
France
After 30 years of development, Thomas de Colmar launched the mechanical calculator industry by starting the manufacturing of a much simplified Arithmometer (invented in 1820). Aside from its clones, which started thirty years later, it was the only calculating machine available anywhere in the world for forty years (Dorr Felt|Dorr E. Felt only sold one hundred comptometers and a few comptographs from 1887 to 1890). Its simplicity made it the most reliable calculator to date. It was a big machine (a 20 digit arithmometer was long enough to occupy most of a desktop). Even though the arithmometer was only manufactured until 1915, twenty European companies manufactured improved clones of its design until the beginning of WWII ; they were Burkhardt, Layton, Saxonia, Gräber, Peerless, Mercedes-Euklid, XxX, Archimedes, etc...

1853
Sweden
To Babbage's delight, the Scheutzes completed the first full-scale difference engine, which they called a Tabulating Machine. It operated on 15-digit numbers and 4th-order differences, and produced printed output just as Babbage's would have. A second machine was later built to the same design by the firm of Bryan Donkin of London.

1858
United States
The first Tabulating Machine (see 1853) was bought by the Dudley Observatory in Albany, New York, and the second by the British government. The Albany machine was used to produce a set of astronomical tables; but the Observatory's director was fired for this extravagant purchase, and the machine never seriously used again, eventually ending up in a museum. The second machine had a long and useful life.

1869
United Kingdom
The first practical logic machine was built by William Stanley Jevons.

1871
United Kingdom
Babbage produced a prototype section of the Analytical Engine's mill and printer.

1875
Sweden
Martin Wiberg produced a reworked difference-engine-like machine intended to prepare logarithmic tables.

1878
Spain
Ramon Verea, living in New York City, invented a calculator with an internal multiplication table; this was much faster than the shifting carriage, or other digital methods of the time. He wasn't interested in putting it into production, however; it seems he just wanted to show that a Spaniard could invent as well as an American.

1879
United Kingdom
A committee investigated the feasibility of completing the Analytical Engine, and concluded that it would be impossible now that Babbage was dead. The project was then largely forgotten, except by a very few; Howard Aiken was a notable exception.

1884
United States
Dorr Felt, of Chicago, developed his Comptometer. This was the first calculator in which operands are entered by pressing keys rather than having to be, for example, dialled in. It was feasible because of Felt's invention of a carry mechanism fast enough to act while the keys return from being pressed. Felt and Tarrant started a partnership to manufacture the comptometer in 1887.

1885
United States, Sweden, Russia

A multiplying calculator more compact than the Arithmometer entered mass production. The design was the independent, and more or less simultaneous, invention of Frank S. Baldwin, of the United States, and Willgodt Theophil Odhner, a Swede living in Russia. Fluted drums were replaced by a "variable-toothed gear" design: a disk with radial pegs that could be made to protrude or retract from it.

1886
United States

Herman Hollerith developed the first version of his tabulating system in the Baltimore Department of Health.

1889
United States

Dorr Felt invented the first printing desk calculator.

1890
United States

The 1880 US census had taken 7 years to complete since all processing had been done by hand from journal sheets. The increasing population suggested that by the 1890 census, data processing would take longer than the 10 years before the next census—so a competition was held to find a better method. It was won by a Census Department employee, Herman Hollerith, who went on to found the Tabulating Machine Company, later to become IBM. He invented the recording of data on a medium that could then be read by a machine. Prior uses of machine readable media had been for control (Automatons, Piano rolls, looms, ...), not data. "After some initial trials with paper tape, he settled on punched cards..." His machines used mechanical relays (and solenoids) to increment mechanical counters. This method was used in the 1890 census and the completed results (62,622,250 people) were "... finished months ahead of schedule and far under budget". The inspiration for this invention was Hollerith's observation of railroad conductors during a trip in the western US; they encoded a crude description of the passenger (tall, bald, male) in the way they punched the ticket.

1892
United States

William S. Burroughs of St. Louis, invented a machine similar to Felt's (see 1884) in 1885 but unlike the comptometer it was a 'key-set' machine which only processed each number after a crank handle was pulled. The true manufacturing of this machine started in 1892 even though Burroughs had started his" American Arithmometer Company" in 1886 (it later became Burroughs Corporation and is now called Unisys).

1896
United States

Herman Hollerith introduced an Integrating Tabulator that could add numbers encoded on punched cards to one of several 7-digit counters. His earlier tabulators simply incremented counters based on whether a hole was punched or not.

1901
United States

The Standard Adding Machine Company released the first 10-key adding machine in between 1901 and 1903. The inventor, William Hopkins, filed his first patent on October 4, 1892. The 10 keys were set on a single row.

1902
United States

Remington advertised the Dalton adding machine as the first 10-key printing adding machine. The 10 keys were set on two rows. Six machines had been manufactured by the end of 1906

1906
United Kingdom

Henry Babbage, Charles's son, with the help of the firm of R. W. Munro, completed the 'mill' from his father's Analytical Engine, to show that it would have worked. It does. The complete machine was not produced.

1906
United States

Vacuum tube (or thermionic valve) invented by Lee De Forest.

1906
United States
Herman Hollerith introduces a tabulator with a plug-board that can be rewired to adapt the machine for different applications. Plugboards were widely used to direct machine calculations until displaced by stored programs in the 1950s.

1919
United Kingdom
William Henry Eccles and F. W. Jordan published the first flip-flop circuit design.

1924
Germany
Walther Bothe built an "'AND'" logic gate - the "co-incidence circuit", for use in physics experiments, for which he received the Nobel Prize in Physics 1954. Digital circuitries of all kinds make heavy use of this technique.

1926
United States
Westinghouse AC Calculating board. A Network analyzer (AC power) used for electrical transmission line simulations up until the 1960s.

1928
United States
IBM standardizes on punched cards with 80 columns of data and rectangular holes. Widely known as IBM Cards, they dominate the data processing industry for almost half a century.

1930
United States
Vannevar Bush built a partly electronic difference engine capable of solving differential equations.

1930
United Kingdom
Welsh physicist C. E. Wynn-Williams<!— (1903-1979) —>, at Cambridge, England, used a ring of thyratron tubes to construct a binary digital counter that counted emitted Alpha particles.

1931
Austria
Kurt Gödel of Vienna University, Austria, published a paper on a universal formal language based on arithmetic operations. He used it to encode arbitrary formal statements and proofs, and showed that formal systems such as traditional mathematics are either inconsistent in a certain sense, or contain unprovable but true statements. This result is often called the fundamental result of theoretical computer science.

1931
United States
IBM introduced the IBM 601 Multiplying Punch, an electromechanical machine that could read two numbers, up to 8 digits long, from a card and punch their product onto the same card.

1934
United States
Wallace Eckert of Columbia University connects an IBM 285 Tabulator, an 016 Duplicating Punch and an IBM 601 Multiplying Punch with a cam-controlled sequencer switch that he designed. The combined system was used to automate the integration of differential equations.

1936
United Kingdom
Alan Turing of Cambridge University, England, published a paper on 'computable numbers' which reformulated Kurt Gödel's results (see related work by Alonzo Church). His paper addressed the famous 'Entscheidungsproblem' whose solution was sought in the paper by reasoning (as a mathematical device) about a simple and theoretical computer, known today as a Turing machine. In many ways, this device was more convenient than Gödel's arithmetics-based universal formal system.

1937
United States
George Stibitz of the Bell Telephone Laboratories (Bell Labs), New York City, constructed a demonstration 1-bit binary adder using relays. This was one of the first binary computers, although at this

stage it was only a demonstration machine; improvements continued leading to the "Complex Number Calculator" of January 1940.

1937
United States
Claude E. Shannon published a paper on the implementation of symbolic logic using relays as his MIT Master's thesis.

1938
Nazi Germany
Konrad Zuse of Berlin, completed the 'Z1', the first mechanical binary programmable computer. It was based on Boolean Algebra and had some of the basic ingredients of modern machines, using the binary system and floating-point arithmetic. Zuse's 1936 patent application (Z23139/GMD Nr. 005/021) also suggested a 'von Neumann' architecture (re-invented about 1945) with program and data modifiable in storage. Originally the machine was called the 'V1' but retroactively renamed after the war, to avoid confusion with the V-1 flying bomb. It worked with floating point numbers (7-bit exponent, 16-bit mantissa, and sign bit). The memory used sliding metal parts to store 16 such numbers, and worked well; but the arithmetic unit was less successful, occasionally suffering from certain mechanical engineering problems. The program was read from holes punched in discarded 35 mm movie film. Data values could have been entered from a numeric keyboard, and outputs were displayed on electric lamps. The machine was not a general purpose computer (i.e., Turing complete) because it lacked loop capabilities.

1939
United States
William Hewlett and David Packard established the Hewlett-Packard Company in Packard's garage in Palo Alto, California with an initial investment of $538; this was considered to be the symbolic founding of Silicon Valley. HP would grow to become one of the largest technology companies in the world today.

1939, Nov
United States
John Vincent Atanasoff and graduate student Clifford Berry of Iowa State College (now the Iowa State University), Ames, Iowa, completed a prototype 16-bit adder. This was the first machine to calculate using vacuum tubes.

1939
Nazi Germany
Konrad Zuse completed the 'Z2' (originally 'V2'), which combined the Z1's existing mechanical memory unit with a new arithmetic unit using relay logic. Like the Z1, the Z2 lacked loop capabilities. The project was interrupted for a year when Zuse was drafted, but continued after he was released.

1939
Nazi Germany
Helmut Schreyer completed a prototype 10-bit adder using vacuum tubes, and a prototype memory using neon lamps.

1940, Jan
United States
At Bell Labs, Samuel Williams and George Stibitz completed a calculator which could operate on complex numbers, and named it the 'Complex Number Calculator'; it was later known as the 'Model I Relay Calculator'. It used telephone switching parts for logic: 450 relays and 10 crossbar switches. Numbers were represented in 'plus 3 BCED'; that is, for each decimal digit, 0 is represented by binary 0011, 1 by 0100, and so on up to 1100 for 9; this scheme requires fewer relays than straight BCED. Rather than requiring users to come to the machine to use it, the calculator was provided with three remote keyboards, at various places in the building, in the form of teletypes. Only one could be used at a time, and the output was automatically displayed on the same one. On 9 September 1940, a teletype was set up at a Dartmouth College in Hanover, New Hampshire, with a connection to New York, and those attending the conference could use the machine remotely.

1940, Apr 1
GER|Nazi
In 1940 Zuse presented the Z2 to an audience of the {{lang|de|"Deutsche Versuchsanstalt für Luftfahrt"}} ("German Laboratory for Aviation") in Berlin-Adlershof.
|}

==1941–1949==
{| class="wikitable sortable"
|-
! Date
! class="unsortable" | Place
! class="unsortable" | Event

1941, May 11
GER|Nazi

Now working with limited backing from the DVL (German Aeronautical Research Institute), Konrad Zuse completed the "Z3" (originally 'V3'): the first operational programmable computer. One major improvement over Charles Babbage's non-functional device is the use of Leibniz's binary system (Babbage and others unsuccessfully tried to build decimal programmable computers). Zuse's machine also featured floating point numbers with a 7-bit exponent, 14-bit mantissa (with a '1' bit automatically prefixed unless the number is 0), and a sign bit. The memory held 64 of these words and therefore required over 1400 relays; there were 1200 more in the arithmetic and control units. It also featured parallel adders. The program, input, and output were implemented as described above for the Z1. Although conditional jumps were not available, it has been shown that Zuse's Z3 is, in principle, capable of functioning as a universal computer. The machine could do 3-4 additions per second, and took 3–5 seconds for a multiplication. The Z3 was destroyed in 1943 during an Allied bombardment of Berlin, and had no impact on computer technology in America and England.

1942, Summer
United States

Atanasoff and Berry completed a special-purpose calculator for solving systems of simultaneous linear equations, later called the 'ABCE' ('Atanasoff–Berry Computer'). This had 60 50-bit words of memory in the form of capacitors (with refresh circuits—the first regenerative memory) mounted on two revolving drums. The clock speed was 60 Hz, and an addition took 1 second. For secondary memory it used punched cards, moved around by the user. The holes were not actually punched in the cards, but burned. The punched card system's error rate was never reduced beyond 0.001%, and this was inadequate.

Atanasoff left Iowa State after the U.S. entered the war, ending his work on digital computing machines.

1942
GER|Nazi

Helmut Hölzer built an analog computer to calculate and simulate V-2 rocket trajectories.

1942
GER|Nazi

Konrad Zuse developed the S1, the world's first process computer, used by Henschel to measure the surface of wings.

1943, Apr
United Kingdom

Max Newman, Wynn-Williams and their team at the secret Government Code and Cypher School ('Station X'), Bletchley Park, Bletchley, England, completed the 'Heath Robinson'. This was a specialized counting machine used for cipher-breaking, not a general-purpose calculator or computer, but a logic device using a combination of electronics and relay logic. It read data optically at 2000 characters per second from 2 closed loops of paper tape, each typically about 1000 characters long. It was significant since it was the forerunner of Colossus. Newman knew Turing from Cambridge (Turing was a student of Newman's), and had been the first person to see a draft of Turing's 1936 paper. Heath Robinson is the name of a British cartoonist known for drawings of comical machines, like the American Rube Goldberg. Two later machines in the series will be named after London stores with 'Robinson' in their names.

1943, Sep
United States

Williams and Stibitz completed the 'Relay Interpolator', later called the 'Model II Relay Calculator'. This was a programmable calculator; again, the program and data were read from paper tapes. An innovative feature was that, for greater reliability, numbers were represented in a biquinary format using 7 relays for each digit, of which exactly 2 should be "on": 01 00001 for 0, 01 00010 for 1, and so on up to 10 10000 for 9. Some of the later machines in this series would use the biquinary notation for the digits of floating-point numbers.

1943, Dec
United Kingdom

The Colossus was built, by Dr Thomas Flowers at The Post Office Research Laboratories in London, to crack the German Lorenz (SZ42) cipher. It contained 2400 vacuum tubes for logic and applied a programmable logical function to a stream of input characters, read from punched tape at a rate of 5000 characters a second. Colossus was used at Bletchley Park during World War II—as a successor to the unreliable Heath Robinson machines. Although 10 were eventually built, most were destroyed immediately after they had finished their work to maintain the secrecy of the work.

1944, Aug 7
United States

The IBM Automatic Sequence Controlled Calculator was turned over to Harvard University, which called it the Harvard Mark I. It was designed by Howard Aiken and his team, financed and built by IBM—it became the second program controlled machine (after Konrad Zuse's). The whole machine was {{convert|51|ft|m}} long, weighed 5 (short) tons (4.5 tonnes), and incorporated 750,000 parts. It used 3304 electromechanical relays as on-off switches, had 72 accumulators (each with its own arithmetic unit), as well as a mechanical register with a capacity of 23 digits plus sign. The arithmetic was fixed-point and decimal, with a control panel setting determining the number of decimal places. Input-output facilities include card readers, a card punch, paper tape readers, and typewriters. There were 60 sets of rotary switches, each of which could be used as a constant register—sort of mechanical read-only memory. The program was read from one paper tape; data could be read from the other tapes, or the card readers, or from the constant registers. Conditional jumps were not available. However, in later years, the machine was modified to support multiple paper tape readers for the program, with the transfer from one to another being conditional, rather like a conditional subroutine call. Another addition allowed the provision of plug-board wired subroutines callable from the tape. Used to create ballistics tables for the US Navy.

1945
GER|Nazi

Konrad Zuse developed Plankalkül, the first higher-level programming language. He also presented the Z4 in March.

1945
United States

Vannevar Bush developed the theory of the memex, a hypertext device linked to a library of books and films.

1945
United States

John von Neumann drafted a report describing the future computer eventually built as the EDVAC (Electronic Discrete Variable Automatic Computer). "First Draft of a Report on the EDVAC" includes the first published description of the design of a stored-program computer, giving rise to the term von Neumann architecture. It directly or indirectly influenced nearly all subsequent projects, especially EDSAC. The design team included John W. Mauchly and J. Presper Eckert.

1946, Feb 14
United States

ENIAC (Electronic Numerical Integrator and Computer): One of the first totally electronic, valve driven, digital, program-controlled computers was unveiled although it was shut down on 9 November 1946 for a refurbishment and a memory upgrade, and was transferred to Aberdeen Proving Ground, Maryland in 1947. Development had started in 1943 at the Ballistic Research Laboratory, USA, by John W. Mauchly and J. Presper Eckert. It weighed 30 tonnes and contained 18,000 electronic valves, consuming around 160 kW of electrical power. It could do 50,000 basic calculations a second. It was used for calculating ballistic trajectories and testing theories behind the hydrogen bomb.

1946, Feb 19
United Kingdom

ACE (Automatic Computing Engine): Alan Turing presented a detailed paper to the National Physical Laboratory (NPL) Executive Committee, giving the first reasonably complete design of a stored-program computer. However, because of the strict and long-lasting secrecy around his wartime work at Bletchley Park, he was prohibited (having signed the Official Secrets Act) from explaining that he knew that his ideas could be implemented in an electronic device.

1946
United Kingdom

The trackball was invented as part of a radar plotting system named Comprehensive Display System (CDS) by Ralph Benjamin when working for the British Royal Navy Scientific Service. Benjamin's project used analog computers to calculate the future position of target aircraft based on several initial input points provided by a user with a joystick. Benjamin felt that a more elegant input device was needed and invented a "ball tracker" system called the "roller ball" for this purpose in 1946. The device was patented in 1947 but only a prototype was ever built and the device was kept as a secret outside military.

1947, Dec 16
United States

Invention of the transistor at Bell Laboratories, USA, by William B. Shockley, John Bardeen and Walter Brattain.

1947
United States

Howard Aiken completed the Harvard Mark II.

1947
United States

The Association for Computing Machinery (ACM), was founded as the world's first scientific and educational computing society. It remains to this day with a membership currently around 78,000. Its headquarters are in New York City.

1948, Jan 27
United States

IBM finished the SSEC (Selective Sequence Electronic Calculator). It was the first computer to modify a stored program. "About 1300 vacuum tubes were used to construct the arithmetic unit and eight very high-speed registers, while 23000 relays were

used in the control structure and 150 registers of slower memory."

1948, Jul 21
United Kingdom

SSEM, Small-Scale Experimental Machine or 'Baby' was built at the University of Manchester. It ran its first program on this date. It was the first computer to store both its programs and data in RAM, as modern computers do. By 1949 the 'Baby' had grown, and acquired a magnetic drum for more permanent storage, and it became the Manchester Mark 1.

1948
United States

ANACOM from Westinghouse was an AC-energized electrical analog computer system used up until the early 1990s for problems in mechanical and structural design, fluidics, and various transient problems.

1948
United States

IBM introduced the '604', the first machine to feature Field Replaceable Units (FRUs), which cut downtime as entire pluggable units can simply be replaced instead of troubleshot.

1948

The first Curta handheld mechanical calculator was sold. The Curta computed with 11 digits of decimal precision on input operands up to 8 decimal digits. The Curta was about the size of a handheld pepper grinder.

1949, Mar
United States

John Presper Eckert and John William Mauchly construct the BINAC for Northrop Corporation|Northrop.

1949, May 6
United Kingdom

This is considered the birthday of modern computing. Maurice Wilkes and a team at Cambridge University executed the first stored program on the EDSAC computer, which used paper tape input-output. Based on ideas from John von Neumann about stored program computers, the EDSAC was the first complete, fully functional von Neumann architecture computer.

1949, Oct
United Kingdom
The Manchester Mark 1 final specification is completed; this machine was notably in being the first computer to use the equivalent of base/index registers, a feature not entering common computer architecture until the second generation around 1955.

1949
Australia
CSIR Mk I (later known as CSIRAC), Australia's first computer, ran its first test program. It was a vacuum tube based electronic general purpose computer. Its main memory stored data as a series of acoustic pulses in 5-foot-long tubes filled with mercury.

1949
United Kingdom
MONIAC (Monetary National Income Analogue Computer) also known as the Phillips Hydraulic Computer, was created in 1949 to model the national economic processes of the United Kingdom. The MONIAC consisted of a series of transparent plastic tanks and pipes. It is thought that twelve to fourteen machines were built.

1949
United States
"Computers in the future may weigh no more than 1.5 tons." "Popular Mechanics", forecasting the relentless march of science.

TIMELINE OF MICROPROCESSORS

Year	Microprocessors
1971	Intel 4004
1972	Fairchild PPS-25; Intel 8008; Rockwell PPS-4
1973	Burroughs Mini-D; National IMP-16; NEC μCOM
1974	General Instrument CP1600; Intel 4040, 8080; Mostek 5065; Motorola 6800; National IMP-4, IMP-8, ISP-8A/500, PACE; Texas Instruments TMS 1000; Toshiba TLCS-12
1975	Fairchild F-8; Hewlett Packard BPC; Intersil 6100; MOS Technology 6502; RCA CDP 1801; Rockwell PPS-8; Signetics 2650
1976	RCA CDP 1802; Signetics 8x300; Texas Instruments TMS9900; Zilog Z-80
1977	Intel 8085
1978	Intel 8086; Motorola 6801, 6809
1979	Intel 8088; Motorola 68000; Zilog Z8000
1980	National Semi 16032; Intel 8087
1981	DEC T-11; Harris 6120; IBM ROMP
1982	Hewlett Packard FOCUS; Intel 80186, 80188,; 80286; Berkeley RISC-I
1983	Stanford MIPS; UC Berkeley RISC-II
1984	Motorola 68020; National Semi 32032; NEC V20
1985	DEC MicroVax II; Harris Novix; Intel 80386; MIPS R2000
1986	NEC V60; Sun SPARC; Zilog Z80000
1987	Acorn ARM2; DEC CVAX 78034; Hitachi Gmicro/200; Motorola 68030; NEC V70
1988	Intel 80386SX, i960; MIPS R3000
1989	DEC VAX DC520 Rigel; Intel 80486, i860
1990	IBM POWER1; Motorola 68040
1991	DEC NVAX; IBM RSC; MIPS R4000
1992	DEC Alpha 21064; Hewlett Packard PA-7100; Sun microSPARC I
1993	IBM POWER2, PowerPC 601; Intel Pentium

1994	DEC Alpha 21064A; Hewlett Packard PA-7100LC, PA-7200; IBM PowerPC 603, PowerPC 604; Motorola 68060; QED R4600
1995	DEC Alpha 21164; HAL Computer SPARC64; Intel Pentium Pro; Sun UltraSPARC
1996	AMD K5; DEC Alpha 21164A; HAL Computer SPARC64 II; Hewlett Packard PA-8000; IBM P2SC; MTI R10000; QED R5000
1997	AMD K6; IBM PowerPC 620, PowerPC 750,; RS64, ES/390 G4; Intel Pentium II; Sun UltraSPARC IIs
1998	DEC Alpha 21264; HAL Computer SPARC64 III; Hewlett Packard PA-8500; IBM POWER3, RS64-II; ES/390 G5; QED RM7000; SGI MIPS R12000
1999	AMD Athlon; IBM RS64-III; Intel Pentium III; Motorola PowerPC 7400
2000	AMD Athlon XP; Duron; Fujitsu SPARC64 IV; IBM RS64-IV; z900; Intel Pentium 4
2001	IBM POWER4; Intel Itanium; Motorola PowerPC 7450; SGI MIPS R14000; Sun UltraSPARC III
2002	Fujitsu SPARC64 V; Intel Itanium 2
2003	AMD Opteron; IBM PowerPC 970; Intel Pentium M
2004	IBM POWER5; PowerPC BGL
2005	AMD Athlon 64 X2; Opteron Athens; IBM PowerPC 970MP; Xenon; Intel Pentium D; Sun UltraSPARC IV; UltraSPARC T1
2006	IBM Cell/B.E.; Intel Core 2; Core Duo; Itanium Montecito
2007	AMD Opteron Barcelona; Fujitsu SPARC64 VI; IBM POWER, PowerPC BGP; Sun UltraSPARC T2; Tilera TILE64
2008	AMD Opteron Shanghai, Phenom; Fujitsu SPARC64 VII; IBM PowerXCell 8i; IBM z10; Intel Atom, Core i7; Tilera TILEPro64
2009	AMD Opteron Istanbul, Phenom II
2010	AMD Opteron Magny-cours; Fujitsu SPARC64 VII+; IBM POWER7; z196; Intel Itanium Tukwila, Westmere; Xeon, Nehalem-EX; Sun SPARC T3
2011	AMD FX Bulldozer, Interlagos, Llano; Fujitsu SPARC64 VIIIfx; Freescale PowerPC e6500; Intel Sandy Bridge, Xeon E7; Oracle SPARC T4
2012	Fujitsu SPARC64 IXfx; IBM POWER7+, zEC12; Intel Itanium Poulson
2013	Fujitsu SPARC64 X; Intel Haswell; Oracle SPARC T5
2014	IBM POWER8

THE PIONEERS OF COMPUTER SCIENCE

al-Khwārizmī	The term "algorithm" is derived from the algorism, the technique of performing arithmetic with Hindu-Arabic numerals developed by al-Khwarizmi. Both "algorithm" and "algorism" are derived from the Latinized forms of al-Khwarizmi's name, Algoritmi and Algorismi, respectively.	c. 780 – c. 850
John Atanasoff	Built the first electronic digital computer, the Atanasoff–Berry Computer, though it was neither programmable nor Turing-complete (1939).	b. 1903 d. 1995
Charles Babbage	Originated the concept of a programmable general-purpose computer. Designed the Analytical Engine and built a prototype for a less powerful mechanical calculator.	1822 1837
John Warner Backus	Invented FORTRAN ("For"mula "Tran"slation), the first practical high-level programming language (1954), and he formulated the Backus–Naur form that described the formal language syntax (1963).	b. 1924 d. 2007
Jean Bartik	One of the first computer programmers, on ENIAC (1946). Worked with John Mauchly toward BINAC (1949), EDVAC (1949), UNIVAC (1951) to develop early "stored program" computers.	b. 1924 d. 2011
Sir Timothy John Berners-Lee	Invented worldwide web. With Robert Cailliau, sent first HTTP communication between client and server (1989).	b. 1955
George Boole	Formalized Boolean algebra (1854) in *The Laws of Thought*, the basis for digital logic and computer science.	b. 1815 d. 1864
Per Brinch Hansen	lDeveloped the RC 4000 multiprogramming system introducing the concept of an operating system kernel and the separation of policy and mechanism (1967). Co-developed the monitor with Tony Hoare (1973), and created the first monitor implementation (1975). Implemented the first form of remote procedure call in the RC 4000 (1978)	l1969
Nikolay Brusentsov	Built ternary computer Setun (1958).	b. 1925 d. 2014
Vannevar Bush	Originator of the Memex concept (1930), which led to the development of Hypertext.	b. 1890 d. 1974
David Caminer	Developed the LEO computer the first business computer with John Pinkerton, for J. Lyons and Co. (1951)	b. 1915 d. 2008

Vinton Gray Cerf	With Bob Kahn, designed the Transmission Control Protocol and Internet Protocol (TCP/IP), the primary data communication protocols of the Internet and other computer networks (1978).	b. 1943
Avrom Noam Chomsky	Responsible for Chomsky hierarchy (1956), a discovery which has directly impacted Programming language theory and other branches of computer science.	b. 1928
Alonzo Church	Contributions to theoretical computer science, specifically for the development of the lambda calculus and the discovery of the undecidability problem within it (1936).	B, 1903 d. 1995
Wesley Allison. Clark	Designed LINC, the first functional computer scaled down and priced for the individual user (1963) with many of its features prototypes of what became essential elements of personal computers.	b. 1927 d. 2016
Edmund Melson Clarke, Jr.	Developed model checking and formal verification of software and hardware with E. Allen Emerson (1981).	b. 1945
Edgar Frank "Ted" Codd	Proposed and formalized the relational model of data management, the theoretical basis of relational databases (1970).	b. 1923 d. 2003
Stephen Cook	Formalized the notion of NP-completeness, inspiring a great deal of research in computational complexity theory (1971).	b. 1939
James William Cooley	Created the Fast Fourier Transform (FFT) with John W. Tukey (1965).	b. 1926 d. 2016
Ole-Johan Dahl	With Kristen Nygaard, invented the proto-object oriented language SIMULA (1962).	b. 1931 d. 2002
Edsger Wybe Dijkstra	Made advances in algorithms (1965), the semaphore, rigor, and pedagogy (1968).	b. 1930 d. 2002
John Adam Presper "Pres" Eckert, Jr.	Designed and built the ENIAC (1943) with John Mauchly, the first all electronic, Turing-complete) computer, and the UNIVAC I, the first commercially available computer (1951).	b. 1919 d. 1995
Enest Allen Emerson II	Developed model checking and formal verification of software and hardware together with Edmund M. Clarke (1981).	b. 1954
Douglas Carl Engelbart	Invented the computer mouse with Bill English (1963); a pioneer of human-computer interaction whose Augment team developed hypertext, networked computers, and precursors to GUIs (1968).	b. 1925 d. 2013

Thomas "Tommy" Harold Flowers	Designed and built the Mark 1 (1943) and the improved Mark 2 (1944) Colossus computers, the world's first programmable, digital, electronic, computing devices.	1943 b. 1905 d. 1998
Friederich Ludwig Gottlob Frege	Developed first-order predicate calculus, a crucial precursor requirement to developing computation theory (1879).	b. 1848 d. 1935
Seymour Ginsburg	Proved "don't-care" circuit minimization does not necessarily yield optimal results (1958); ALGOL programming language is context-free (thus linking formal language theory to the problem of compiler writing. 1966); invented AFL Theory (1967).	b. 1927 d. 2004
Kurt Friedrich Gödel	Proved Peano axiomatized arithmetic could not be both logically consistent and complete in first-order predicate calculus (1931). Church, Kleene, and Turing developed the foundations of computation theory based on corollaries to Gödel's work.	b. 1906 d. 1978
Lois Haibt	Was a member of the ten person team that invented Fortran (1954) and among the first women to play a crucial role in the development of computer science.	b. 1934
Margaret Heafield Hamilton	Credited with coining the phrase "Software engineering"; developed the concepts of asynchronous software, priority scheduling, end-to-end testing, and human-in-the-loop decision capability which became the foundation for ultra reliable software design (1965).	b. 1936
Sir Charles Anthony Richard "C.A.R." Hoare	Developed the formal language Communicating Sequential Processes (CSP) and Quicksort (1960).	b. 1934
Herman Hollerith	Widely regarded as the father of modern machine data processing. Invented punched card evaluating machine (1889) which began the era of automatic data processing systems.	b. 1860 d. 1929
Grace Brewster Murray Hopper	Pioneered work on the necessity for high-level programming languages, which she termed "automatic programming"; her A-O compiler (written in 1952) influenced the COBOL language.	b. 1906 d. 1992
Cuthbert Corwin Hurd	Helped the International Business Machines Corporation develop its first general-purpose computer, the IBM 701 (1952).	b. 1911 d. 1996
Kenneth Eugene Iverson	Assisted in establishing and taught the first graduate course in computer science at Harvard (1955); invented the APL programming language (1962); made contributions to interactive computing.	b. 1920 d. 2004

Joseph Marie Jacquard	Built and demonstrated the Jacquard loom, a programmable mechanized loom controlled by punch cards (1801).	b. 1752 d. 1834
Maurice Karnaugh	Inventor of the Karnaugh map, used for logic function minimization (1953).	b. 1924
Jacek Karpiński	Developed the first differential analyzer that used transistors; one of the first machine learning algorithms for character and image recognition; inventor of one of the first minicomputers, the K-202 (1973)	b. 1927 d. 2010
Alan Curtis Kay	Pioneered ideas at the root of object-oriented programming languages (1966), led the team that developed Smalltalk (1976), and made fundamental contributions to personal computing.	b. 1940
Stephen Cole Kleene	Pioneered work with Alonzo Church on the Lambda Calculus that first laid down the foundations of computation theory (1935).	b. 1909 d. 1994
Donald Ervin Knuth	Wrote *The Art of Computer Programming* (1968) and created TeX (1978).	1968 b. 1938
Leslie B. Lamport	Formulated algorithms to solve fundamental problems in distributed systems (e.g. the bakery algorithm. 1974); logical clock (1978), enabling synchronization between distributed entities based on the events through which they communicate; created LaTeX (1985)	b. 1941
	Developed the concept of a.	1978
Sergei Alexeyevich Lebedev	Independently designed the first electronic computer in the Soviet Union, MESM, in Kiev, Ukraine (1951).	b. 1902 d. 1974
Gottfried Wilhelm Leibniz	Advances in symbolic logic, such as the Calculus ratiocinator, heavily influential on Gottlob Frege. Made developments in first-order predicate calculus, crucial for the theoretical foundations of computer science. (1670s)	b. 1646 d. 1716
Joseph Carl Robnett "J. C. R." Licklider	Began the investigation of human-computer interaction (1960), leading to many advances in computer interfaces as well as in cybernetics and artificial intelligence.	b. 1915 d. 1990
Ramon Llull	Pioneer of computational theory whose notions of symbolic representation and manipulation to produce knowledge were major influences on Leibniz.	b. c.1232 d. c. 1315-1316

Ada Lovelace (Augusta Ada King-Noel, Countess of Lovelace (*née* Byron)	Began the study of scientific computation, analyzing Babbage's work in her "Sketch of the Analytical Engine" (1842). Namesake for the Ada programming language.	b. 1815 d. 1852
John William Mauchly		
	Designed and built the ENIAC (1946) with J. Presper Eckert (all electronic, Turing-complete) computer. Also worked on BINAC (1949), EDVAC (1949), UNIVAC (1951) with Grace Hopper and Jean Bartik, to develop early "Stored program" computers.	b. 1907 d. 1980
John McCarthy	Invented LISP, a functional programming language (1958).	b. 1927 d. 2011
Marvin Lee Minsky	Co-founder of Artificial Intelligence Lab at Massachusetts Institute of Technology (1963), author of several texts on AI and philosophy.	b. 1927 d. 2016
Peter Naur	Edited the ALGOL 60 Revised Report, introducing Backus-Naur form (1960).	b. 1928 d. 2016
Maxwell Herman Alexander "Max" Neumann	Instigated the production of the Colossus computers at Bletchley Park (1941). Established the Computing Machine Laboratory at the University of Manchester (1946) where the Manchester Small-Scale Experimental Machine was invented (1948).	b. 1897 d. 1984
John von Neumann	Formulated the von Neumann architecture upon which most modern computers are based (1945).	b. 1903 d. 1957
Kristen Nygaard	With Ole-Johan Dahl, invented the proto-object oriented language SIMULA (1962).	b. 1926 d. 2002
Blaise Pascal	Invented the mechanical calculator (1642).	b. 1623 d. 1662
Emil Leon Post	Developed the Post machine (1936). Known also for developing truth tables, the Post correspondence problem (1946) and Post's theorem.	1936 b. 1897 d. 1954
Dennis MacAlistair Ritchie	Created the C programming language with Ken Thompson and the Unix computer operating system at Bell Labs (1970s).	b. 1941 d. 2011
Saul Rosen	Designed the software of the first transistor-based computer (1957). Also influenced the ALGOL programming language (1958–1960).	b, 1922 d., 1991
Bertrand Arthur Russell	Worked on Mathematical logic (example: Truth function). Introduced the notion of Type theory. Introduced Type system with Alfred North Whitehead in *Principia Mathematica* (1910).	b. 1872 d. 1979

Gerard Salton	Pioneer of automatic information retrieval (1960s); proposed the vector space model and the inverted index (1975).	b. 1927 d. 1995
Claude Elwood Shannon	Founded practical digital circuit design (1937) and information theory (1948).	b. 1916 d. 2001
Herbert Alexander Simon	A political scientist and economist who pioneered artificial intelligence. Co-creator of the Logic Theory Machine with Allen Newell (1956) and the General Problem Solver with J. C. Shaw and Allen Newell (1959).	1956 b. 1916 d. 2001
Ivan Edward Sutherland	Author of Sketchpad (1963), precursor to modern computer-aided drafting (CAD) programs and an early example of object-oriented programming.	b. 1938
John Wilder Tukey	Created the Fast Fourier Transform (FFT algorithm), with James Cooley (1965).	b. 1915 d. 2000
Alan Mathison Turing	Made several founding contributions to computer science: the Turing machine and the high-speed ACE design (1936). Widely considered as the father of Computer Science and Artificial Intelligence.	b. 1912 d. 1954
Willis Howard Ware	Co-designer of JOHNNIAC (1955). Chaired committee that developed the Code of Fair Information Practice (1960s) and led to the Privacy Act of 1974. Vice-chair of the Privacy Protection Study Commission (1974).	b. 1920 d. 2013
Adriaan van Wijngaarden	Developer of the W-grammar first used in the definition of ALGOL 68 (1968)	b. 1916 d. 1987
Sir Maurice Vincent Wilkes	Built the first practical stored program computer (EDSAC) to be completed (1949) and for being credited with the ideas of several high-level programming language constructs.	b. 1913 d. 2010
Sophie Wilson	Wrote BBC Basic programming language (1981). Also designed the ARM architecture (1985) and Firepath (2001) processsors	
Niklaus Wirth	Designed the Pascal (1970), Modula-2 (1978) and Oberon (1986) programming languages.	b. 1934
Konrad Zuse	Built the first digital freely programmable computer, the Z1 (1938). Built the first functional tape-stored program-controlled computer, the Z3 (1941), proven to be Turing-complete in 1998. Produced the world's first commercial computer, the Z4. Designed the first high-level programming language, Plankalkül (1943–1945).	b. 1910 d. 1995

GLOSSARY

1-bit watermarking: a type of digital watermark that embeds one bit of binary data in the signal to be transmitted; also called "0-bit watermarking."

3-D rendering: the process of creating a 2-D animation using 3-D models.

3D Touch: a feature that senses the pressure with which users exert upon Apple touch screens.

abstraction: a technique used to reduce the structural complexity of programs, making them easier to create, understand, maintain, and use.

accelerometer IC: an integrated circuit that measures acceleration.

access level: in a computer security system, a designation assigned to a user or group of users that allows access a predetermined set of files or functions.

actuator: a motor designed to control the movement of a device or machine by transforming potential energy into kinetic energy.

Additive White Gaussian Noise (AWGN): a model used to represent imperfections in real communication channels.

address space: the amount of memory allocated for a file or process on a computer.

adware: software that generates advertisements to present to a computer user.

affinity chromatography: a technique for separating a particular biochemical substance from a mixture based on its specific interaction with another substance.

agile software development: an approach to software development that addresses changing requirements as they arise throughout the process, with programming, implementation, and testing occurring simultaneously.

algorithm: a set of step-by-step instructions for performing computations.

American National Standards Institute (ANSI): a nonprofit organization that oversees the creation and use of standards and certifications such as those offered by CompTIA.

amplifier: a device that strengthens the power, voltage, or current of a signal.

analog signal: a continuous signal whose values or quantities vary over time.

analytic combinatorics: a method for creating precise quantitative predictions about large sets of objects.

Android Open Source Project: a project undertaken by a coalition of mobile phone manufacturers and other interested parties, under the leadership of Google. The purpose of the project is to develop the Android platform for mobile devices.

animation variables (avars): defined variables used in computer animation to control the movement of an animated figure or object.

anthropomorphic: resembling a human in shape or behavior; from the Greek words *anthropos* (human) and *morphe* (form).

app: an abbreviation for "application," a program designed to perform a particular task on a computer or mobile device.

application program interface (API): the code that defines how two pieces of software interact, particularly a software application and the operating system on which it runs.

application suite: a set of programs designed to work closely together, such as an office suite that includes a word processor, spreadsheet, presentation creator, and database application.

application-level firewalls: firewalls that serve as proxy servers through which all traffic to and from applications must flow.

application-specific GUI: a graphical interface designed to be used for a specific application.

artificial intelligence: the intelligence exhibited by machines or computers, in contrast to human, organic, or animal intelligence.

asymmetric-key encryption: a process in which data is encrypted using a public encryption key but can only be decrypted using a different, private key.

attenuation: the loss of intensity from a signal being transmitted through a medium.

attributes: the specific features that define an object's properties or characteristics.

audio codec: a program that acts as a "coder-decoder" to allow an audio stream to be encoded for storage or transmission and later decoded for playback.

authentication: the process by which the receiver of encrypted data can verify the identity of the sender or the authenticity of the data.

automatic sequential control system: a mechanism that performs a multistep task by triggering a series of actuators in a particular sequence.

automaton: a machine that mimics a human but is generally considered to be unthinking.

autonomic components: self-contained software or hardware modules with an embedded capacity for self-management, connected via input/outputs to other components in the system.

autonomous: able to operate independently, without external or conscious control.

autonomous agent: a system that acts on behalf of another entity without being directly controlled by that entity.

backdoor: a hidden method of accessing a computer system that is placed there without the knowledge of the system's regular user in order to make it easier to access the system secretly.

base-16: a number system using sixteen symbols, 0 through 9 and A through F.

base-2 system: a number system using the digits 0 and 1.

BCD-to-seven-segment decoder/driver: a logic gate that converts a four-bit binary-coded decimal (BCD) input to decimal numerals that can be output to a seven-segment digital display.

behavioral marketing: advertising to users based on their habits and previous purchases.

binder jetting: the use of a liquid binding agent to fuse layers of powder together.

bioinformatics: the scientific field focused on developing computer systems and software to analyze and examine biological data.

bioinstrumentation: devices that combine biology and electronics in order to interface with a patient's body and record or monitor various health parameters.

biomarker: short for "biological marker"; a measurable quality or quantity (e.g., internal temperature, amount of iron dissolved in blood) that serves as an indicator of an organism's health, or some other biological phenomenon or state.

biomaterials: natural or synthetic materials that can be used to replace, repair, or modify organic tissues or systems.

biomechanics: the various mechanical processes such as the structure, function, or activity of organisms.

bioMEMS: short for "biomedical micro-electromechanical system"; a microscale or nanoscale self-contained device used for various applications in health care.

biometrics: measurements that can be used to distinguish individual humans, such as a person's height, weight, fingerprints, retinal pattern, or genetic makeup.

bionics: the use of biologically based concepts and techniques to solve mechanical and technological problems.

biosignal processing: the process of capturing the information the body produces, such as heart rate, blood pressure, or levels of electrolytes, and analyzing it to assess a patient's status and to guide treatment decisions.

bit: a single binary digit that can have a value of either 0 or 1.

bit rate: the amount of data encoded for each second of video; often measured in kilobits per second (kbps) or kilobytes per second (Kbps).

bit width: the number of bits used by a computer or other device to store integer values or other data.

black-box testing: a testing technique in which function is analyzed based on output only, without knowledge of structure or coding.

blotting: a method of transferring RNA, DNA, and other proteins onto a substrate for analysis.

bootstrapping: a self-starting process in a computer system, configured to automatically initiate other processes after the booting process has been initiated.

bridge: a connection between two or more networks, or segments of a single network, that allows the computers in each network or segment to communicate with one another.

broadcast: an audio or video transmission sent via a communications medium to anyone with the appropriate receiver.

building information modeling (BIM): the creation of a model of a building or facility that accounts for its function, physical attributes, cost, and other characteristics.

butterfly effect: an effect in which small changes in a system's initial conditions lead to major, unexpected changes as the system develops.

byte: a group of eight bits.

caching: the storage of data, such as a previously accessed web page, in order to load it faster upon future access.

carrier signal: an electromagnetic frequency that has been modulated to carry analog or digital information.

cathode ray tube (CRT): a vacuum tube used to create images in devices such as older television and computer monitors.

central processing unit (CPU): electronic circuitry that provides instructions for how a computer handles processes and manages data from applications and programs.

channel capacity: the upper limit for the rate at which information transfer can occur without error.

character: a unit of information that represents a single letter, number, punctuation mark, blank space, or other symbol used in written language.

chatterbot: a computer program that mimics human conversation responses in order to interact with people through text; also called "talkbot," "chatbot," or simply "bot."

circular wait: a situation in which two or more processes are running and each one is waiting for a resource that is being used by another; one of the necessary conditions for deadlock.

class: a collection of independent objects that share similar properties and behaviors.

class-based inheritance: a form of code reuse in which attributes are drawn from a preexisting class to create a new class with additional attributes.

clinical engineering: the design of medical devices to assist with the provision of care.

clock speed: the speed at which a microprocessor can execute instructions; also called "clock rate."

coding theory: the study of codes and their use in certain situations for various applications.

combinatorial design: the study of the creation and properties of finite sets in certain types of designs.

command line: a text-based computer interface that allows the user to input simple commands via a keyboard.

command-line interpreter: an interface that interprets and carries out commands entered by the user.

commodities: consumer products, physical articles of trade or commerce.

communication architecture: the design of computer components and circuitry that facilitates the rapid and efficient transmission of signals between different parts of the computer.

communication devices: devices that allow drones to communicate with users or engineers in remote locations.

compliance: adherence to standards or specifications established by an official body to govern a particular industry, product, or activity.

component-based development: an approach to software design that uses standardized software components to create new applications and new software.

compressed data: data that has been encoded such that storing or transferring the data requires fewer bits of information.

computational linguistics: a branch of linguistics that uses computer science to analyze and model language and speech.

Computer Fraud and Abuse Act (CFAA): a 1986 legislative amendment that made accessing a protected computer without authorization, or exceeding one's authorized level of access, a federal offense.

computer technician: a professional tasked with the installation, repair, and maintenance of computers and related technology.

constraints: limitations on values in computer programming that collectively identify the solutions to be produced by a programming problem.

context switch: a multitasking operating system shifting from one task to another; for example, after formatting a print job for one user, the computer might switch to resizing a graphic for another user.

context switching: pausing and recording the progress of a thread or process such that the process or thread can be executed at a later time.

control characters: units of information used to control the manner in which computers and other devices process text and other characters.

control unit design: describes the part of the CPU that tells the computer how to perform the instructions sent to it by a program.

converter: a device that expands a system's range of reception by bringing in and adapting signals that the system was not originally designed to process.

cookies: small data files that allow websites to track users.

cooperative multitasking: an implementation of multitasking in which the operating system will not initiate a context switch while a process is running in order to allow the process to complete.

core voltage: the amount of power delivered to the processing unit of a computer from the power supply.

counter: a digital sequential logic gate that records how many times a certain event occurs in a given amount of time.

coupling: the degree to which different parts of a program are dependent upon one another.

cracker: a criminal hacker; one who finds and exploits weak points in a computer's security system to gain unauthorized access for malicious purposes.

crippleware: software programs in which key features have been disabled and can only be activated after registration or with the use of a product key.

crosstalk: interference of the signals on one circuit with the signals on another, caused by the two circuits being too close together.

cybercrime: crime that involves targeting a computer or using a computer or computer network to commit a crime, such as computer hacking, digital piracy, and the use of malware or spyware.

data granularity: the level of detail with which data is collected and recorded.

data integrity: the degree to which collected data is and will remain accurate and consistent.

data source: the origin of the information used in a computer model or simulation, such as a database or spreadsheet.

data width: a measure of the amount of data that can be transmitted at one time through the computer bus, the specific circuits and wires that carry data from one part of a computer to another.

datapath design: describes how data flows through the CPU and at what points instructions will be decoded and executed.

declarative language: language that specifies the result desired but not the sequence of operations needed to achieve the desired result.

deep learning: an emerging field of artificial intelligence research that uses neural network algorithms to improve machine learning.

delta debugging: an automated method of debugging intended to identify a bug's root cause while eliminating irrelevant information.

deterministic algorithm: an algorithm that when given a particular input will always produce the same output.

device: equipment designed to perform a specific function when attached to a computer, such as a scanner, printer, or projector.

device fingerprinting: information that uniquely identifies a particular computer, component, or piece of software installed on the computer. This can be used to find out precisely which device accessed a particular online resource.

dexterity: finesse; skill at performing delicate or precise tasks.

digital commerce: the purchase and sale of goods and services via online vendors or information technology systems.

digital legacy (digital remains): the online accounts and information left behind by a deceased person.

digital literacy: familiarity with the skills, behaviors, and language specific to using digital devices to access, create, and share content through the Internet.

digital native: an individual born during the digital age or raised using digital technology and communication.

direct manipulation interfaces: computer interaction format that allows users to directly manipulate graphical objects or physical shapes that are automatically translated into coding.

direct-access storage: a type of data storage in which the data has a dedicated address and location on the storage device, allowing it to be accessed directly rather than sequentially.

directed energy deposition: a process that deposits wire or powdered material onto an object and then melts it using a laser, electron beam, or plasma arc.

distributed algorithm: an algorithm designed to run across multiple processing centers and so is capable of directing a concentrated action between several computer systems.

domain: the range of values that a variable may take on, such as any even number or all values less than −23.7.

domain-dependent complexity: a complexity that results from factors specific to the context in which the computational problem is set.

DRAKON chart: a flowchart used to model algorithms and programmed in the hybrid DRAKON computer language.

dual-mesh network: a type of mesh network in which all nodes are connected both through wiring and wirelessly, thus increasing the reliability of the system. A node is any communication intersection or end-point in the network (e.g., computer or terminal).

dynamic balance: the ability to maintain balance while in motion.

dynamic random-access memory (DRAM): a form of RAM in which the device's memory must be refreshed on a regular basis, or else the data it contains will disappear.

dynamic testing: a testing technique in which software input is tested and output is analyzed by executing the code.

edit decision list (EDL): a list that catalogs the reel or time code data of video frames so that the frames can be accessed during video editing.

electromagnetic spectrum: the complete range of electromagnetic radiation, from the longest wavelength and lowest frequency (radio waves) to the shortest wavelength and highest frequency (gamma rays).

Electronic Communications Privacy Act: a 1986 law that extended restrictions on wiretapping to cover the retrieval or interception of information transmitted electronically between computers or through computer networks.

electronic interference: the disturbance generated by a source of electrical signal that affects an electrical circuit, causing the circuit to degrade or malfunction. Examples include noise, electrostatic discharge, and near-field and far-field interference.

embedded systems: computer systems that are incorporated into larger devices or systems to monitor performance or to regulate system functions.

entanglement: the phenomenon in which two or more particles' quantum states remain linked even if the particles are later separated and become part of distinct systems.

enumerative combinatorics: a branch of combinatorics that studies the number of ways that certain patterns can be formed using a set of objects.

Environmental Protection Agency (EPA): US government agency tasked with combating environmental pollution.

e-waste: short for "electronic waste"; computers and other digital devices that have been discarded by their owners.

false match rate: the probability that a biometric system incorrectly matches an input to a template contained within a database.

fault detection: the monitoring of a system in order to identify when a fault occurs in its operation.

fiber: a small thread of execution using cooperative multitasking.

field programmable gate array: an integrated circuit that can be programmed in the field and can therefore allow engineers or users to alter a machine's programming without returning it to the manufacturer.

filter: in signal processing, a device or procedure that takes in a signal, removes certain unwanted elements, and outputs a processed signal.

firewall: a virtual barrier that filters traffic as it enters and leaves the internal network, protecting internal resources from attack by external sources.

fixed point: a type of digital signal processing in which numerical data is stored and represented with a fixed number of digits after (and sometimes before) the decimal point, using a minimum of sixteen bits.

fixed-point arithmetic: a calculation involving numbers that have a defined number of digits before and after the decimal point.

flash memory: nonvolatile computer memory that can be erased or overwritten solely through electronic signals, i.e. without physical manipulation of the device.

flashing: a process by which the flash memory on a motherboard or an embedded system is updated with a newer version of software.

floating point: a type of digital signal processing in which numerical data is stored and represented in the form of a number (called the mantissa, or significand) multiplied by a base number (such as base-2) raised to an exponent, using a minimum of thirty-two bits.

floating-point arithmetic: a calculation involving numbers that have a decimal point that can be placed anywhere through the use of exponents, as is done in scientific notation.

flooding: sending information to every other node in a network to get the data to its appropriate destination.

force-sensing touch technology: touch display that can sense the location of the touch as well as the amount of pressure the user applies, allowing for a wider variety of system responses to the input.

four-dimensional building information modeling (4-D BIM): the process of creating a 3-D model that incorporates time-related information to guide the manufacturing process.

Fourier transform: a mathematical operator that decomposes a single function into the sum of multiple sinusoidal (wave) functions.

frames per second (FPS): a measurement of the rate at which individual video frames are displayed.

free software: software developed by programmers for their own use or for public use and distributed without charge; it usually has conditions attached that prevent others from acquiring it and then selling it for their own profit.

function: instructions read by a computer's processor to execute specific events or operations.

functional programming: a theoretical approach to programming in which the emphasis is on applying mathematics to evaluate functional relationships that are, for the most part, static.

functional requirement: a specific function of a computer system, such as calculating or data processing.

game loop: the main part of a game program that allows the game's physics, artificial intelligence, and graphics to continue to run with or without user input.

gateway: a device capable of joining one network to another that has different protocols.

genetic modification: direct manipulation of an organism's genome, often for the purpose of engineering useful microbes or correcting for genetic disease.

genome-wide association study: a type of genetic study that compares the complete genomes of individuals within a population to find which genetic markers, if any, are associated with various traits, most often diseases or other health problems.

gestures: combinations of finger movements used to interact with multitouch displays in order to accomplish various tasks. Examples include tapping the finger on the screen, double-tapping, and swiping the finger along the screen.

glyph: a specific representation of a grapheme, such as the letter A rendered in a particular typeface.

graph theory: the study of graphs, which are diagrams used to model relationships between objects.

grapheme: the smallest unit used by a writing system, such as alphabetic letters, punctuation marks, or Chinese characters.

graphical user interface (GUI): an interface that allows users to control a computer or other device by interacting with graphical elements such as icons and windows.

graphical user interface (GUI): an interface that allows a user to interact with pictures instead of requiring the user to type commands into a text-only interface.

Green Electronics Council: a US nonprofit organization dedicated to promoting green electronics.

hacking: the use of technical skill to gain unauthorized access to a computer system; also, any kind of advanced tinkering with computers to increase their utility.

hamming distance: a measurement of the difference between two characters or control characters that effects character processing, error detection, and error correction.

hardware: the physical parts that make up a computer. These include the motherboard and processor, as well as input and output devices such as monitors, keyboards, and mice.

hardware interruption: a device attached to a computer sending a message to the operating system to inform it that the device needs attention, thereby "interrupting" the other tasks that the operating system was performing.

Harvard architecture: a computer design that has physically distinct storage locations and signal routes for data and for instructions.

hash function: an algorithm that converts a string of characters into a different, usually smaller, fixed-length string of characters that is ideally impossible either to invert or to replicate.

hashing algorithm: a computing function that converts a string of characters into a different, usually smaller string of characters of a given length, which is ideally impossible to replicate without knowing both the original data and the algorithm used.

Health Insurance Portability and Accountability Act (HIPAA): a 1996 law that established national standards for protecting individuals' medical records and other personal health information.

heavy metal: one of several toxic natural substances often used as components in electronic devices.

heterogeneous scalability: the ability of a multiprocessor system to scale up using parts from different vendors.

hexadecimal: a base-16 number system that uses the digits 0 through 9 and the letters A, B, C, D, E, and F as symbols to represent numbers.

hibernation: a power-saving state in which a computer shuts down but retains the contents of its random-access memory.

hidden Markov model: a type of model used to represent a dynamic system in which each state can only be partially seen.

hierarchical file system: a directory with a treelike structure in which files are organized.

high-fidelity wireframe: an image or series of images that represents the visual elements of the website in great detail, as close to the final product as possible, but still does not permit user interaction.

homebrew: software that is developed for a device or platform by individuals not affiliated with the device manufacturer; it is an unofficial or "homemade" version of the software that is developed to provide additional functionality not included or not permitted by the manufacturer.

hopping: the jumping of a data packet from one device or node to another as it moves across the network. Most transmissions require each packet to make multiple hops.

horizontal scaling: the addition of nodes (e.g., computers or servers) to a distributed system.

host-based firewalls: firewalls that protect a specific device, such as a server or personal computer, rather than the network as a whole.

HTML editor: a computer program for editing web pages encoded in hypertext markup language (HTML).

Human Brain Project: a project launched in 2013 in an effort at modeling a functioning brain by 2023; also known as HBP.

humanoid: resembling a human.

hybrid cloud: a cloud computing model that combines public cloud services with a private cloud platform linked through an encrypted connection.

identifiers: measurable characteristics used to identify individuals.

imitation game: Alan Turing's name for his proposed test, in which a machine would attempt to respond to questions in such a way as to fool a human judge into thinking it was human.

immersive mode: a full-screen mode in which the status and navigation bars are hidden from view when not in use.

imperative language: language that instructs a computer to perform a particular sequence of operations.

imperative programming: programming that produces code that consists largely of commands issued to the computer, instructing it to perform specific actions.

in-circuit emulator: a device that enables the debugging of a computer system embedded within a larger system.

information hierarchy: the relative importance of the information presented on a web page.

information technology: the use of computers and related equipment for the purpose of processing and storing data.

infrastructure as a service: a cloud computing platform that provides additional computing resources by linking hardware systems through the Internet; also called "hardware as a service."

inheritance: a technique that reuses and repurposes sections of code.

Initiative for Software Choice (ISC): a consortium of technology companies founded by CompTIA, with the goal of encouraging governments to allow competition among software manufacturers.

input/output instructions: instructions used by the central processing unit (CPU) of a computer when information is transferred between the CPU and a device such as a hard disk.

integration testing: a process in which multiple units are tested individually and when working in concert.

interaction design: the practice of designing a user interface, such as a website, with a focus on how to facilitate the user's experience.

interface: the function performed by the device driver, which mediates between the hardware of the peripheral and the hardware of the computer.

interface metaphors: linking computer commands, actions, and processes with real-world actions, processes, or objects that have functional similarities.

interference: anything that disrupts a signal as it moves from source to receiver.

interferometry: a technique for studying biochemical substances by superimposing light waves, typically one reflected from the substance and one reflected from a reference point, and analyzing the interference.

internal-use software: software developed by a company for its own use to support general and administrative functions, such as payroll, accounting, or personnel management and maintenance.

Internet of things: a wireless network connecting devices, buildings, vehicles, and other items with network connectivity.

interpolation: a process of estimating intermediate values when nearby values are known; used in image editing to "fill in" gaps by referring to numerical data associated with nearby points.

interrupt vector table: a chart that lists the addresses of interrupt handlers.

intrusion detection system: a system that uses hardware, software, or both to monitor a computer or network in order to determine when someone attempts to access the system without authorization.

inverter: a logic gate whose output is the inverse of the input; also called a NOT gate.

iOS: Apple's proprietary mobile operating system, installed on Apple devices such as the iPhone, iPad, and iPod touch.

IRL relationships: relationships that occur "in real life," meaning that the relationships are developed or sustained outside of digital communication.

jailbreak: the removal of restrictions placed on a mobile operating system to give the user greater control over the mobile device.

jailbreaking: the process of removing software restrictions within iOS that prevent a device from running certain kinds of software.

JavaScript: a flexible programming language that is commonly used in website design.

jQuery: a free, open-source JavaScript library.

kernel: the central component of an operating system.

keyframing: a part of the computer animation process that shows, usually in the form of a drawing, the position and appearance of an object at the beginning of a sequence and at the end.

language interface packs: programs that translate interface elements such as menus and dialog boxes into different languages.

learner-controlled program: software that allows a student to set the pace of instruction, choose which content areas to focus on, decide which areas to explore when, or determine the medium or difficulty level of instruction; also known as a "student-controlled program."

learning strategy: a specific method for acquiring and retaining a particular type of knowledge, such as memorizing a list of concepts by setting the list to music.

learning style: an individual's preferred approach to acquiring knowledge, such as by viewing visual stimuli, reading, listening, or using one's hands to practice what is being taught.

lexicon: the total vocabulary of a person, language, or field of study.

linear predictive coding: a popular tool for digital speech processing that uses both past speech samples and a mathematical approximation of a human vocal tract to predict and then eliminate certain vocal frequencies that do not contribute to meaning. This allows speech to be processed at a faster bit rate without significant loss in comprehensibility.

livelock: a situation in which two or more processes constantly change their state in response to one another in such a way that neither can complete.

local area network (LAN): a network that connects electronic devices within a limited physical area.

logic implementation: the way in which a CPU is designed to use the open or closed state of combinations of circuits to represent information.

logical copy: a copy of a hard drive or disk that captures active data and files in a different configuration from the original, usually excluding free space and artifacts such as file remnants; contrasts with a physical copy, which is an exact copy with the same size and configuration as the original.

logotype: a company or brand name rendered in a unique, distinct style and font; also called a "wordmark."

lossless compression: data compression that allows the original data to be compressed and reconstructed without any loss of accuracy.

lossy compression: a method of reducing the size of an audio file while sacrificing some of the quality of the original file.

lossy compression: a method of decreasing image file size by discarding some data, resulting in some image quality being irreversibly sacrificed.

low-energy connectivity IC: an integrated circuit that enables wireless Bluetooth connectivity while using little power.

LZW compression: a type of lossless compression that uses a table-based algorithm.

magnetic storage: a device that stores information by magnetizing certain types of material.

main loop: the overarching process being carried out by a computer program, which may then invoke subprocesses.

main memory: the primary memory system of a computer, often called "random access memory" (RAM), accessed by the computer's central processing unit (CPU).

mastering: the creation of a master recording that can be used to make other copies for distribution.

Material Design: a comprehensive guide for visual, motion, and interaction design across Google platforms and devices.

material extrusion: a process in which heated filament is extruded through a nozzle and deposited in layers, usually around a removable support.

material jetting: a process in which drops of liquid photopolymer are deposited through a printer head and heated to form a dry, stable solid.

Medical Device Innovation Consortium: a nonprofit organization established to work with the US Food and Drug Administration on behalf of medical device manufacturers to ensure that these devices are both safe and effective.

medical imaging: the use of devices to scan a patient's body and create images of the body's internal structures to aid in diagnosis and treatment planning.

memory dumps: computer memory records from when a particular program crashed, used to pinpoint and address the bug that caused the crash.

memristor: a memory resistor, a circuit that can change its own electrical resistance based on the resistance it has used in the past and can respond to familiar phenomena in a consistent way.

mesh network: a type of network in which each node relays signal through the network. A node is any communication intersection or endpoint in the network (e.g., computer or terminal).

meta-complexity: a complexity that arises when the computational analysis of a problem is compounded by the complex nature of the problem itself.

metadata: data that contains information about other data, such as author information, organizational information, or how and when the data was created.

method: a procedure that describes the behavior of an object and its interactions with other objects.

microcontroller: a tiny computer in which all of the essential parts of a computer are united on a single microchip—input and output channels, memory, and a processor.

microcontroller: an integrated circuit that contains a very small computer.

micron: a unit of measurement equaling one millionth of a meter; typically used to measure the width of a core in an optical figure or the line width on a microchip.

microwaves: electromagnetic radiation with a frequency higher than that of radio wave but lower than that of visible light.

middle computing: computing that occurs at the application tier and involves intensive processing of data that will subsequently be presented to the user or another, intervening application.

million instructions per second (MIPS): a unit of measurement used to evaluate computer performance or the cost of computing resources.

mixer: a component that converts random input radio frequency (RF) signal into a known, fixed frequency to make processing the signal easier.

mixing: the process of combining different sounds into a single audio recording.

modeling: the process of creating a 2-D or 3-D representation of the structure being designed.

modulator: a device used to regulate or adjust some aspect of an electromagnetic wave.

morphology: a branch of linguistics that studies the forms of words.

multi-agent system: a system consisting of multiple separate agents, either software or hardware systems, that can cooperate and organize to solve problems.

multibit watermarking: a watermarking process that embeds multiple bits of data in the signal to be transmitted.

multicast: a network communications protocol in which a transmission is broadcast to multiple recipients rather than to a single receiver.

multimodal monitoring: the monitoring of several physical parameters at once in order to better evaluate a patient's overall condition, as well as how different parameters affect one another or respond to a given treatment.

multiplexing: combining multiple data signals into one in order to transmit all signals simultaneously through the same medium.

multiplier-accumulator: a piece of computer hardware that performs the mathematical operation of multiplying two numbers and then adding the result to an accumulator.

multiprocessing: the use of more than one central processing unit to handle system tasks; this requires an operating system capable of dividing tasks between multiple processors.

multitasking: in the mobile phone environment, allowing different apps to run concurrently, much like the ability to work in multiple open windows on a PC.

multitasking: in computing, the process of executing multiple tasks concurrently on an operating system (OS).

multitenancy: a software program that allows multiple users to access and use the software from different locations.

multi-terminal configuration: a computer configuration in which several terminals are connected to a single computer, allowing more than one person to use the computer.

multitouch gestures: combinations of finger movements used to interact with touch-screen or other touch-sensitive displays in order to accomplish various tasks. Examples include double-tapping and swiping the finger along the screen.

multiuser: capable of being accessed or used by more than one user at once.

mutual exclusion: a rule present in some database systems that prevents a resource from being accessed by more than one operation at a time; one of the necessary conditions for deadlock.

near-field communication: a method by which two devices can communicate wirelessly when in close proximity to one another.

near-field communications antenna: an antenna that enables a device to communicate wirelessly with a nearby compatible device.

negative-AND (NAND) gate: a logic gate that produces a false output only when both inputs are true

nervous (neural) system: the system of nerve pathways by which an organism senses changes in itself and its environment and transmits electrochemical signals describing these changes to the brain so that the brain can respond.

network: two or more computers being linked in a way that allows them to transmit information back and forth.

network firewalls: firewalls that protect an entire network rather than a specific device.

networking: the use of physical or wireless connections to link together different computers and computer networks so that they can communicate with one another and collaborate on computationally intensive tasks.

neural network: in computing, a model of information processing based on the structure and function of biological neural networks such as the human brain.

neuroplasticity: the capacity of the brain to change as it acquires new information and forms new neural connections.

nibble: a group of four bits.

node: any point on a computer network where communication pathways intersect, are redistributed, or end (i.e., at a computer, terminal, or other device).

noise: interferences or irregular fluctuations affecting electrical signals during transmission.

noise-tolerant signal: a signal that can be easily distinguished from unwanted signal interruptions or fluctuations (i.e., noise).

nondestructive editing: a mode of image editing in which the original content of the image is not destroyed because the edits are made only in the editing software.

nonfunctional requirement: also called extrafunctional requirement; an attribute of a computer system, such as reliability, scalability, testability, security, or usability, that reflects the operational performance of the system.

nongraphical: not featuring graphical elements.

nonlinear editing: a method of editing video in which each frame of video can be accessed, altered, moved, copied, or deleted regardless of the order in which the frames were originally recorded.

nonvolatile memory: computer storage that retains its contents after power to the system is cut off, rather than memory that is erased at system shutdown.

nonvolatile random-access memory (NVRAM): a form of RAM in which data is retained even when the device loses access to power.

normalization: a process that ensures that different code points representing equivalent characters will be recognized as equal when processing text.

nuclear magnetic resonance (NMR) spectroscopy: a technique for studying the properties of atoms or molecules by applying an external magnetic field to atomic nuclei and analyzing the resulting difference in energy levels.

object: an element with a unique identity and a defined set of attributes and behaviors.

object-oriented programming: a type of programming in which the source code is organized into objects, which are elements with a unique identity that have a defined set of attributes and behaviors.

object-oriented user interface: an interface that allows users to interact with onscreen objects as they would in real-world situations, rather than selecting objects that are changed through a separate control panel interface.

open-source software: software that makes its source code available to the public for free use, study, modification, and distribution.

operating system (OS): a specialized program that manages a computer's functions.

operating system shell: a user interface used to access and control the functions of an operating system.

optical storage: storage of data by creating marks in a medium that can be read with the aid of a laser or other light source.

optical touchscreens: touchscreens that use optical sensors to locate the point where the user touches before physical contact with the screen has been made.

packet filters: filters that allow data packets to enter a network or block them on an individual basis.

packet forwarding: the transfer of a packet, or unit of data, from one network node to another until it reaches its destination.

packet sniffers: a program that can intercept data or information as it moves through a network.

packet switching: a method of transmitting data over a network by breaking it up into units called packets, which are sent from node to node along the network until they reach their destination and are reassembled.

parallel processing: the division of a task among several processors working simultaneously, so that the task is completed more quickly.

parameter: a measurable element of a system that affects the relationships between variables in the system.

PATRIOT Act: a 2001 law that expanded the powers of federal agencies to conduct surveillance and intercept digital information for the purpose of investigating or preventing terrorism.

pedagogy: a philosophy of teaching that addresses the purpose of instruction and the methods by which it can be achieved.

peer-to-peer (P2P) network: a network in which all computers participate equally and share coordination of network operations, as opposed to a client-server network, in which only certain computers coordinate network operations.

pen/trap: short for pen register or trap-and-trace device, devices used to record either all numbers called from a particular telephone (pen register) or all numbers making incoming calls to that phone (trap and trace); also refers to the court order that permits the use of such devices.

peripheral: a device that is connected to a computer and used by the computer but is not part of the computer, such as a printer, scanner, external storage device, and so forth.

personal area network (PAN): a network generated for personal use that allows several devices to connect to one another and, in some cases, to other networks or to the Internet.

personally identifiable information (PII): information that can be used to identify a specific individual.

pharmacogenomics: the study of how an individual's genome influences his or her response to drugs.

phased software development: an approach to software development in which most testing occurs after the system requirements have been implemented.

phishing: the use of online communications in order to trick a person into sharing sensitive personal information, such as credit card numbers or social security numbers.

Phonebloks: a concept devised by Dutch designer Dave Hakkens for a modular mobile phone intended to reduce electronic waste.

phoneme: a sound in a specified language or dialect that is used to compose words or to distinguish between words.

pipelined architecture: a computer design where different processing elements are connected in a series, with the output of one operation being the input of the next.

piracy: in the digital context, unauthorized reproduction or use of copyrighted media in digital form.

planned obsolescence: a design concept in which consumer products are given an artificially limited lifespan, therefore creating a perpetual market.

platform: the specific hardware or software infrastructure that underlies a computer system; often refers to an operating system, such as Windows, Mac OS, or Linux.

platform: the underlying computing system on which applications can run, which may be the computer's hardware or its operating system.

platform as a service: a category of cloud computing that provides a virtual machine for users to develop, run, and manage web applications.

plug-in: an application that is easily installed (plugged in) to add a function to a computer system.

polymerase chain reaction (PCR) machine: a machine that uses polymerase chain reaction to amplify segments of DNA for analysis; also called a thermal cycler.

polymorphism: the ability to maintain the same method name across subclasses even when the method functions differently depending on its class.

port scanning: the use of software to probe a computer server to see if any of its communication ports have been left open or vulnerable to an unauthorized connection that could be used to gain control of the computer.

portlet: an independently developed software component that can be plugged into a website to generate and link to external content.

postproduction: the period after a model has been designed and an image has been rendered, when the architect may manipulate the created image by adding effects or making other aesthetic changes.

powder bed fusion: the use of a laser to heat layers of powdered material in a movable powder bed.

precoding: a technique that uses the diversity of a transmission by weighting an information channel.

preemptive multitasking: an implementation of multitasking in which the operating system will initiate a context switch while a process is running, usually on a schedule so that switches between tasks occur at precise time intervals.

Pretty Good Privacy: a data encryption program created in 1991 that provides both encryption and authentication.

principle of least privilege: a philosophy of computer security that mandates users of a computer or network be given, by default, the lowest level of privileges that will allow them to perform their jobs. This way, if a user's account is compromised, only a limited amount of data will be vulnerable.

printable characters: characters that can be written, printed, or displayed in a manner that can be read by a human.

printed circuit board: a flat copper sheet shielded by fiberglass insulation in which numerous lines have been etched and holes have been punched, allowing various electronic components to be connected and to communicate with one another and with external components via the exposed copper traces.

printed circuit board (PCB): a component of electronic devices that houses and connects many smaller components.

Privacy Incorporated Software Agents (PISA): a project that sought to identify and resolve privacy problems related to intelligent software agents.

process: the execution of instructions in a computer program.

processor coupling: the linking of multiple processors within a computer so that they can work together to perform calculations more rapidly. This can be characterized as loose or tight, depending on the degree to which processors rely on one another.

processor symmetry: multiple processors sharing access to input and output devices on an equal basis and being controlled by a single operating system.

programmable oscillator: an electronic device that fluctuates between two states that allows user modifications to determine mode of operation.

programming languages: sets of terms and rules of syntax used by computer programmers to create instructions for computers to follow. This code is then compiled into binary instructions for a computer to execute.

projected capacitive touch: technology that uses layers of glass etched with a grid of conductive material that allows for the distortion of voltage flowing through the grid when the user touches the surface; this distortion is measured and used to determine the location of the touch.

proprietary software: software owned by an individual or company that places certain restrictions on its use, study, modification, or distribution and typically withholds its source code.

protocol processor: a processor that acts in a secondary capacity to the CPU, relieving it from some of the work of managing communication protocols that are used to encode messages on the network.

prototypal inheritance: a form of code reuse in which existing objects are cloned to serve as prototypes.

prototype: an early version of software that is still under development, used to demonstrate what the finished product will look like and what features it will include.

proxy: a dedicated computer server that functions as an intermediary between a user and another server.

proxy server: a computer through which all traffic flows before reaching the user's computer.

pseudocode: a combination of a programming language and a spoken language, such as English, that is used to outline a program's code.

public-key cryptography: a system of encryption that uses two keys, one public and one private, to encrypt and decrypt data.

push technology: a communication protocol in which a messaging server notifies the recipient as soon as the server receives a message, instead of waiting for the user to check for new messages.

quantum bit (qubit): a basic unit of quantum computation that can exist in multiple states at the same time, and can therefore have multiple values simultaneously.

quantum logic gate: a device that alters the behavior or state of a small number of qubits for the purpose of computation.

radio waves: low-frequency electromagnetic radiation, commonly used for communication and navigation.

random access memory (RAM): memory that the computer can access very quickly, without regard to where in the storage media the relevant information is located.

ransomware: malware that encrypts or blocks access to certain files or programs and then asks users to pay to have the encryption or other restrictions removed.

rapid prototyping: the process of creating physical prototype models that are then tested and evaluated.

raster: a means of storing, displaying, and editing image data based on the use of individual pixels.

read-only memory (ROM): type of nonvolatile data storage that is typically either impossible to erase and alter or requires special equipment to do so; it is usually used to store basic operating information needed by a computer.

real-time monitoring: a process that grants administrators access to metrics and usage data about a software program or database in real time.

real-time operating system: an operating system that is designed to respond to input within a set amount of time without delays caused by buffering or other processing backlogs.

receiver: a device that reads a particular type of transmission and translates it into audio or video.

recursive: describes a method for problem solving that involves solving multiple smaller instances of the central problem.

remote monitoring: a platform that reviews the activities on software or systems that are located off-site.

render farm: a cluster of powerful computers that combine their efforts to render graphics for animation applications.

rendering: the process of transforming one or more models into a single image; the production of a computer image from a 2-D or 3-D computer model; the process of selecting and displaying glyphs.

resistive touchscreens: touchscreens that can locate the user's touch because they are made of several layers of conductive material separated by small spaces; when the user touches the screen, the layers touch each other and complete a circuit.

resource allocation: a system for dividing computing resources among multiple, competing requests so that each request is eventually fulfilled.

resource distribution: the locations of resources available to a computing system through various software or hardware components or networked computer systems.

resource holding: a situation in which one process is holding at least one resource and is requesting further resources; one of the necessary conditions for deadlock.

retriggerable single shot: a monostable multivibrator (MMV) electronic circuit that outputs a single pulse when triggered but can identify a new trigger during an output pulse, thus restarting its pulse time and extending its output.

reversible data hiding: techniques used to conceal data that allow the original data to be recovered in its exact form with no loss of quality.

RGB: a color model that uses red, green, and blue to form other colors through various combinations.

routing: selecting the best path for a data packet to take in order to reach its destination on the network.

Sarbanes-Oxley Act (SOX): a 2002 law that requires all business records, including electronic records and electronic messages, to be retained for a certain period of time.

scareware: malware that attempts to trick users into downloading or purchasing software or applications to address a computer problem.

Scientific Working Group on Digital Evidence (SWGDE): an American association of various academic and professional organizations interested in the development of digital forensics systems, guidelines, techniques, and standards.

screen-reading program: a computer program that converts text and graphics into a format accessible to visually impaired, blind, learning disabled, or illiterate users.

script: a group of written signs, such as Latin or Chinese characters, used to represent textual information in a writing system.

scrubbing: navigating through an audio recording repeatedly in order to locate a specific cue or word.

search engine optimization (SEO): techniques used to increase the likelihood that a website will appear among the top results in certain search engines.

self-star properties: a list of component and system properties required for a computing system to be classified as an autonomic system.

self-star properties: a list of component and system properties required for a computing system to be classified as an autonomic system.

semantics: a branch of linguistics that studies the meanings of words and phrases.

semiconductor intellectual property (SIP) block: a quantity of microchip layout design that is owned by a person or group; also known as an "IP core."

sensors: devices capable of detecting, measuring, or reacting to external physical properties.

separation of concerns: a principle of software engineering in which the designer separates the computer program's functions into discrete elements.

shadow RAM: a form of RAM that copies code stored in read-only memory into RAM so that it can be accessed more quickly.

sheet lamination: a process in which thin layered sheets of material are adhered or fused together and then extra material is removed with cutting implements or lasers.

shell: an interface that allows a user to operate a computer or other device.

Short Message Service (SMS): the technology underlying text messaging used on cell phones.

signal-to-noise ratio (SNR): the power ratio between meaningful information, referred to as "signal," and background noise.

silent monitoring: listening to the exchanges between an incoming caller and an employee, such as a customer service agent.

simulation: a computer model executed by a computer system.

software: the sets of instructions that a computer follows in order to carry out tasks. Software may be stored on physical media, but the media is not the software.

software as a service: a software service system in which software is stored at a provider's data center and accessed by subscribers.

software patches: updates to software that correct bugs or make other improvements.

software-defined antennas: reconfigurable antennas that transmit or receive radio signals according to software instructions.

solid modeling: the process of creating a 3-D representation of a solid object.

solid-state storage: computer memory that stores information in the form of electronic circuits, without the use of disks or other read/write equipment.

sound card: a peripheral circuit board that enables a computer to accept audio input, convert signals between analog and digital formats, and produce sound.

source code: the set of instructions written in a programming language to create a program.

speaker independent: describes speech software that can be used by any speaker of a given language.

special character: a character such as a symbol, emoji, or control character.

spyware: software installed on a computer that allows a third party to gain information about the computer user's activity or the contents of the user's hard drive.

state: a technical term for all of the stored information, and the configuration thereof, that a program or circuit can access at a given time; a complete description of a physical system at a specific point in time, including such factors as energy, momentum, position, and spin.

stateful filters: filters that assess the state of a connection and allow or disallow data transfers accordingly.

static random-access memory (SRAM): a form of RAM in which the device's memory does not need to be regularly refreshed but data will still be lost if the device loses power.

static testing: a testing technique in which software is tested by analyzing the program and associated documentation, without executing its code.

substitution cipher: a cipher that encodes a message by substituting one character for another.

subtyping: a relation between data types where one type is based on another, but with some limitations imposed.

supercomputer: an extremely powerful computer that far outpaces conventional desktop computers.

superposition: the principle that two or more waves, including waves describing quantum states, can be combined to give rise to a new wave state with unique properties. This allows a qubit to potentially be in two states at once.

surface capacitive technology: a glass screen coated with an electrically conductive film that draws current across the screen when it is touched; the flow of current is measured in order to determine the location of the touch.

symmetric-key cryptography: a system of encryption that uses the same private key to encrypt and decrypt data.

syntax: a branch of linguistics that studies how words and phrases are arranged in sentences to create meaning.

syntax: rules that describe how to correctly structure the symbols that comprise a language.

system: a set of interacting or interdependent component parts that form a complex whole; a computer's combination of hardware and software resources that must be managed by the operating system.

system agility: the ability of a system to respond to changes in its environment or inputs without failing altogether.

system identification: the study of a system's inputs and outputs in order to develop a working model of it.

system software: the basic software that manages the computer's resources for use by hardware and other software.

tableless web design: the use of style sheets rather than HTML tables to control the layout of a web page.

TCO certification: a credential that affirms the sustainability of computers and related devices.

telecom equipment: hardware that is intended for use in telecommunications, such as cables, switches, and routers.

telemedicine: health care provided from a distance using communications technology, such as video chats, networked medical equipment, smartphones, and so on.

telemetry: automated communication process that allows a machine to identify its position relative to external environmental cues.

temporal synchronization: the alignment of signals from multiple devices to a single time standard, so that, for example, two different devices that record the same event will show the event happening at the exact same time.

terminals: a set of basic input devices, such as a keyboard, mouse, and monitor, that are used to connect to a computer running a multi-user operating system.

third-party data center: a data center service provided by a separate company that is responsible for maintaining its infrastructure.

thread: the smallest part of a programmed sequence that can be managed by a scheduler in an OS.

time-sharing: a strategy used by multi-user operating systems to work on multiple user requests by switching between tasks in very small intervals of time.

time-sharing: the use of a single computing resource by multiple users at the same time, made possible by the computer's ability to switch rapidly between users and their needs.

topology: the way a network is organized, including nodes and the links that connect nodes.

toxgnostics: a subfield of personalized medicine and pharmacogenomics that is concerned with whether an individual patient is likely to suffer a toxic reaction to a specific medication.

trace impedance: a measure of the inherent resistance to electrical signals passing through the traces etched on a circuit board.

transactional database: a database management system that allows a transaction— a sequence of operations to achieve a single, self-contained task—to be undone, or "rolled back," if it fails to complete properly.

transistor: a computing component generally made of silicon that can amplify electronic signals or work as a switch to direct electronic signals within a computer system.

transmission medium: the material through which a signal can travel.

transmitter: a device that sends a signal through a medium to be picked up by a receiver.

transparent monitoring: a system that enables employees to see everything that the managers monitoring them can see.

transposition cipher: a cipher that encodes a message by changing the order of the characters within the message.

trusted platform module (TPM): a standard used for designing cryptoprocessors, which are special chips that enable devices to translate plain text into cipher text and vice versa.

tuning: the process of making minute adjustments to a computer's settings in order to improve its performance.

Turing complete: a programming language that can perform all possible computations.

typography: the art and technique of arranging type to make language readable and appealing.

ubiquitous computing: an approach to computing in which computing activity is not isolated in a desktop, laptop, or server, but can occur everywhere and at any time through the use of microprocessors embedded in everyday devices.

undervolting: reducing the voltage of a computer system's central processing unit to decrease power usage.

unmanned aerial vehicle (UAV): an aircraft that does not have a pilot onboard but typically operates through remote control, automated flight systems, or preprogrammed computer instructions.

user-centered design: design based on a perceived understanding of user preferences, needs, tendencies, and capabilities.

utility program: a type of system software that performs one or more routine functions, such as disk partitioning and maintenance, software installation and removal, or virus protection.

variable: a symbol representing a quantity with no fixed value.

vat photopolymerization: a process in which a laser hardens layers of light-sensitive material in a vat.

vector: a means of storing, displaying, and editing image data based on the use of defined points and lines.

vertical scaling: the addition of resources to a single node (e.g., a computer or server) in a system.

vibrator: an electronic component that vibrates.

video scratching: a technique in which a video sequence is manipulated to match the rhythm of a piece of music.

virtual device driver: a type of device driver used by the Windows operating system that handles communications between emulated hardware and other devices.

virtual memory: memory used when a computer configures part of its physical storage (on a hard drive, for example) to be available for use as additional RAM. Information is copied from RAM and moved into virtual memory whenever memory resources are running low.

virtual reality: the use of technology to create a simulated world into which a user may be immersed through visual and auditory input.

vision mixing: the process of selecting and combining multiple video sources into a single video.

visual programming: a form of programming that allows a programmer to create a program by dragging and dropping visual elements with a mouse instead of having to type in text instructions.

voice over Internet Protocol (VoIP): a set of parameters that make it possible for telephone calls to be transmitted digitally over the Internet, rather than as analog signals through telephone wires.

volatile memory: memory that stores information in a computer only while the computer has power; when the computer shuts down or power is cut, the information is lost.

wardriving: driving around with a device such as a laptop that can scan for wireless networks that may be vulnerable to hacking.

web application: an application that is downloaded either wholly or in part from the Internet each time it is used.

white-box testing: a testing technique that includes complete knowledge of an application's coding and structure.

widget: an independently developed segment of code that can be plugged into a website to display information retrieved from other sites.

widgets: small, self-contained applications that run continuously without being activated like a typical application.

wireframe: a schematic or blueprint that represents the visual layout of a web page, without any interactive elements.

word prediction: a software feature that recognizes words that the user has typed previously and offers to automatically complete them each time the user begins typing.

worm: a type of malware that can replicate itself and spread to other computers independently; unlike a computer virus, it does not have to be attached to a specific program.

zombie computer: a computer that is connected to the Internet or a local network and has been compromised such that it can be used to launch malware or virus attacks against other computers on the same network.

BIBLIOGRAPHY

"3D Printing Processes: The Free Beginner's Guide." *3D Printing Industry*. 3D Printing Industry, 2015. Web. 6 Jan. 2016.

"About Additive Manufacturing." *Additive Manufacturing Research Group*. Loughborough U, 2015. Web. 6 Jan. 2016.

"About ANSI." *ANSI*. American Natl. Standards Inst., 2016. Web. 31 Jan. 2016.

"About Threads and Processes." *MSDN.Microsoft*. Windows, n.d. Web. 15 Mar 2016.

Abramovich, Sergei, ed. *Computers in Education*. 2 vols. New York: Nova, 2012. Print.

"A Catalog of Published Genome-Wide Association Studies." *National Human Genome Research Institute*. Natl. Insts. of Health, 16 Sept. 2015. Web. 23 Dec. 2015.

Adams, Ty, and Stephen A. Smith. *Communication Shock: The Rhetoric of New Technology*. Newcastle upon Tyne: Cambridge Scholars, 2015. Print.

Adee, Sally. "Thanks for the Memories." *IEEE Spectrum*. IEEE, 1 May 2009. Web. 10 Mar. 2016.

Adelman, L. M. Molecular computation of solutions to combinatorial problems, *Science* 226, 1021-1024 (1994).

Agrawal, Manindra, S. Barry Cooper, and Angsheng Li, eds. *Theory and Applications of Models of Computation: 9th Annual Conference, TAMC 2012, Beijing, China, May 16–21, 2012*. Berlin: Springer, 2012. Print.

Agrawal, Manish. *Business Data Communications*. Hoboken: Wiley, 2011. Print.

"A History of Technology in the Architecture Office." *Architizer*. Architizer, 23 Dec. 2014. Web. 31 Jan. 2016.

"A History of Windows." *Microsoft*. Microsoft, 2016. Web. 2 Jan. 2016.

Alcorn, Paul. "SMR (Shingled Magnetic Recording) 101." *Tom's IT Pro*. Purch, 10 July 2015. Web. 7 Mar. 2016.

Allanwood, Gavin, and Peter Beare. *User Experience Design: Creating Designs Users Really Love*. New York: Fairchild, 2014. Print.

Alton, Larry. "Email Security in 2016: What You Need to Know." *Inc*. Mansueto Ventures, 18 Feb. 2016. 21 Feb. 2016.

Amadeo, Ron. "The History of Android." *Ars Technica*. Condé Nast, 15 June 2014. Web. 2 Jan. 2016.

Ambainis, Andris. "What Can We Do with a Quantum Computer?" *Institute Letter* Spring 2014: 6–7. *Institute for Advanced Study*. Web. 24 Mar. 2016.

Ambinder, Marc. "What's Really Limiting Advances in Computer Tech." *Week*. The Week, 2 Sept. 2014. Web. 4 Mar. 2016.

Ambrose, Gavin, Paul Harris, and Sally Stone. *The Visual Dictionary of Architecture*. Lausanne: AVA, 2008. Print.

Amer. Standards Assn. *American Standard Code for Information Interchange*. Amer. Standards Assn., 17 June 1963. Digital file.

"An Introduction to Public Key Cryptography and PGP." *Surveillance Self-Defense*. Electronic Frontier Foundation, 7 Nov. 2014. Web. 4 Feb. 2016.

Anderson, Deborah. "Global Linguistic Diversity for the Internet." *Communications of the ACM* Jan. 2005: 27. PDF file.

Anderson, Gary H. *Video Editing and Post Production: A Professional Guide*. Woburn: Focal, 1999. Print.

Anderson, James A, Edward Rosenfeld, and Andras Pellionisz. *Neurocomputing*. Cambridge, Mass: MIT Press, 1990. Print.

Anderson, Nate. "CompTIA Backs Down; Past Certs Remain Valid for Life." *Ars Technica*. Condé Nast, 26 Jan. 2010. Web. 31 Jan. 2016.

Anderson, Thomas, and Michael Dahlin. *Operating Systems: Principles and Practice*. West Lake Hills: Recursive, 2014. Print.

Andrews, Jean. *A+ Guide to Hardware: Managing, Maintaining, and Troubleshooting*. 6th ed. Boston: Course Tech., 2014. Print.

Andrews, Jean. *A+ Guide to Managing and Maintaining Your PC*. 8th ed. Boston: Course Tech., 2014. Print.

"Android: A Visual History." *Verge*. Vox Media, 7 Dec. 2011. Web. 2 Jan. 2016.

"ANSI Accredits Four Personnel Certification Programs." *ANSI*. American Natl. Standards Inst., 8 Apr. 2008. Web. 31 Jan. 2016.

Anthes, Gary. "Back to Basics: Algorithms." *Computerworld*. Computerworld, 24 Mar. 2008. Web. 19 Jan. 2016.

Ao, Sio-Iong, and Len Gelman, eds. *Electrical Engineering and Intelligent Systems*. New York: Springer, 2013. Print.

"A Quantum Leap in Computing." *NOVA*. WGBH/PBS Online, 21 July 2011. Web. 24 Mar. 2016.

Aravind, Alex A., and Sibsankar Haldar. *Operating Systems*. Upper Saddle River: Pearson, 2010. Print.

Ascher, Steven. *The Filmmaker's Handbook: A Comprehensive Guide for the Digital Age*. New York: Plume, 2012. Print.

"A Short History of US Internet Legislation: Privacy on the Internet." *ServInt*. ServInt, 17 Sept. 2013. Web. 28 Feb. 2016.

"A Tutorial on Data Representation: Integers, Floating-Point Numbers, and Characters." *NTU. edu*. Nanyang Technological U, Jan. 2014. Web. 20 Feb. 2016.

Australian National University. *Binary Representation and Computer Arithmetic*. Australian National U, n.d. Digital file.

Aziz, Scott. "With Regulation Looming, It's Time for Industry to Raise the Bar for Software Quality." *Wired*. Condé Nast, 28 Aug. 2014. Web. 31 Mar. 2016.

Badilescu, Simona, and Muthukumaran Packirisamy. *BioMEMS: Science and Engineering Perspectives*. Boca Raton: CRC, 2011. Print.

Bajarin, Tim. "Google Is at a Major Crossroads with Android and Chrome OS." *PCMag*. Ziff Davis, 21 Dec. 2015. Web. 4 Jan. 2016.

Bajo, Javier, et al., eds. *Highlights of Practical Applications of Agents, Multi-Agent Systems, and Sustainability*. Proc. of the International Workshops of PAAMS 2015, June 3–4, 2015, Salamanca, Spain. Cham: Springer, 2015. Print.

Baldé, C. P., et al. *E-waste Statistics: Guidelines on Classification, Reporting and Indicators, 2015*. Bonn: United Nations U, 2015. *United Nations University*. Web. 9 Feb. 2016.

Ball, Phillip. "The Truth about the Turing Test." *Future*. BBC, 24 July 2015. Web. 18 Dec. 2015.

Ballew, Joli, and Ann McIver McHoes. *Operating Systems DeMYSTiFieD*. New York: McGraw, 2012. Print.

Balovich, David. "Sarbanes-Oxley Document Retention and Best Practices." *Creditworthy News*. 3JM Company, 5 Sept. 2007. Web. 1 Apr. 2016.

Banga, Cameron, and Josh Weinhold. *Essential Mobile Interaction Design: Perfecting Interface Design in Mobile*

Apps. Upper Saddle River: Addison-Wesley, 2014. Print.

Baptiste, Philippe, Claude Le Pape, and Wim Nuijten. *Constraint-Based Scheduling: Applying Constraint Programming to Scheduling Problems*. New York: Springer, 2013. Print.

Barski, Conrad. *Land of Lisp. Learn to Program LISP One Game at a Time*. San Francisco, CA: No Starch Press, 2011. Print.

Basagni, Stefano, et al., eds. *Mobile Ad Hoc Networking: Cutting Edge Directions*. 2nd ed. Hoboken: Wiley, 2013. Print.

Bass, Len, Paul Clements, and Rick Kazman. *Software Architecture in Practice*. 3rd ed. Upper Saddle River: Addison, 2012. Print.

Batchelor, Bruce. *Intelligent Image Processing in PROLOG*. London, UK: Springer-Verlag, 1991. Print.

Beattie, Andrew. "Cloud Computing: Why the Buzz?" *Techopedia*. Techopedia, 30 Nov. 2011. Web. 21 Jan. 2016.

Beeler, Robert A., *How to Count: An Introduction to Combinatorics*. New York: Springer, 2015. Print.

Bell, Tim, et al. "Algorithms." *Computer Science Field Guide*. U of Canterbury, 3 Feb. 2015. Web. 19 Jan. 2016.

Bell, Tom. *Programming: A Primer; Coding for Beginners*. London: Imperial Coll. P, 2016. Print.

Belton, Padraig. "Coding the Future: What Will the Future of Computing Look Like?" *BBC News*. BBC, 15 May 2015. Web. 24 Feb. 2016.

Bembenik, Robert, Łukasz Skonieczny, Henryk Rybiński, Marzena Kryszkiewicz, and Marek Niezgódka, eds. *Intelligent Tools for Building a Scientific Information Platform: Advanced Architectures and Solutions*. New York: Springer, 2013. Print.

Ben-Ari, M. *Principles of Concurrent and Distributed Programming*. 2nd ed. New York: Addison, 2006. Print.

Bergin, Michael S. "History of BIM." *Architecture Research Lab*. Architecture Research Lab, 21 Aug. 2011. Web. 31 Jan. 2016.

Berman, P. Gennady, Garu D. Doolen, Ronnie Mainieri, and Vladimir Tsifrinovich. (1998) *Introduction to Quantum Computers*. River Edge, NJ: World Scientific Publishing. Print.

Bernal, Paul. *Internet Privacy Rights: Rights to Protect Autonomy*. New York: Cambridge UP, 2014. Print.

Bernhardt, Chris. *Turing's Vision: The Birth of Computer Science*. MIT Press, 2016.

Bernier, Samuel N., Bertier Luyt, Tatiana Reinhard, and Carl Bass. *Design for 3D Printing: Scanning, Creating, Editing, Remixing, and Making in Three Dimensions*. San Francisco: Maker Media, 2014. Print.

Berns, Andrew, and Sukumar Ghosh. "Dissecting Self-Properties." *SASO 2009: Third IEEE International Conference on Self-Adaptive and Self-Organizing Systems*. Los Alamitos: IEEE, 2009. 10–19. *Andrew Berns: Homepage*. Web. 20 Jan. 2016.

Berry, John D. *Language Culture Type: International Type Design in the Age of Unicode*. New York: Graphis, 2002. Print.

Bessière, Pierre, et al. *Bayesian Programming*. Boca Raton: CRC, 2014. Print.

Bibby, Joe. "Robonaut: Home." *Robonaut*. NASA, 31 May 2013. Web. 21 Jan. 2016.

Biere, Armin, Amir Nahir, and Tanja Vos, eds. *Hardware and Software: Verification and Testing*. New York: Springer, 2013. Print.

Binh, Le Nguyen. *Digital Processing: Optical Transmission and Coherent Receiving Techniques*. Boca Raton: CRC, 2013. Print.

Biometric Center of Excellence. Federal Bureau of Investigation, 2016. Web. 21 Jan. 2016.

"Biomedical Engineers." *Occupational Outlook Handbook, 2016–2017 Edition*. Bureau of Labor Statistics, US Dept. of Labor, 17 Dec. 2015. Web. 23 Jan. 2016.

"Biotechnology." *ACS*. Amer. Chemical Soc., n.d. Web. 27 Jan. 2016.

Bird, Steven, Ewan Klein, and Edward Loper. *Natural Language Processing with Python, 2nd ed*. O'Reilly Media, 2017.

Bishop, Owen (2011) *Electronics. Circuits and Systems* 4th ed., New York, NY: E;sevier. Print.

Bishop, Todd. "Microsoft Exec Admits New Reality: Market Share No Longer 90%— It's 14%." *Geekwire*. GeekWire, 14 July 2014. Web. 30 Jan. 2015.

Black, Jeremy. *The Power of Knowledge: How Information and Technology Made the Modern World*. New Haven: Yale UP, 2014. Print.

Blanco, Xiomara. "Top Tablets with Expandable Storage." *CNET*. CBS Interactive, 25 Jan. 2016. Web. 7 Mar. 2016.

Boashash, Boualem, ed. *Time-Frequency Signal Analysis and Processing: A Comprehensive Reference*. 2nd ed. San Diego: Academic, 2016. Print.

Bone, Simon, and Matias Castro. "A Brief History of Quantum Computing." *SURPRISE* May–June 1997: n. pag. *Department of Computing, Imperial College London*. Web. 24 Mar. 2016.

Booch, Grady, et al. *Object-Oriented Analysis and Design with Applications*. 3rd ed. Upper Saddle River: Addison, 2007. Print.

"Book—Understanding Biometrics." *Griaule Biometrics*. Griaule Biometrics, 2008. Web. 21 Jan. 2016.

Borkar, Shekhar, and Andrew A. Chien. "The Future of Microprocessors." *Communications of the ACM*. ACM, May 2011. Web. 3 Mar. 2016.

Boyle, David. *Alan Turing: Unlocking the Enigma*. CreateSpace Independent Publishing Platform, 2014.

Boyle, Randall, and Raymond R. Panko. *Corporate Computer Security*. 4th ed. Boston: Pearson, 2015. Print.4 Science Reference Center™ Computer Security

Bradley, Tony. "Experts Pick the Top 5 Security Threats for 2015." *PCWorld*. IDG Consumer & SMB, 14 Jan. 2015. Web. 12 Mar. 2016.

Brandom, Russell. "Google Survey Finds More than Five Million Users Infected with Adware." *The Verge*. Vox Media, 6 May 2015. Web. 12 Mar. 2016.

Brauer, Max. *Logic Programming With Prolog*. New York, NY: Springer, 2005. Print.

Briassouli, Alexia, Jenny Benois-Pineau, and Alexander Hauptmann, eds. *Health Monitoring and Personalized Feedback Using Multimedia Data*. Cham: Springer, 2015. Print.

Bright, Peter. "Locking the Bad Guys Out with Asymmetric Encryption." *Ars Technica*. Condé Nast, 12 Feb. 2013. Web. 23 Feb. 2016.

Brindley, Keith (2011) *Starting Electronics* 4th ed., New York, NY: Elsevier. Print.

Brooks, R. R. *Introduction to Computer and Network Security: Navigating Shades of Gray*. Boca Raton: CRC, 2014. Digital file.

Brown, Adrian. *Graphics File Formats*. Kew: Natl. Archives, 2008. PDF file. Digital Preservation Guidance Note 4.

Brownlee, John. "The History of Web Design Explained in 9 GIFs." *Co. Design*. Fast Co., 5 Dec. 2014. Web. 22 Feb. 2016.

Bryden, Douglas. *CAD and Rapid Prototyping for Product Design.* London: King, 2014. Print.

Bucchi, Massimiano, and Brian Trench, eds. *Routledge Handbook of Public Communication of Science and Technology.* 2nd ed. New York: Routledge, 2014. Print.

Burger, John R. *Brain Theory from a Circuits and Systems Perspective: How Electrical Science Explains Neuro-Circuits, Neuro-Systems, and Qubits.* New York: Springer, 2013. Print.

Busch, David D. *Mastering Digital SLR Photography.* 2nd ed. Boston: Thompson Learning, 2008. Print.

Bwalya, Kelvin J., Nathan M. Mnjama, and Peter M. I. I. M. Sebina. *Concepts and Advances in Information Knowledge Management: Studies from Developing and Emerging Economies.* Boston: Elsevier, 2014. Print.

Calude, C. S., and Gheorge Paun, *Computing with Cells and Atoms.* (New York: Taylor & Francis, 2001. Print.

Calude, Cristian S., ed. *The Human Face of Computing.* London: Imperial Coll. P, 2016. Print.

Campbell-Kelly, Martin, William Aspray, Nathan Ensmenger, and Jeffrey R. Yost. *Computer: A History of the Information Machine.* Boulder: Westview, 2014. Print.

Carlson, Wayne. "A Critical History of Computer Graphics and Animation." *Ohio State University.* Ohio State U, 2003. Web. 31 Jan. 2016.

Case, Meredith A., et al. "Accuracy of Smartphone Applications and Wearable Devices for Tracking Physical Activity Data." *Journal of the American Medical Association* 313.6 (2015): 625–26. Web. 2 Mar. 2016.

Cassell, Eric J. *The Nature of Healing: The Modern Practice of Medicine.* New York: Oxford UP, 2013. Print.

"Categories of Free and Nonfree Software." *GNU Operating System.* Free Software Foundation, 1 Jan. 2016. Web. 31 Mar. 2016.

Cavanagh, Joseph (2013) *X86 Assembly Language and C Fundamentals* Boca Raton, FL: CRC Press. Print.

Ceberio, Martine, and Vladik Kreinovich. *Constraint Programming and Decision Making.* New York: Springer, 2014. Print.

Celada, Laura. "What Are the Most Common Graphics File Formats." *FESPA.* FESPA, 27 Mar. 2015. Web. 11 Feb. 2016.

Cervantes, Humberto, and Rick Kazman. *Designing Software Architectures: A Practical Approach.* Upper Saddle River: Addison, 2016. Print.

Cezzar, Juliette. "What Is Graphic Design?" *AIGA.* Amer. Inst. of Graphic Arts, 2016. Web. 16 Mar. 2016.

Cha, Ariana Eunjung. "Health and Data: Can Digital Fitness Monitors Revolutionise Our Lives?" *Guardian.* Guardian News and Media, 19 May 2015. Web. 26 Feb. 2016.

Chan, Melanie. *Virtual Reality: Representations in Contemporary Media.* New York: Bloomsbury, 2014. Print.

Chao, Loretta. "Tech Partnership Looks beyond the Bar Code with Digital Watermarks." *Wall Street Journal.* Dow Jones, 12 Jan. 2016. Web. 14 Mar. 2016.

Cheever, Erik. "Representation of Numbers." *Swarthmore College.* Swarthmore College, n.d. Web. 20 Feb. 2016.

Chen, Yufeng, and Zhiwu Li. *Optimal Supervisory Control of Automated Manufacturing Systems.* Boca Raton: CRC, 2013. Print.

Chivers, Ian, and Jane Sleightholme. *Introduction to Programming With Fortran, with Coverage of Fortran 90, 95, 2003, 2008 and 77.* 3rd ed., New York, NY: Springer, 2015. Print.

"Choose a Website Builder: 14 Top Tools." *Creative Bloq.* Future, 8 Feb. 2016. Web. 22 Feb. 2016.

Christiano, Marie. "What Are Integrated Development Environments?" *All about Circuits.* EETech Media, 3 Aug. 2015. Web. 23 Feb. 2016.

Chua, Chee Kai, Kah Fai Leong, and Chu Sing Lim. *Rapid Prototyping: Principles and Applications.* Hackensack: World Scientific, 2010. Print.

Clark, Alexander, Chris Fox, and Shalom Lappin (eds.) *The Handbook of Computational Linguistics and Natural Language Processing.* Wiley-Blackwell, 2012.

Clark, Don. "Smart-Home Gadgets Still a Hard Sell." *Wall Street Journal.* Dow Jones, 5 Jan. 2016. Web. 12 Mar. 2016.

Clarke, Dave, James Noble, and Tobias Wrigstad, eds. *Aliasing in Object-Oriented Programming: Types, Analysis and Verification.* Berlin: Springer, 2013. Print.

Clements, Alan (2006) *Principles of Computer Hardware* 4th ed., New York, NY: Oxford University Press. Print.

Clements, Paul, et al. *Documenting Software Architectures: Views and Beyond.* 2nd ed. Upper Saddle River: Addison, 2011. Print.

Cline, Hugh F. *Information Communication Technology and Social Transformation: A Social and Historical Perspective.* New York: Routledge, 2014. Print.

Coelho, Helder, and José C. Cotta. *Prolog by Example: How to Learn, Teach and Use It.* New York, NY: Springer, 1996. Print.

Coleman, E Gabriella. *Coding Freedom: The Ethics and Aesthetics of Hacking.* Princeton: Princeton UP, 2013. Print.

Collins, Lauren, and Scott Ellis. *Mobile Devices: Tools and Technologies.* Boca Raton: CRC, 2015. Print.

Collins, Mike. *Pro Tools 11: Music Production, Recording, Editing, and Mixing.* Burlington: Focal, 2014. Print.

"Combinatorics." *Mathigon.* Mathigon, 2015. Web. 10 Feb. 2016.

Comer, Douglas E. *Computer Networks and Internets.* 6th ed. Boston: Pearson, 2015. Print.

"CompTIA A+." *CompTIA.* Computing Technology Industry Assn., 2015. Web. 31 Jan. 2016.

"Computer Crime Laws." *Frontline.* WGBH Educ. Foundation, 2014. Web. 28 Mar. 2016.

"Computer-Aided Design (CAD) and Computer-Aided Manufacturing (CAM)." *Inc.* Mansueto Ventures, n.d. Web. 31 Jan. 2016.

"Computer—Networking." *Tutorials Point.* Tutorials Point, 2016. Web. 22 Feb. 2016.

"Computer Fraud and Abuse Act (CFAA)." *Internet Law Treatise.* Electronic Frontier Foundation, 24 Apr. 2013. Web. 31 Mar. 2016.

Comstock, Jonah. "Eight Years of Fitbit News Leading Up to Its Planned IPO." *MobiHealthNews.* HIMSS Media, 11 May 2015. Web. 28 Feb. 2016.

Connolly, Thomas M., and Carolyn E. Begg. *Database Systems: A Practical Approach to Design, Implementation, and Management.* 6th ed. Boston: Pearson, 2015. Print.

Cooper, Stephen. "Motherboard Design Process." *MBReview.com.* Author, 4 Sept. 2009. Web. 14 Mar. 2016.

Corbet, Jonathan, Alessandro Rubini, and Greg Kroah-Hartman. *Linux Device Drivers.* 3rd ed. Cambridge: O'Reilly, 2005. Print.

Cormen, Thomas H. *Algorithms Unlocked.* Cambridge: MIT P, 2013. Print.

Cormen, Thomas H., et al. *Introduction to Algorithms.* 3rd ed. Cambridge: MIT P, 2009. Print.

Costello, Vic, Susan Youngblood, and Norman E. Youngblood. *Multimedia Foundations: Core Concepts for Digital Design.* New York: Focal, 2012. Print.

Counihan, Martin *Fortran 95.* London, UK: University College Press. 1996. Print.

Couts, Andrew. "Drones 101: A Beginner's Guide to Taking Flight, No License Needed." *Digital Trends.* Designtechnica, 16 Nov. 2013. Web. 27 Jan. 2015.

Cox, Ingemar J., Jessica Fridrich, Matthew L. Miller, Jeffrey A. Bloom, and Ton Kalker. "Practical Dirty-Paper Codes." *Digital Watermarking and Steganography.* 2nd ed. Amsterdam: Elsevier, 2008. 183–212. Digital file.

Cross, Mark. *Audio Post Production for Film and Television.* Boston: Berklee, 2013. Print.

Crothers, Brooke. "Microsoft Explains Quantum Computing So Even You Can Understand." *CNET.* CBS Interactive, 25 July 2014. Web. 24 Mar. 2016.

Dale, Nell, and John Lewis. *Computer Science Illuminated.* 6th ed. Burlington: Jones, 2016. Print.

Dancyger, Ken. *The Technique of Film and Video Editing: History, Theory, and Practice.* Burlington: Focal, 2013. Digital file.

Dastbaz, Mohammad, Colin Pattinson, and Bakbak Akhgar, eds. *Green Information Technology: A Sustainable Approach.* Waltham: Elsevier, 2015. Print.

Davies, Alan. *An Introduction to Applied Linguistics: From Practice to Theory.* 2nd ed. Edinburgh: Edinburgh UP, 2007. Print.

Davis, Stephen R. (2015) *Beginning Programming with C++ for Dummies* 2nd ed. Joboken, NJ: John Wiley & Sons. Print.

"DCN—Computer Network Types." *Tutorials Point.* Tutorials Point, 2016. Web. 22 Feb. 2016.

De Filippi, Primavera. "It's Time to Take Mesh Networks Seriously (and Not Just for the Reasons You Think)." *Wired.* Condé Nast, 2 Jan. 2014. Web. 22 Feb. 2016.

De George, Andy. "How to Autoscale an Application." *Microsoft Azure.* Microsoft, 7 Dec. 2015. Web. 17 Feb. 2016.

Deitel, H.M. And Deitel, P.J. (2009) *C++ for Programmers* Upper Saddle River, NJ: Pearson Education Incorporated. Print.

Delforge, Pierre. "America's Data Centers Consuming and Wasting Growing Amounts of Energy." *NRDC.* Natural Resources Defense Council, 6 Feb. 2015. Web. 17 Mar. 2016.

Delfs, Hans, and Helmut Knebl. *Introduction to Cryptography: Principles and Applications.* 3rd ed. Berlin: Springer, 2015. Print.

Dennis, Alan, Barbara Haley Wixom, and David Tegarden. *Systems Analysis and Design: An Object-Oriented Approach with UML.* 5th ed. Hoboken: Wiley, 2015. Print.

Deransart, P., A. Ed-Dbali, and L. Cervoni. *PROLOG: The Standard Reference Manual.* New York, NY: Springer, 1996. Print.

"Design and Technology: Manufacturing Processes." *GCSE Bitesize.* BBC, 2014. Web. 31 Jan. 2016.

"Designing for Screen Reader Compatibility." *WebAIM.* Center for Persons with Disabilities, Utah State U, 19 Nov. 2014. Web. 22 Feb. 2016.

Devroye, N., P. Mitran, and V. Tarokh. "Limits on Communications in a Cognitive Radio Channel." *IEEE Communication Magazine* 44.6 (2006): 4449. *Inspec.* Web. 9 Mar. 2016.

Devroye, Natasha, Patrick Mitran and Vahid Tarokh. *On Cognitive Graphs: Decomposing Wireless Networks.* New York: Wiley Interscience, 2006. Print.

Dey, Pradip, and Manas Ghosh. *Computer Fundamentals and Programming in C.* 2nd ed. New Delhi: Oxford UP, 2013. Print.

Dice, Pete. *Quick Boot: A Guide for Embedded Firmware Developers.* Hillsboro: Intel, 2012. Print.

"Differences between Multithreading and Multitasking for Programmers." *NI.* National Instruments, 20 Jan. 2014. Web. 15 Mar 2016.

"Digital Evidence and Forensics." *National Institute of Justice.* Office of Justice Programs, 28 Oct. 2015. Web. 12 Feb. 2016.

Dillinger, Markus, Kambiz Madani, and Nancy Alonistioti. *Software Defined Radio: Architectures, Systems, and Functions.* Hoboken: Wiley, 2003. Print.

Doeppner, Thomas W. *Operating Systems in Depth.* Hoboken: Wiley, 2011. Print.

Dor, Daniel. *The Instruction of Imagination: Language as a Social Communication Technology.* New York: Oxford UP, 2015. Print.

Drake, Joshua J. *Android Hacker's Handbook.* Indianapolis: Wiley, 2014. Print.

Duncan, Geoff. "Can the Government Regulate Internet Privacy?" *Digital Trends.* Designtechnica, 21 Apr. 2014. Web. 28 Mar. 2016.

Duntemann, Jeff (2011) *Assembly Language Programming Step-by-Step: Programming with Linux* 3rd ed., Hoboken, NJ: John Wiley & Sons. Print.

Dutson, Phil. *Responsive Mobile Design: Designing for Every Device.* Upper Saddle River: Addison-Wesley, 2015. Print.

Dyson, George. *Turing's Cathedral: The Origins of the Digital Universe.* Vintage, 2012.

Edwards, Benj. "From Paper Tape to Data Sticks: The Evolution of Removable Storage." *PCWorld.* IDG Consumer & SMB, 7 Feb. 2010. Web. 7 Mar. 2016.

Edwards, Jim. "Proof That Android Really Is for the Poor." *Business Insider.* Business Insider, 27 June 2014. Web. 4 Jan. 2016.

Edwards, Paul N. *A Vast Machine: Computer Models, Climate Data, and the Politics of Global Warming.* Cambridge: MIT P, 2010. Print.

El-Rewini, Hesham, and Mostafa Abd-El-Barr. *Advanced Computer Architecture and Parallel Processing.* Hoboken: Wiley, 2005. Print.

Elahi, Ata, and Mehran Elahi. *Data, Network, and Internet Communications Technology.* Clifton Park: Thomson, 2006. Print.

Elenkov, Nikolay. *Android Security Internals: An In-Depth Guide to Android's Security Architecture.* San Francisco: No Starch, 2015. Print.

Enderle, John Denis, and Joseph D. Brozino. *Introduction to Biomedical Engineering.* 3rd ed. Burlington: Elsevier, 2012. Print.

Englander, Irv. *The Architecture of Computer Hardware, Systems Software, & Networking: An Information Technology Approach.* 5th ed. Hoboken: Wiley, 2014. Print.

Erben, Tony, Ruth Ban, and Martha E. Castañeda. *Teaching English Language Learners through Technology.* New York: Routledge, 2009. Print.

Esslinger, Bernhard, et al. *The CrypTool Script: Cryptography, Mathematics, and More.* 11th ed. Frankfurt: CrypTool, 2013. *CrypTool Portal.* Web. 2 Mar. 2016.

Evans, Kirk. "Autoscaling Azure—Virtual Machines." *.NET from a Markup Perspective.* Microsoft, 20 Feb. 2015. Web. 17 Feb. 2016.

Ewing, Martin. *The ABCs of Software Defined Radio.* Hartford: Amer. Radio Relay League, 2012. Print.

"Examples and Explanations of BME." *Biomedical Engineering Society.* Biomedical Engineering Soc., 2012–14. Web. 23 Jan. 2016.

Fasano, Philip. *Transforming Health Care: The Financial Impact of Technology, Electronic Tools and Data Mining.* Hoboken: Wiley, 2013. Print.

Faticoni, Theodore G., *Combinatorics: An Introduction.* New York: Wiley, 2014. Digital file.

Feng, Wu-chun, ed. *The Green Computing Book: Tackling Energy Efficiency at Large Scale.* Boca Raton: CRC, 2014. Print.

"Fitbit Flex Teardown." *iFixit.* iFixit, 2013. Web. 28 Feb. 2016.

Firtman, Maximiliano R. *Programming the Mobile Web.* Sebastopol: O'Reilly Media, 2013. Print.

Fischer, Eric. *The Evolution of Character Codes, 1874–1968.* N.p.: Fischer, n.d. *Trafficways.org.* Web. 22 Feb. 2016.

Flach, Peter. *Machine Learning: The Art and Science of Algorithms that Make Sense of Data.* Cambridge University Press, 2012.

Follin, Steve. "Preparing for IT Infrastructure Autonomics." *IndustryWeek.* Penton, 19 Nov. 2015. Web. 20 Jan. 2016.

Foote, Steven. *Learning to Program.* Upper Saddle River: Pearson, 2015. Print.

Fountain, T J. *Parallel Computing: Principles and Practice.* Cambridge: Cambridge University Press, 1994. Print.

Fowler, Geoffrey, A. "The Drones on Autopilot That Follow Your Lead (Usually)." *Wall Street Journal.* Dow Jones, 23 Dec. 2014. Web. 20 Jan. 2016.

Fox, Richard. *Information Technology: An Introduction for Today's Digital World.* Boca Raton: CRC, 2013. Print.

France, Anna Kaziunas, comp. *Make: 3D Printing—The Essential Guide to 3D Printers.* Sebastopol: Maker Media, 2013. Print.

Franceschi-Bicchierai, Lorenzo. "Love Bug: The Virus That Hit 50 Million People Turns 15." *Motherboard.* Vice Media, 4 May 2015. Web. 16 Mar. 2016.

Freedman, Jeri. *Software Development.* New York: Cavendish Square, 2015. Print.

Freeman, Michael. *Digital Image Editing & Special Effects: Quickly Master the Key Techniques of Photoshop & Lightroom.* New York: Focal, 2013. Print.

"Frequently Asked Questions." *Digital Watermarking Alliance.* DWA, n.d. Web. 11 Mar. 2016.

Frenzel, Louis E., Jr. *Electronics Explained: The New Systems Approach to Learning Electronics.* Burlington: Elsevier, 2010. Print.

Friedman, Daniel P., and Mitchell Wand. *Essentials of Programming Languages.* Cambridge: MIT P, 2006. Print.

Fuchs, Christian, and Marisol Sandoval, eds. *Critique, Social Media and the Information Society.* New York: Routledge, 2014. Print.

Gaffney, Alexander. "FDA Confirms It Won't Regulate Apps or Devices Which Store Patient Data." *Regulatory Affairs Professionals Society.* Regulatory Affairs Professional Soc., 6 Feb. 2015. Web. 31 Mar. 2016.

Galer, Mark, and Philip Andrews. *Photoshop CC Essential Skills: A Guide to Creative Image Editing.* New York: Focal, 2014. Print.

Gallagher, Sean. "Cortana for All: Microsoft's Plan to Put Voice Recognition behind Anything." *Ars Technica.* Condé Nast, 15 May 2015. Web. 21 Mar. 2016.

Gallagher, Sean. "'Locky' Crypto-Ransomware Rides In on Malicious Word Document Macro." *Ars Technica.* Condé Nast, 17 Feb. 2016. Web. 16 Mar. 2016.

Gallagher, Sean. "Though 'Barely an Operating System,' DOS Still Matters (to Some People)." *Ars Technica.* Condé Nast, 14 July 2014. Web. 31 Jan. 2016.

Galushkin, Alexander I. *Neural Network Theory.* New York, NY: Springer, 2007. Print.

Gancarz, Mike. *The UNIX Philosophy.* Woburn: Butterworth, 1995. Print.

Garrido, José M., Richard Schlesinger, and Kenneth E. Hoganson. *Principles of Modern Operating Systems.* 2nd ed. Burlington: Jones, 2013. Print.

Garza, George. "Working with the Cons of Object Oriented Programming." Ed. Linda Richter. *Bright Hub.* Bright Hub, 19 May 2011. Web. 6 Feb. 2016.

Gaudin, Sharon. "Quantum Computing May Be Moving out of Science Fiction." *Computerworld.* Computerworld, 15 Dec. 2015. Web. 24 Mar. 2016.

Gee, James Paul. *Unified Discourse Analysis: Language, Reality, Virtual Worlds, and Video Games.* New York: Routledge, 2015. Print.

Gehrke, Wilhelm. *Fortran 90 Language Guide.* New York, NY: Springer, 1995. Print.

"Ghana: Digital Dumping Ground." *Frontline.* PBS, 23 June 2009. Web. 29 Jan. 2016.

Gibbs, Samuel. "From Windows 1 to Windows 10: 29 Years of Windows Evolution." *Guardian*. Guardian News and Media, 2 Oct. 2014. Web. 2 Jan. 2016.

Gibbs, Samuel. "Google's Massive Humanoid Robot Can Now Walk and Move without Wires." *Guardian*. Guardian News and Media, 21 Jan. 2015. Web. 21 Jan. 2016.

Gibbs, W. Wayt. "Autonomic Computing." *Scientific American*. Nature Amer., 6 May 2002. Web. 20 Jan. 2016.

Gibson, J.R. (2011) *Electronic Logic Circuits* 3rd ed., New York, NY: Routledge. Print.

Gibson, Jerry D., ed. *Mobile Communications Handbook*. 3rd ed. Boca Raton: CRC, 2013. Print.

Gilder, Jules H. (1986) *Apple IIc and IIe Assembly Language* New York, NY: Chapman and Hall. Print.

Gillam, Richard. *Unicode Demystified: A Practical Programmer's Guide to the Encoding Standard*. Boston: Addison-Wesley, 2002. Print.

Gillespie, Tarleton, Pablo J. Boczkowski, and Kirsten A. Foot, eds. *Media Technologies: Essays on Communication, Materiality, and Society*. Cambridge: MIT P, 2014. Print.

Glanz, James. "Power, Pollution and the Internet." *New York Times*. New York Times, 22 Sept. 2012. Web. 28 Feb. 2016.

Glaser, Anton. *History of Binary and Other Nondecimal Numeration*. Rev. ed. Los Angeles: Tomash, 1981. Print.

Glaser, J. D. *Secure Development for Mobile Apps: How to Design and Code Secure Mobile Applications with PHP and JavaScript*. Boca Raton: CRC, 2015. Print.

Glaubitz, John Paul Adrian. "Modern Consumerism and the Waste Problem." *ArXiv.org*. Cornell U, 4 June 2012. Web. 9 Feb. 2016.

Glink, Ilyce. "10 Smart Home Features Buyers Actually Want." *CBS News*. CBS Interactive, 11 Apr. 2015. Web. 12 Mar. 2016.

Godbey, W. T. *An Introduction to Biotechnology: The Science, Technology and Medical Applications*. Waltham: Academic, 2014. Print.

Goelker, Klaus. *Gimp 2.8 for Photographers: Image Editing with Open Source Software*. Santa Barbara: Rocky Nook, 2013. Print.

Gogolin, Greg. *Digital Forensics Explained*. Boca Raton: CRC, 2013. Print.

Goldsborough, Reid. "Android on the Rise." *Tech Directions* May 2014: 12. *Academic Search Complete*. Web. 2 Jan. 2016.

Gonzalez, Teofilo, and Jorge Díaz-Herrera, eds. *Computing Handbook: Computer Science and Software Engineering*. 3rd ed. Boca Raton: CRC, 2014. Print.

Goode, Lauren. "Fitbit Hit with Class-Action Suit over Inaccurate Heart Rate Monitoring." *Verge*. Vox Media, 6 Jan. 2016. Web. 28 Feb. 2016.

Goodman, Robert, and Patrick McGrath. *Editing Digital Video: The Complete Creative and Technical Guide*. New York: McGraw, 2003. Print.

Goodwins, Rupert. "The Future of Storage: 2015 and Beyond." *ZDNet*. CBS Interactive, 1 Jan. 2015. Web. 7 Mar. 2016.

Goriunova, Olga, ed. *Fun and Software: Exploring Pleasure, Paradox, and Pain in Computing*. New York: Bloomsbury, 2014. Print.

Govindjee, S. *Internal Representation of Numbers*. Dept. of Civil and Environmental Engineering, U of California Berkeley, Spring 2013. Digital File.

Graham W. Seed (2012) *An Introduction to Object-Oriented Programming in C++ with Applications in Computer Graphics* New York, NY: Springer Science+Business Media. Print.

Graham, Tony. *Unicode: A Primer*. Foster City: M&T, 2000. Print.

"Graphical User Interface (GUI)." *Techopedia*. Techopedia, n.d. Web. 5 Feb. 2016.

Graupe, Daniel. *Principles of Artificial Neural Networks*. 2nd ed. Hackensack, NJ: World Scientific, 2007. Print.

Grayver, Eugene. *Implementing Software Defined Radio*. New York: Springer, 2012. Print.

Greenberg, Saul, et al. *Sketching User Experiences: The Workbook*. Waltham: Morgan, 2012. Print. 4 Science Reference Center™ Wireframes

Griffiths, Devin C. *Virtual Ascendance: Video Games and the Remaking of Reality*. Lanham: Rowman, 2013. Print.

Guichard, David. "An Introduction to Combinatorics and Graph Theory." *Whitman*. Whitman Coll., 4 Jan 2016. Web. 10 Feb. 2016.

Gulchak, Daniel J. "Using a Mobile Handheld Computer to Teach a Student with an Emotional and Behavioral Disorder to Self-Monitor Attention." *Education and Treatment of Children* 31.4 (2008): 567–81. PDF file.

Gupta, Shalene. "For the Disabled, Smart Homes Are Home Sweet Home." *Fortune*. Fortune, 1 Feb. 2015. Web. 15 Mar. 2016.

Gupta, Siddarth, and Vagesh Porwal. "Recent Digital Watermarking Approaches, Protecting

Multimedia Data Ownership." *Advances in Computer Science* 4.2 (2015): 21–30. Web. 14 Mar. 2016.

Habiballah, N., M. Qjani, A. Arbaoui, and J. Dumas. "Effect of a Gaussian White Noise on the Charge Density Wave Dynamics in a One Dimensional Compound." *Journal of Physics and Chemistry of Solids* 75.1 (2014): 153–56. *Inspec.* Web. 9 Mar. 2016.

Haerens, Margaret, and Lynn M. Zott, eds. *Hacking and Hackers.* Detroit: Greenhaven, 2014. Print.

Hagan, Martin T., Howard B. Demuth, , Mark H. Beale, and Orlando de Jesús. *Neural Network Design.* 2nd ed. Martin Hagan, 2014. Print.

Hagen, Rebecca, and Kim Golombisky. *White Space Is Not Your Enemy: A Beginner's Guide to Communicating Visually through Graphic, Web & Multimedia Design.* 2nd ed. Burlington: Focal, 2013. Print.

Hamburg, Margaret A., and Francis S. Collins. "The Path to Personalized Medicine." *New England Journal of Medicine* 363.4 (2010): 301–4. Web. 23 Dec. 2015.

Hamm, Matthew J. *Wireframing Essentials: An Introduction to User Experience Design.* Birmingham: Packt, 2014. Print.

Harbour, Jonathan S. *Beginning Game Programming.* 4th ed. Boston: Cengage, 2015. Print.

Harel, Jacob. "SynthOS and Task-Oriented Programming." *Embedded Computing Design.* Embedded Computing Design, 2 Feb. 2016. Web. 7 Feb. 2016.

Harper, Robert. *Practical Foundations for Programming Languages.* Cambridge: Cambridge UP, 2013. Print.

Harris, David Money, and Sarah L. Harris. *Digital Design and Computer Architecture.* 2nd ed. Waltham: Morgan, 2013. Print.

Harrison, Virginia, and Jose Pagliery. "Nearly 1 Million New Malware Threats Released Every Day." *CNNMoney.* Cable News Network, 14 Apr. 2015. Web. 16 Mar. 2016.

Hart, Archibald D., and Sylvia Hart Frejd. *The Digital Invasion: How Technology Is Shaping You and Your Relationships.* Grand Rapids: Baker, 2013. Print.

Harth, Andreas, Katja Hose, and Ralf Schenkel, eds. *Linked Data Management.* Boca Raton: CRC, 2014. Print.

Haug, Hartmut, and Stephan W. Koch. *Quantum Theory of the Optical and Electronic Properties of Semiconductors.* New Jersey [u.a.]: World Scientific, 2009. Print.

Havens, John C. *Hacking Happiness: Why Your Personal Data Counts and How Tracking It Can Change the World.* New York: Tarcher, 2014. Print.

"Health Information Privacy." *HHS.gov.* Dept. of Health and Human Services, n.d. Web. 28 Mar. 2016.

Heisler, Yoni. "The History and Evolution of iOS, from the Original iPhone to iOS 9." *BGR.* BGR Media, 12 Feb. 2016. Web. 26 Feb. 2016.

Henz, Martin. *Objects for Concurrent Constraint Programming.* New York: Springer, 1998. Print.

Herlihy, Maurice, and Nir Shavit. *The Art of Multiprocessor Programming.* New York: Elsevier, 2012. Print.

Herrman, John. "How to Get Started: 3D Modeling and Printing." *Popular Mechanics.* Hearst Digital Media, 15 Mar. 2012. Web. 31 Jan. 2016.

Hey, Tony, and Gyuri Pápay. *The Computing Universe: A Journey through a Revolution.* New York: Cambridge UP, 2015. Print.

Higginbotham, Stacey. "5 Reasons Why the 'Smart Home' Is Still Stupid." *Fortune.* Fortune, 19 Aug. 2015. Web. 12 Mar. 2016.

Highfield, Roger. "Fast Forward to Cartoon Reality." *Telegraph.* Telegraph Media Group, 13 June 2006. Web. 31 Jan. 2016.

Hill, Mark D. "What Is Scalability?" *Scalable Shared Memory Multiprocessors.* Ed. Michel Dubois and Shreekant Thakkar. New York: Springer, 1992. 89–96. Print.

Hillis, W. Daniel. "Richard Feynman and the Connection Machine." *Phys. Today Physics Today* 42.2 (1989): 78. Web.

Hirvensalo, Mike. *Quantum Computing.* New York, NY: Springer, 2001. Print.

"History and Timeline." *Open Group.* Open Group, n.d. Web. 28 Feb. 2016.

History of Cryptography: An Easy to Understand History of Cryptography. N.p.: Thawte, 2013. *Thawte.* Web. 4 Feb. 2016.

Hodges, Andrew. *Alan Turing: The Enigma: The Book That Inspired the Film "The Imitation Game."* Princeton University Press, 2014.

Hof, Robert. "How Fitbit Survived as a Hardware Startup." *Forbes.* Forbes.com, 4 Feb. 2014. Web. 28 Feb. 2016.

Hoffer, Jeffrey A., V. Ramesh, and Heikki Topi. *Modern Database Management.* 12th ed. Boston: Pearson, 2016. Print.

Hoffstein, Jeffrey, Jill Pipher, and Joseph H. Silverman. *An Introduction to Mathematical Cryptography.* 2nd ed. New York: Springer, 2014. Print.

Hofstedt, Petra. *Multiparadigm Constraint Programming Languages.* New York: Springer, 2013. Print.

Holcombe, Jane, and Charles Holcombe. *Survey of Operating Systems.* New York: McGraw, 2015. Print.

Holland, Bill. "Software Testing: A History." *SitePoint.* SitePoint, 15 Feb. 2012. Web. 9 Feb. 2016.

Holleley, Douglas. *Photo-Editing and Presentation: A Guide to Image Editing and Presentation for Photographers and Visual Artists.* Rochester: Clarellen, 2009. Print.

Holt, Thomas J., Adam M. Bossler, and Kathryn C. Seigfried-Spellar. *Cybercrime and Digital Forensics: An Introduction.* New York: Routledge, 2015. Print.

Hopfield, J. J. "Neural Networks and Physical Systems with Emergent Collective Computational Abilities." *Proceedings of the National Academy of Sciences* 79.8 (1982): 2554-558. Web.

Horspool, Nigel, and Nikolai Tillmann. *Touchdevelop: Programming on the Go.* New York: Apress, 2013. Print.

Horvath, Joan. *Mastering 3D Printing: Modeling, Printing, and Prototyping with Reprap- Style 3D Printers.* Berkeley: Apress, 2014. Print.

Hoskins, Stephen. *3D Printing for Artists, Designers and Makers.* London: Bloomsbury, 2013. Print.

"How Firewalls Work." *Boston University Information Services and Technology.* Boston U, n.d. Web. 28 Feb. 2016.

"How Speech-Recognition Software Got So Good." *Economist.* Economist Newspaper, 22 Apr 2014. Web. 21 Mar. 2016.

Hsu, John Y. (2002) *Computer Logic Design Principles and Applications* New York, NY: Springer. Print.

"HTML Tutorial." *Tutorials Point.* Tutorials Point, 2016. Web. 22 Feb. 2016.

Hughes, Cameron, and Tracey Hughes. *Parallel and Distributed Programming Using C++.* Boston: Addison-Wesley, 2004. Print.

Hughes, John F. *Computer Graphics: Principles and Practice.* Upper Saddle River: Addison, 2014. Print.

Human Brain Project. Human Brain Project, 2013. Web. 16 Feb. 2016.

Hutchinson, Lee. "Home 3D Printers Take Us on a Maddening Journey into Another Dimension." *Ars Technica.* Condé Nast, 27 Aug. 2013. Web. 6 Jan. 2016.

Huth, Alexa, and James Cebula. *The Basics of Cloud Computing.* N.p.: Carnegie Mellon U and US Computer Emergency Readiness Team, 2011. PDF file.

Hyde, Randall (2010) *The Art of Assembly Language* 2nd ed., San Francisco, CA: No Starch Press. Print.

Hyde, Randall. *Write Great Code: Understanding the Machine.* Vol. 1. San Francisco: No Starch, 2005. Print.

Information Resources Management Association, ed. *Assistive Technologies: Concepts, Methodologies, Tools, and Applications.* Vol. 1. Hershey: Information Science Reference, 2014. Print.

Ingersoll, Grant S., Thomas S. Morton, and Drew Farris. *Taming Text: How to Find, Organize, and Manipulate It.* Manning Publications, 2013.

Ingham, Kenneth, and Stephanie Forrest. *A History and Survey of Network Firewalls.* Albuquerque: U of New Mexico, 2002. PDF file.

Iniewski, Krzysztof. *Embedded Systems: Hardware, Design, and Implementation.* Hoboken: Wiley, 2013. Print.

"Internet Privacy." *ACLU.* American Civil Liberties Union, 2015. Web. 28 Feb. 2016.

"Intro to Algorithms." *Khan Academy.* Khan Acad., 2015. Web. 19 Jan. 2016.

"Introduction to Biometrics." *Biometrics.gov.* Biometrics.gov, 2006. Web. 21 Jan. 2016.

"Introduction to Biotechnology." *Center for Bioenergy and Photosynthesis.* Arizona State U, 13 Feb. 2006. Web. 20 Jan 2016.

"Introduction to Image Files Tutorial." *Boston University Information Services and Technology.* Boston U, n.d. Web. 11 Feb. 2016.

Intro to Biotechnology: Techniques and Applications. Cambridge: NPG Educ., 2010. *Scitable.* Web. 20 Jan. 2016.

"iOS: A Visual History." *Verge.* Vox Media, 16 Sept. 2013. Web. 24 Feb. 2016.

ITL Education Solutions. *Introduction to Information Technology.* 2nd ed. Delhi: Pearson, 2012. Print.

Iversen, Jakob, and Michael Eierman. *Learning Mobile App Development: A Hands-On Guide to Building Apps with iOS and Android.* Upper Saddle River: Addison-Wesley, 2014. Print.

Ives, Mike. "Boom in Mining Rare Earths Poses Mounting Toxic Risks." *Environment 360*. Yale U, 28 Jan. 2013. Web. 28 Feb. 2016.

Jacobson, Douglas, and Joseph Idziorek. *Computer Security Literacy: Staying Safe in a Digital World*. Boca Raton: CRC, 2013. Print.

Jadhav, S.S. (2008) *Advanced Computer Architecture & Computing*. Pune, IND: Technical Publishers, 2008. Print.

Jain, Anil K., Arun A. Ross, and Karthik Nandakumar. *Introduction to Biometrics*. New York: Springer, 2011. Print.

Janert, Philipp K. *Feedback Control for Computer Systems*. Sebastopol: O'Reilly, 2014. Print.

Jeannot, Emmanuel, and J. Žilinskas. *High Performance Computing on Complex Environments*. Hoboken: Wiley, 2014. Print.

Jennings, Tom. "An Annotated History of Some Character Codes." *World Power Systems*. Tom Jennings, 29 Oct. 2004. Web. 16 Feb. 2016.

Johnson, C. Richard, and William A. Sethares. *Telecommunications Breakdown: Concepts of Communication Transmitted via Software-Defined Radio*. New York: Prentice, 2003. Print.

Johnson, Jeff. *Designing with the Mind in Mind*. 2nd ed. Waltham: Morgan, 2014. Print.

Jones, M. Tim, *Artificial Intelligence: A Systems Approach*. Sudbury, MA: Jones and Bartlett, 2009. Print.

Jorgensen, Paul C. *Software Testing: A Craftsman's Approach*. 4th ed. Boca Raton: CRC, 2014. Print.

Jurafsky, Daniel, and James H. Martin. *Speech and Language Processing: An Introduction to Natural Language Processing, Computational Linguistics, and Speech Recognition*, PEL, 2008.

Kale, Vivek. *Guide to Cloud Computing for Business and Technology Managers*. Boca Raton: CRC, 2015. Print.

Kaptelinin, Victor, and Mary P. Czerwinski, eds. *Beyond the Desktop Metaphor: Designing Integrated Digital Work Environments*. Cambridge: MIT P, 2007. Print.

Katoh, Shigeo, Jun-ichi Horiuchi, and Fumitake Yoshida. *Biochemical Engineering: A Textbook for Engineers, Chemists and Biologists*. 2nd rev. and enl. ed. Weinheim: Wiley, 2015. Print.

Katz, Jonathan, and Yehuda Lindell. *Introduction to Modern Cryptography*. 2nd ed. Boca Raton: CRC, 2015. Print.

Kay, Roger. "Behind Apple's Siri Lies Nuance's Speech Recognition." *Forbes*. Forbes. com, 24 Mar. 2014. Web. 21 Mar. 2016.

Kaye, Phillip, Raymond Laflamme, and Michele Mosea. *An Introduction to Quantum Computing*. New York, NU: Oxford University Press, 2007. Print.

Kefauver, Alan P., and David Patschke. *Fundamentals of Digital Audio*. Middleton: A-R Editions, 2007. Print.

Kelly, Gordon, "Apple iOS 9: 11 Important New Features." *Forbes*. Forbes.com, 16 Sept. 2015. Web. 28 Feb. 2016.

Kemeny, John G. and Thomas E. Kurtz. *Back To BASIC: The History, Corruption, and Future of the Language*. Boston, MA: Addison-Wesley, 1985. Print.

Kernighan, Brian W. and Dennis M. Ritchie. *The C Programming Language*. Englewood Cliffs, NJ: Prentice-Hall, 1978. Print.

Khan, Gul N., and Krzysztof Iniewski, eds. *Embedded and Networking Systems: Design, Software, and Implementation*. Boca Raton: CRC, 2014. Print.

Khan, Shafiullah, and Al-Sakib Khan Pathan, eds. *Wireless Networks and Security: Issues, Challenges and Research Trends*. Berlin: Springer, 2013. Print.

Kilkelly, Michael. "Which Architectural Software Should You Be Using?" *ArchDaily*. ArchDaily, 4 May 2015. Web. 31 Jan. 2016.

Kilper, Daniel C., and Tucker, Rodney S. "Energy-Efficient Telecommunications." *Optical Fiber Telecommunications*. 6th ed. N.p.: Elsevier, 2013. 747–91. Digital file.

Kim, Chang-Hun, et al. *Real-Time Visual Effects for Game Programming*. Singapore: Springer, 2015. Print.

Kirk, David B, and Wen-mei Hwu. *Programming Massively Parallel Processors: A Hands-on Approach*. Burlington, Massachusetts: Morgan Kaufmann Elsevier, 2013. Print.

Kirkwood, Patricia Elaine, and Necia T. Parker-Gibson. *Informing Chemical Engineering Decisions with Data, Research, and Government Resources*. San Rafael: Morgan, 2013. Digital file.

Kizza, Joseph Migga. *Ethical and Social Issues in the Information Age*. 5th ed. London: Springer, 2013. Print.

Kizza, Joseph Migga. *Guide to Computer Network Security*. 3rd ed. London: Springer, 2015. Print.

Klimczak, Erik. *Design for Software: A Playbook for Developers.* Hoboken: Wiley, 2013. Print.

"Knowledge Base: Technologies in 3D Printing." *DesignTech.* DesignTech Systems, n.d. Web. 6 Jan. 2016.

Köhler, Anna, and Heinz Bässler. *Electronic Processes in Organic Semiconductors: An Introduction.* Wiley-VHC Verlag, 2015. Print.

Kojić, Miloš, et al. *Computer Modeling in Bioengineering: Theoretical Background, Examples and Software.* Hoboken: Wiley, 2008. Print.

Könenkamp, Rolf. *Photoelectric Properties and Applications of Low-Mobility Semiconductors.* Berlin: Springer, 2000. Print.

Korpela, Jukka K. *Unicode Explained.* Sebastopol: O'Reilly Media, 2006. Print.

Kosky, Philip, et al. *Exploring Engineering: An Introduction to Engineering and Design.* 4th ed. Waltham: Academic, 2016. Print.

Kramer, Bill. *The Autocadet's Guide to Visual LISP.* Laurence, KS: CMP Books, 2002. Print.

Krar, Steve, Arthur Gill, and Peter Smid. *Computer Numerical Control Simplified.* New York: Industrial, 2001. Print.

Krug, Steve. *Don't Make Me Think, Revisited: A Common Sense Approach to Web Usability.* 3rd ed. Berkeley: New Riders, 2014. Print.

Kruk, Robert. "Public, Private and Hybrid Clouds: What's the Difference?" *Techopedia.* Techopedia, 18 May 2012. Web. 21 Jan. 2016.

Kshemkalyani, Ajay D., and Mukesh Singhal. *Distributed Computing: Principles, Algorithms, and Systems.* New York: Cambridge UP, 2008. Print.

Kulisch, Ulrich. *Computer Arithmetic and Validity: Theory, Implementation, and Applications.* 2nd ed. Boston: De Gruyter, 2013. Print.

Kumar, Ela. *Natural Language Processing.* I K International Publishing House, 2011.

Kumari, Ramesh (2005) *Computers and Their Applications to Chemistry.* 2nd ed. Oxford, UK: Alpha Science International. Print.

Kuo, Sen M., Bob H. Lee, and Wenshun Tian. *Real-Time Digital Signal Processing: Fundamentals, Implementations and Applications.* 3rd ed. Hoboken: Wiley, 2013. Print.

Kupferschmid, Michael. *Classical Fortran Programming for Engineering and Scientific Applications.* Boca Raton, FL: CRC Press, 2009. Print.

Kurose, James F., and Keith W. Ross. *Computer Networking: A Top-Down Approach.* 6th ed. Boston: Pearson, 2013. Print.

Lackey, Ella Deon, et al. "Introduction to Public-Key Cryptography." *Mozilla Developer Network.* Mozilla, 21 Mar. 2015. Web. 4 Feb. 2016.

Lafferty, Edward L, Marion C. Michaud, and Myra J. Prelle. *Parallel Computing: An Introduction.* Park Ridge: Noyes Data Corporation, 1993. Print.

Lakhtakia, A., and R. J. Martín-Palma. *Engineered Biomimicry.* Amsterdam: Elsevier, 2013. Print.

Lalanda, Philippe, Julie A. McCann, and Ada Diaconescu, eds. *Autonomic Computing: Principles, Design and Implementation.* London: Springer, 2013. Print.

Lande, Daniel R. "Development of the Binary Number System and the Foundations of Computer Science." *Mathematics Enthusiast* 1 Dec. 2014: 513–40. Print.

Lander, Steve. "Disadvantages or Problems for Implementing Wi-Fi Technology." *Small Business—Chron.com.* Hearst Newspapers, n.d. Web. 22 Feb. 2016.

Langer, Arthur M. *Guide to Software Development: Designing and Managing the Life Cycle.* New York: Springer, 2012. Print.

Lardinois, Frederic. "Ukrainian Students Develop Gloves That Translate Sign Language into Speech." *TechCrunch.* AOL, 9 July 2012. Web. 19 Jan 2016.

Laszlo, Arp. "Why Is Website Design So Important?" *Sunrise Pro Websites & SEO.* Sunrise Pro Websites, 22 Jan. 2016. Web. 22 Feb. 2016.

Lathi, B. P. *Linear Systems and Signals.* 2nd rev. ed. New York: Oxford UP, 2010. Print.

Law, Averill M. *Simulation Modeling and Analysis.* 5th ed. New York: McGraw, 2015. Print.

Lee, John D., and Alex Kirlik, eds. *The Oxford Handbook of Cognitive Engineering.* New York: Oxford UP, 2013. Print.

Lee, Kent D. *Foundations of Programming Languages.* Cham: Springer, 2014. Print.

Lee, Roger Y., ed. *Applied Computing and Information Technology.* New York: Springer, 2014. Print.

Li, Han-Xiong, and XinJiang Lu. *System Design and Control Integration for Advanced Manufacturing.* Hoboken: Wiley, 2015. Print.

Liang, Hualou, Joseph D. Bronzino, and Donald R. Peterson, eds. *Biosignal Processing: Principles and Practices.* Boca Raton: CRC, 2012. Print.

Lien, Tracey. "Virtual Reality Isn't Just for Video Games." *Los Angeles Times.* Tribune, 8 Jan. 2015. Web. 23 Mar. 2016.

Lipiansky, Ed. *Electrical, Electronics, and Digital Hardware Essentials for Scientists and Engineers.* Hoboken: Wiley, 2013. Print.

Lippman, Stanley B, Josee Lajoie, and Barbara E. Moo. *C++ Primer* 5th ed. Upper Saddle River, NJ: Addison-Wesley, 2003. Print.

Lipson, Hod, and Melba Kurman. *Fabricated: The New World of 3D Printing.* Indianapolis: Wiley, 2013. Print.

Lipton, R. J., and E. B. Baum, eds., DNA Based Computers, DIMACS Series in Discrete Mathematics, and Theoretical Computer *Science,* 27, American Mathematical Society (1995).

Liu, Shih-Chii, Tobi Delbruck, Giacomo Indiveri, Adrian Whatley, and Rodney Douglas. *Event-Based Neuromorphic Systems.* Chichester: Wiley, 2015. Print.

Livingston, Steven, and Gregor Walter-Drop, eds. *Bits and Atoms: Information and Communication Technology in Areas of Limited Statehood.* New York: Oxford UP, 2014. Print.

Lohr, Steve. "Humanizing Technology: A History of Human-Computer Interaction." *New York Times: Bits.* New York Times, 7 Sept. 2015. Web. 31 Jan. 2016.

Loo, Alfred Waising, ed. *Distributed Computing Innovations for Business, Engineering, and Science.* Hershey: Information Science Reference, 2013. Print.

Luenendonk, Martin. "Top Programming Languages Used in Web Development." *Cleverism.* Cleverism, 21 June 2015. Web. 22 Feb. 2016.

MacLennan, Bruce J. *Principles of Programming Languages: Design, Evaluation, and Implementation.* Oxford: Oxford UP, 1999. Print.

Madhav, Sanjay. *Game Programming Algorithms and Techniques: A Platform-Agnostic Approach.* Upper Saddle River: Addison, 2014. Print.

Malcolme-Lawes, D.J. (1969) *Programming – ALGOL* London, UK: Pergamon Press. Print.

Mallick, Pradeep Kumar, ed. *Research Advances in the Integration of Big Data and Smart Computing.* Hershey: Information Science Reference, 2016. Print.

Malvik, Callie. "Graphic Design vs. Web Design: Which Career Is Right for You?" *Rasmussen College.* Rasmussen Coll., 25 July 2013. Web. 16 Mar. 2016.

Mandl, H., and A. Lesgold, eds. *Learning Issues for Intelligent Tutoring Systems.* New York: Springer, 1988. Print.

Mangan, Dan. "There's a Hack for That: Fitbit User Accounts Attacked." *CNBC.* CNBC, 8 Jan. 2016. Web. 28 Feb. 2016.

Manjoo, Farhad. "Now You're Talking!" *Slate.* Slate Group, 6 Apr. 2011. Web. 21 Mar. 2016.

Manjoo, Farhad. "Planet Android's Shaky Orbit." *New York Times* 28 May 2015: B1. Print.

Mara, Wil. *Software Development: Science, Technology, and Engineering.* New York: Children's, 2016. Print.

Marble, Scott, ed. *Digital Workflows in Architecture.* Basel: Birkhäuser, 2012. Print.

Marchant, Ben. "Game Programming in C and C++." *Cprogramming.com.* Cprogramming.com, 2011. Web. 16 Mar. 2016.

Marchewka, Jack T. *Information Technology Project Management.* 5th ed. Hoboken: Wiley, 2015. Print.

Margush, Timothy S. (2012) *Some Assembly Required. Assembly Language Programming with the AVR Microcontroller* Boca Raton, FL: CRC Press. Print.

Marsh, Joel. *UX for Beginners: A Crash Course in 100 Short Lessons.* Sebastopol: O'Reilly, 2016. Print.

Marshall, Gary. "The Story of Fitbit: How a Wooden Box Became a $4 Billion Company." *Wareable.* Wareable, 30 Dec. 2015. Web. 28 Feb. 2016.

Mason, Paul. *Understanding Computer Search and Research.* Chicago: Heinemann, 2015. Print.

Mathews, Clive. *An Introduction to Natural Language Processing Through Prolog* New York, NY: Routledge, 2014. Print.

Matsudaira, Kate. "Scalable Web Architecture and Distributed Systems." *The Architecture of Open Source Applications.* Ed. Amy Brown and Greg Wilson. Vol. 2. N.p.: Lulu, 2012. PDF file.

Matulka, Rebecca. "How 3D Printers Work." *Energy. gov.* Dept. of Energy, 19 June 2014. Web. 6 Jan. 2016.

McCauley, Renée, et al. "Debugging: A Review of the Literature from an Educational Perspective." *Computer Science Education* 18.2 (2008): 67–92. Print.

McConnell, Robert, James Haynes, and Richard Warren. "Understanding ASCII Codes." *NADCOMM.* NADCOMM, 14 May 2011. Web. 16 Feb. 2016.

McCracken, Harry. "Ten Momentous Moments in DOS History." *PCWorld.* IDG Consumer, n.d. Web. 31 Jan. 2016.

McDonald, Nicholas G. "Past, Present, and Future Methods of Cryptography and Data Encryption." *SpaceStation.* U of Utah, 2009. Web. 4 Feb. 2016.

McFedries, Paul. *Fixing Your Computer: Absolute Beginner's Guide.* Indianapolis: Que, 2014. Print.

McGrath, Mike (2015) *C++ Programming in Easy Steps* 4th ed. Leamington Spa, UK: Easy Steps Limited. Print.

McKay, Sinclair. *The Secret Lives of Codebreakers: The Men and Women Who Cracked the Enigma Code at Bletchley Park.* Plume, 2012.

McLellan, Charles. "The History of Windows: A Timeline." *ZDNet.* CBS Interactive, 14 Apr. 2014. Web. 15 Feb. 2016.

McMillan, Robert. "IBM Bets $3B That the Silicon Microchip Is Becoming Obsolete." *Wired.* Condé Nast, 9 July 2014. Web. 10 Mar. 2016.

McMillan, Robert. "Siri Will Soon Understand You a Whole Lot Better." *Wired.* Condé Nast, 30 June 2014. Web. 21 Mar. 2016.

McMullan, Dawn. "What Is Personalized Medicine?" *Genome* Spring 2014: n. pag. Web. 23 Dec. 2015.

McNeill, Dwight. *Using Person-Centered Health Analytics to Live Longer: Leveraging Engagement, Behavior Change, and Technology for a Healthy Life.* Upper Saddle River: Pearson, 2015. Print.

McNeill, Erin. "Even 'Digital Natives' Need Digital Training." *Education Week.* Editorial Projects in Education, 20 Oct 2015. Web. 26 Jan. 2016.

McNicoll, Arion. "Phonebloks: The Smartphone for the Rest of Your Life." *CNN.* Cable News Network, 19 Sept. 2013. Web. 29 Jan. 2016.

Méndez, Luis Argüelles. *A Practical Introduction to Fuzzy Logic Using LISP.* New York, NY: Springer, 2016. Print.

Menezes, Alfred J., Paul C. van Oorschot, and Scott A. Vanstone. *Handbook of Applied Cryptography.* Boca Raton: CRC, 1996. Print.

Mercer, Christina. "5 Best Storage Devices for Startups: Removable, SSD or Cloud, What Type of Storage Should Your Business Use?" *Techworld.* IDG UK, 22 Dec. 2015. Web. 7 Mar. 2016.

Mermin, David N. *Quantum Computer Science: An introduction.* New York, NY: Cambridge University Press, 2007. Print.

Metcalf, Michael, and John Reid. *The F Programming Language.* New York, NY: Oxford University Press, 1996. Print.

Metz, Sandi. *Practical Object-Oriented Design in Ruby: An Agile Primer.* Upper Saddle River: Addison, 2012. Print.

"Microprocessors: Explore the Curriculum." *Intel.* Intel Corp., 2015. Web. 11 Mar. 2016.

"Microprocessors." *MIT Technology Review.* MIT Technology Review, 2016. Web. 11 Mar. 2016.

Mihalcea, Rada and Radev Dragomir. *Graph-based Natural Language Processing and Information Retrieval.* Cambridge University Press, 2011.

"Milestones of Innovation." *American Institute for Medical and Biological Engineering.* Amer. Inst. for Medical and Biological Engineering, 2016. Web. 25 Jan. 2016.

Mili, Ali, and Fairouz Tchier. *Software Testing: Concepts and Operations.* Hoboken: Wiley, 2015. Print.

Miller, Charles, and Aaron Doering. *The New Landscape of Mobile Learning: Redesigning Education in an App-Based World.* New York: Routledge, 2014. Print.

Miller, Michael J. "The Rise of DOS: How Microsoft Got the IBM PC OS Contract." *PCMag.com.* PCMag Digital Group, 10 Aug. 2011. Web. 31 Jan. 2016.

Miller, Michelle D. *Minds Online: Teaching Effectively with Technology.* Cambridge: Harvard UP, 2014. Print.

Miniman, Stuart. "Hyperscale Invades the Enterprise." *Network Computing.* UBM, 13 Jan. 2014. Web. 8 Mar. 2016.

Mir, Nader F. *Computer and Communication Networks.* 2nd ed. Upper Saddle River: Prentice, 2015. Print.

Mishra, Umesh. *Semiconductor Device Physics and Design.* Place of publication not identified: Springer, 2014. Print.

Miszczak, Jarosław Adam. *High-Level Structures for Quantum Computing.* Williston, VT: Morgan and Claypool, 2012. Print.

Mitchell, Jamie L., and Rex Black. *Advanced Software Testing.* 2nd ed. 3 vols. Santa Barbara: Rocky Nook, 2015. Print.

Mitra, Tilak. *Practical Software Architecture: Moving from System Context to Deployment.* Indianapolis: IBM, 2015. Print.

Modi, Shimon K. *Biometrics in Identity Management.* Boston: Artech House, 2011. Print.

Mooallem, Jon. "The Afterlife of Cellphones." *New York Times Magazine.* New York Times, 13 Jan. 2008. Web. 9 Feb. 2016.

Morreale, Patricia, and Kornel Terplan, eds. *The CRC Handbook of Modern Telecommunications.* 2nd ed. Boca Raton: CRC, 2009. Print.

Morrison, Foster. *The Art of Modeling Dynamic Systems: Forecasting for Chaos, Randomness, and Determinism.* 1991. Mineola: Dover, 2008. Print.

Morselli, Carlo, ed. *Crime and Networks.* New York: Routledge, 2014. Print.

Moss, Frank. *The Sorcerers and Their Apprentices: How the Digital Magicians of the MIT Media Lab Are Creating the Innovative Technologies That Will Transform Our Lives.* New York: Crown Business, 2011. Print.

Moynihan, Tim. "Things Will Get Messy If We Don't Start Wrangling Drones Now." *Wired.* Condé Nast, 30 Jan. 2015. Web. 30 Jan. 2016.

"MS-DOS: A Brief Introduction." *Linux Information Project.* Linux Information Project, 30 Sept. 2006. Web. 31 Jan. 2016.

Mueller, Scott. *Upgrading and Repairing PCs.* 22nd ed. Indianapolis: Que, 2015. Print.

Murphy, Chris. "How to Save $150 Billion: Move All App Dev and Testing to the Cloud." *Forbes.* Forbes.com, 3 Feb. 2016. Web. 10 Feb. 2016.

Murphy, Michael P., and Metin Sitti. "Waalbot: Agile Climbing with Synthetic Fibrillar Dry Adhesives." *2009 IEEE International Conference on Robotics and Automation.* Piscataway: IEEE, 2009. *IEEE Xplore.* Web. 21 Jan. 2016.

Myers, Courtney Boyd, ed. *The AI Report. Forbes.* Forbes.com, 22 June 2009. Web. 18 Dec. 2015.

Myers, Glenford J., Tom Badgett, and Corey Sandler. *The Art of Software Testing.* Hoboken: Wiley, 2012. Print.

Naraine, Ryan. "Metasploit's H. D. Moore Releases 'War Dialing' Tools." *ZDNet.* CBS Interactive, 6 Mar. 2009. Web. 15 Mar. 2016.

Nayeem, Sk. Md. Abu, Jyotirmoy Mukhopadhyay, and S. B. Rao, eds. *Mathematics and Computing: Current Research and Developments.* New Delhi: Narosa, 2013. Print.

Neapolitan, Richard E. *Foundations of Algorithms.* 5th ed. Burlington: Jones, 2015. Print.

Neiderreiter, Harald, and Chaoping Xing. *Algebraic Geometry in Coding Theory and Cryptography.* Princeton: Princeton UP, 2009. Print.

Netzley, Patricia D. *How Serious a Problem Is Computer Hacking?* San Diego: ReferencePoint, 2014. Print.

Neuburg, Matt. *Programming iOS 8: Dive Deep into Views, View Controllers, and Frameworks.* Sebastopol: O'Reilly Media, 2014. Print.

Newman, Jared. "Android Laptops: The $200 Price Is Right, but the OS May Not Be." *PCWorld.* IDG Consumer & SMB, 26 Apr. 2013. Web. 27 Jan. 2016.

Newman, Jared. "With Android Lollipop, Mobile Multitasking Takes a Great Leap Forward." *Fast Company.* Mansueto Ventures, 6 Nov. 2014. Web. 27 Jan. 2016.

Nicolelis, Miguel A. and Ronald M. Cicurel. *The Relativistic Brain: How it works and why it cannot be simulated by a Turing machine.* CreateSpace Independent Publishing Platform, 2015.

Nielsen, Jakob. "F-Shaped Pattern for Reading Web Content." *Nielsen Norman Group.* Nielsen Norman Group, 17 Apr. 2006. Web. 4 Mar. 2016.

Noergaard, Tammy. *Embedded Systems Architecture: A Comprehensive Guide for Engineers and Programmers.* 2nd ed. Boston: Elsevier, 2013. Print.

Norman, Don. *The Design of Everyday Things.* Rev. and expanded ed. New York: Basic, 2013. Print.

Northrup, Tony. "Firewalls." *TechNet.* Microsoft, n.d. Web. 28 Feb. 2016.

Nystrom, Robert. *Game Programming Patterns.* N.p.: Author, 2009–14. Web. 16 Mar. 2016.

"Online Privacy: Using the Internet Safely." *Privacy Rights Clearinghouse.* Privacy Rights Clearinghouse, Jan. 2016. Web. 28 Feb. 2016.

O'Regan, Gerard. *A Brief History of Computing* 2nd ed., New York, NY: Springer-Verlag, 2012. Print.

Ohanian, Thomas. *Digital Nonlinear Editing: Editing Film and Video on the Desktop.* Woburn: Focal, 1998. Print.

Olley, Allan. "Can Machines Think Yet? A Brief History of the Turing Test." *Bubble Chamber.* U of Toronto's Science Policy Working Group, 23 June 2014. Web. 18 Dec. 2015.

Oloruntoba, Samuel. "SOLID: The First 5 Principles of Object Oriented Design." *Scotch.* Scotch.io, 18 Mar. 2015. Web. 1 Feb. 2016.

Openshaw, Stan, and Ian Turton. *High Performance Computing and the Art of Parallel Programming: An*

Introduction for Geographers, Social Scientists, and Engineers. London: Routledge, 2005. Print.

Oppy, Graham, and David Dowe. "The Turing Test." *Stanford Encyclopedia of Philosophy (Spring 2011 Edition).* Ed. Edward N. Zalta. Stanford U, 26 Jan. 2011. Web. 18 Dec. 2015.

Organick, E.I., Forsythe, A.I. and Plummer, R.P. (1978) *Programming Language Structures* New York, NY: Academic Press. Print.

Orwick, Penny, and Guy Smith. *Developing Drivers with the Windows Driver Foundation.* Redmond: Microsoft P, 2007. Print.

"Our Story." *CompTIA.* Computing Technology Industry Assn., n. d. Web. 31 Jan. 2016.

"Our Story." *Pixar.* Pixar, 2016. Web. 31 Jan. 2016.

Owen, Mark. *Practical Signal Processing.* New York: Cambridge UP, 2012. Print.

Paar, Christof, and Jan Pelzi. *Understanding Cryptography: A Textbook for Students and Practitioners.* Heidelberg: Springer, 2010. Print.

Paddock, Catharine. "How Self-Monitoring Is Transforming Health." *Medical News Today.* MediLexicon Intl., 15 Aug. 2013. Web. 26 Feb. 2016.

Pandolfi, Luciano. *Distributed Systems with Persistent Memory: Control and Moment Problems.* New York: Springer, 2014. Print.

Pandzu, Abhujit S. and Robert B. Macy. *Pattern Recognition with Neural Networks in C++.* Boca Raton, FL: CRC Press, 1996. Print.

Parashar, Manish, and Salim Hariri, eds. *Autonomic Computing: Concepts, Infrastructure, and Applications.* Boca Raton: CRC, 2007. Print.

Parent, Rick. *Computer Animation: Algorithms and Techniques.* Waltham: Elsevier, 2012. Print.

Parisi, Tony. *Learning Virtual Reality: Developing Immersive Experiences and Applications for Desktop, Web, and Mobile.* Sebastopol: O'Reilly, 2015. Print.

Parker, Jason, "The Continuing Evolution of iOS." *CNET.* CBS Interactive, 7 May 2014. Web. 26 Feb. 2016.

"Part Two: Communicating with Computers—The Operating System." *Computer Programming for Scientists.* Oregon State U, 2006. Web. 31 Jan. 2016.

Patel, Ruchika, and Parth Bhatt. "A Review Paper on Digital Watermarking and Its Techniques." *International Journal of Computer Applications* 110.1 (2015): 10–13. Web. 14 Mar. 2016.

Patel, Shuchi, and Avani Kasture. "E (Electronic) Waste Management Using Biological Systems—Overview." *International Journal of Current Microbiology and Applied Sciences* 3.7 (2014): 495–504. Web. 9 Feb. 2016.

Pathak, Parth H., and Rudra Dutta. *Designing for Network and Service Continuity in Wireless Mesh Networks.* New York: Springer, 2013. Print.

Patrizio, Andy. "The History of Visual Development Environments: Imagine There's No IDEs. It's Difficult If You Try." *Mendix.* Mendix, 4 Feb. 2013. Web. 23 Feb. 2016.

Patterson, David A., and John L. Hennessy. *Computer Organization and Design: The Hardware/Software Interface.* 5th ed. Waltham: Morgan, 2013. Print.

Pavel, M., et al. "The Role of Technology and Engineering Models in Transforming Healthcare." *IEEE Reviews in Biomedical Engineering.* IEEE, 2013. Web. 25 Jan. 2016.

"Personalized Medicine and Pharmacogenomics." *Mayo Clinic.* Mayo Foundation for Medical Education and Research, 5 June 2015. Web. 23 Dec. 2015.

Pele, Maria, and Carmen Cimpeanu. *Biotechnology: An Introduction.* Billerica: WIT, 2012. Print.

Pelleau, Marie, and Narendra Jussien. *Abstract Domains in Constraint Programming.* London: ISTE, 2015. Print.

Peterson, James L. *Computer Organization and Assembly Language Programming* New York, NY: Academic Press, 1978. Print.

Peterson, Larry L., and Bruce S. Davie. *Computer Networks: A Systems Approach.* 5th ed. Burlington: Morgan, 2012. Print.

Petzold, Charles. *The Annotated Turing: A Guided Tour Through Alan Turing's Historic Paper on Computability and the Turing Machine.* Wiley, 2008.

Pinch, T. J., and Karin Bijsterveld. *The Oxford Handbook of Sound Studies.* New York: Oxford UP, 2013. Print.

Pinola, Melanie. "Speech Recognition through the Decades: How We Ended Up with Siri." *PCWorld.* IDG Consumer & SMB, 2 Nov. 2011. Web. 21 Mar. 2016.

"Planned Obsolescence: A Weapon of Mass Discarding, or a Catalyst for Progress?" *ParisTech Review.* ParisTech Rev., 27 Sept. 2013. Web. 9 Feb. 2016.

Plumb, Charles. "Drones in the Workplace." *EmployerLINC.* McAfee and Taft, 14 Dec. 2015. Web. 21 Jan. 2016.

Pollitt, Mark. "A History of Digital Forensics." *Advances in Digital Forensics VI.* Ed. Kam-Pui Chow and Sujeet Shenoi. Berlin: Springer, 2010. 3–15. Print.

Pourhashemi, Ali, ed. *Chemical and Biochemical Engineering: New Materials and Developed Components.* Rev. Gennady E. Zaikov and A. K. Haghi. Oakville: Apple Acad., 2015. Print.

Prandoni, Paolo, and Martin Vetterli. *Signal Processing for Communications.* Boca Raton: CRC, 2008. Print.

Prasad, Bhanu, and S.R. Mahadeva Prasanna, eds. *Speech, Audio, Image and Biomedical Signal Processing Using Neural Networks.* New York, NY: Springer, 2008. Print.

"Precision (Personalized) Medicine." US Food and Drug Administration. Dept. of Health and Human Services, 18 Nov. 2015. Web. 23 Dec. 2015.

Priddy, Kevin L. and Keller, Paul E. *Artificial Neural Networks, An Introduction.* Bellingham, WA: SPIE Press, 2005. Print.

Proakis, John G., and Dimitris G. Manolakis. *Digital Signal Processing: Principles, Algorithms, and Applications.* 4th ed. Upper Saddle River: Prentice, 2007. Print.

Prokopenko, Mikhail. *Advances in Applied Self-Organizing Systems.* London: Springer, 2013. Print.

"Pros & Cons of Website Templates." Entheos. Entheos, n.d. Web. 22 Feb. 2016

"Protect Your Privacy on the Internet." Safety and Security Center. Microsoft, n. d. Web. 28 Feb. 2016.

Protalinski, Emil. "Windows 10 Ends 2015 under 10% Market Share." *VentureBeat.* VentureBeat, 1 Jan. 2016. Web. 26 Feb. 2016.

Pu, Di, and Alexander M. Wyglinski. *Digital Communication Systems Engineering with Software-Defined Radio.* London: Artech, 2013. Print.

Pullen, John Patrick. "This Is How Drones Work." *Time.* Time, 3 Apr. 2015. Web. 27 Jan. 2016.

Puryear, Martin. "Programming Trends to Look for This Year." *TechCrunch.* AOL, 13 Jan. 2016. Web. 7 Feb. 2016.

Quian, Quiroga R., and Stefano Panzeri. *Principles of Neural Coding.* Boca Raton: CRC, 2013. Print.

Rahimi, Saeed K., and Frank S. Haug. *Distributed Database Management Systems: A Practical Approach.* Hoboken: Wiley, 2010. Print.

Rajaraman, V. *Computer Programming in Fortran 77.* 4th ed., New Delhi, IND: Prentice-Hall of India Pvt., 2006. Print.

Rajasekaran, Sanguthevar. *Multicore Computing: Algorithms, Architectures, and Applications.* Boca Raton: CRC, 2013. Print.

Ramasubbu, Suren. "How Technology Can Help Language Learning." *Huffington Post.* TheHuffingtonPost.com, 3 June 2015. Web. 19 Jan. 2016.

Rao, M. Ananda, and J. Srinavas. *Neural Networks. Algorithms and Applications.* Pangbourne, UK: Alpha Science International, 2003. Print.

Rasmussen, Nicolas. *Gene Jockeys: Life Science and the Rise of Biotech Enterprise.* Baltimore: Johns Hopkins UP, 2014. Print.

Ravindranath, Mohana. "PCs Lumber towards the Technological Graveyard." *Guardian.* Guardian News and Media, 11 Feb. 2014. Web. 10 Mar. 2016.

Rawls, Rod. R., Paul F. Richard and Mark A. Hagen. *Visual LISP Programming: Principles and Techniques* Tinley Park, IL: Goodheart-Willcox, 2007. Print.

Raymond, Eric S. "Origins and History of Unix, 1969–1995." *The Art of UNIX Programming.* Boston: Pearson Education, 2004. Print.

Reed, Jeffrey H. *Software Radio: A Modern Approach to Radio Engineering.* New York: Prentice, 2002. Print.

Reimer, Jeremy. "A History of the GUI." *Ars Technica.* Condé Nast, 5 May 2005. Web. 31 Jan. 2016.

Ribble, Mike. *Digital Citizenship in Schools: Nine Elements All Students Should Know.* Eugene: Intl. Soc. for Technology in Education, 2015. Print.

Rice, Daniel M. *Calculus of Thought: Neuromorphic Logistic Regression in Cognitive Machines.* Waltham: Academic, 2014. Print.

Rieffel, Eleanor, and Wolfgang Polak. *Quantum Computing: A Gentle Introduction.* Cambridge, MA: MIT Press, 2011. Print.

Roberts, Fred S., and Barry Tesman. *Applied Combinatorics.* 2nd ed. Boca Raton: Chapman, 2012. Print.

Roberts, Richard M. *Computer Service and Repair.* 4th ed. Tinley Park: Goodheart, 2015. Print.

Roblyer, M. D., and Aaron H. Doering. *Integrating Educational Technology into Teaching.* 6th ed. Boston: Pearson, 2013. Print.

Rockett, Angus. *The Materials Science of Semiconductors.* New York, NY: Springer, 2010. Print.

Rogers, Joey. *Object-Oriented Neural Networks in C++.* New York, NY: Academic Press, 1997. Print.

Rogers, Scott. *Swipe This!: The Guide to Great Touchscreen Game Design.* Chichester: Wiley, 2012. Print.

Roosta, Seyed H. *Parallel Processing and Parallel Algorithms: Theory and Computation.* New York: Springer, 2013. Print.

Rountree, Derrick, and Ileana Castrillo. *The Basics of Cloud Computing.* Waltham: Elsevier, 2014. Print.

Rouse, Margaret. "Sarbanes-Oxley Act (SOX)." *TechTarget.* TechTarget, June 2014. Web. 31 Mar. 2014.

Rozanski, Nick, and Eoin Woods. *Software Systems Architecture: Working with Stakeholders Using Viewpoints and Perspectives.* 2nd ed. Upper Saddle River: Addison, 2012. Print.

Rubin, Michael. *Nonlinear—A Field Guide to Digital Video and Film Editing.* Gainesville: Triad, 2000, Print.

Rutishauer, Heinz (1967) "Description of ALGOL-60" Chapter in Bauer, F.L., Householder, I.S., Olver, F.W.J., Rutishauer, H., Samelson, K. and Stiefel, E., eds. *Handbook for Automatic Computation* Berlin, GER: Springer-Verlag. Print.

Ryan, Janel. "Five Basic Things You Should Know about Cloud Computing." *Forbes.* Forbes.com, 30 Oct. 2013. Web. 30 Oct. 2013.

Sabbatini, Renato M. E. "Imitation of Life: A History of the First Robots." *Brain & Mind* 9 (1999): n. pag. Web. 21 Jan. 2016.

Saffer, Dan. *Designing Gestural Interfaces.* Beijing: O'Reilly, 2008. Print.

Saltman, Dave. "Tech Talk: Turning Digital Natives into Digital Citizens." *Harvard Education Letter* 27.5 (2011): n. pag. *Harvard Graduate School of Education.* Web. 27 Jan. 2016.

Saltzman, Steven. *Music Editing for Film and Television: The Art and the Process.* Burlington: Focal, 2015. Print.

Saltzman, W. Mark. "Lecture 1—What Is Biomedical Engineering?" *BENG 100: Frontiers of Biomedical Engineering.* Yale U, Spring 2008. Web. 23 Jan. 2016.

Salz, Peggy Anne, and Jennifer Moranz. *The Everything Guide to Mobile Apps: A Practical Guide to Affordable Mobile App Development for Your Business.* Avon: Adams Media, 2013. Print.

Sammons, John. *The Basics of Digital Forensics: The Primer for Getting Started in Digital Forensics.* Waltham: Syngress, 2012. Print.

Sandberg, Bobbi. *Networking: The Complete Reference.* 3rd ed. New York: McGraw, 2015. Print.

Sanders, James. "Hybrid Cloud: What It Is, Why It Matters." *ZDNet.* CBS Interactive, 1 July 2014. Web. 10 Jan. 2016.

Sarkar, Jayanta. *Computer Aided Design: A Conceptual Approach.* Boca Raton: CRC, 2015. Print.

Savage, Terry Michael, and Karla E. Vogel. *An Introduction to Digital Multimedia.* 2nd ed. Burlington: Jones, 2014. 256–58. Print.

Savischenko, Nikolay V. *Special Integral Functions Used in Wireless Communications Theory.* N.p.: World Scientific, 2014. Digital file.

Schmidt, Richard F. *Software Engineering: Architecture-driven Software Development.* Waltham: Morgan, 2013. Print.

Schmidt, Silke, and Otto Rienhoff, eds. *Interdisciplinary Assessment of Personal Health Monitoring.* Amsterdam: IOS, 2013. Print.

Schmitt, Christopher. *Designing Web & Mobile Graphics: Fundamental Concepts for Web and Interactive Projects.* Berkeley: New Riders, 2013. Print.

Schou, Corey, and Steven Hernandez. *Information Assurance Handbook: Effective Computer Security and Risk Management Strategies.* New York: McGraw, 2015. Print.

Schwartz, John. "In the Lab: Robots That Slink and Squirm." *New York Times.* New York Times, 27 Mar. 2007. Web. 21 Jan. 2016.

Scott, Michael L. *Programming Language Pragmatics.* 4th ed., Waltham, MA: Morgan Kaufmann, 2016. Print.

Segall, Richard S., Jeffrey S. Cook, and Qingyu Zhang, eds. *Research and Applications in Global Supercomputing.* Hershey: Information Science Reference, 2015. Print.

Seibel, Peter. *Practical Common LISP* New York, NY: Apress/Springer-Verlag. Print.

Seidl, Martina, et al. *UML@Classroom: An Introduction to Object-Oriented Modeling.* Cham: Springer, 2015. Print.

Serpanos, Dimitrios N., and Tilman Wolf. *Architecture of Network Systems.* Burlington: Morgan, 2011. Print.

"Shadow RAM Basics." *Microsoft Support.* Microsoft, 4 Dec. 2015. Web. 10 Mar. 2016.

Shahani, Aarthi. "Biometrics May Ditch the Password, But Not the Hackers." *All Things Considered.* NPR, 26 Apr. 2015. Web. 21 Jan. 2016.

Shakarian, Paulo, Jana Shakarian, and Andrew Ruef. *Introduction to Cyber-Warfare: A Multidisciplinary Approach.* Waltham: Syngress, 2013. Print.

Shen, Jialie, John Shepherd, Bin Cui, and Ling Liu. *Intelligent Music Information Systems: Tools and Methodologies.* Hershey: IGI Global, 2008. Print.

Shenoi, Belle A. *Introduction to Digital Signal Processing and Filter Design.* Hoboken: Wiley, 2006. Print.

Shinder, Deb. "So You Want to Be a Computer Forensics Expert." *TechRepublic.* CBS Interactive, 27 Dec. 2010. Web. 2 Feb. 2016.

Shustek, Len. "Microsoft MS-DOS Early Source Code." *Computer History Museum.* Computer History Museum, 2013. Web. 31 Jan. 2016.

Silberschatz, Abraham, Peter B. Galvin, and Greg Gagne. *Operating Systems Concepts.* 9th ed. Hoboken: Wiley, 2012. Print.

Silberschatz, Abraham, Peter B. Galvin, and Greg Gagne. *Operating System Concepts Essentials.* 2nd ed. Wiley, 2014. Print.

Silver, H. Ward. "Digital Code Basics." *Qst* 98.8 (2014): 58–59. PDF file.

Simpson, James, ed. *The Routledge Handbook of Applied Linguistics.* New York: Routledge, 2011. Print.

Singh, Amandeep. "Top 13 Tips for Writing Effective Test Cases for Any Application." *Quick Software Testing.* QuickSoftwareTesting, 23 Jan. 2014. Web. 11 Feb. 2016.

Sinnen, Oliver. *Task Scheduling for Parallel Systems.* Hoboken: Wiley, 2007. Print.

Sito, Tom. *Moving Innovation: A History of Computer Animation.* Cambridge: MIT P, 2013. Print.

Smith, Bud E. *Green Computing: Tools and Techniques for Saving Energy, Money, and Resources.* Boca Raton: CRC, 2014. Print.

Smith, Catharine. "Tim Cook Issues Apology for Apple Maps." *Huffpost Tech.* TheHuffingtonPost.com, 28 Sept. 2012. Web. 11 Feb. 2016.

Smith, Eric N. *Workplace Security Essentials: A Guide for Helping Organizations Create Safe Work Environments.* Oxford: Butterworth-Heinemann, 2014. Print.

Smith, Matt. "Wi-Fi vs. Ethernet: Has Wireless Killed Wired?" *Digital Trends.* Designtechnica, 18 Jan. 2013. Web. 22 Feb. 2016.

Snoke, David W. *Electronics: A Physical Approach.* Boston: Addison, 2014. Print.

Soare, Robert I. *Turing Computability: Theory and Applications (Theory and Applications of Computability).* Springer, 2016.

Soares, Marcelo M., and Francisco Rebelo. *Advances in Usability Evaluation.* Boca Raton: CRC, 2013. Print.

Solnon, Christine. *Ant Colony Optimization and Constraint Programming.* Hoboken: Wiley, 2010. Print.

Solomon, Nancy B., ed. *Architecture: Celebrating the Past, Designing the Future.* New York: Visual Reference, 2008. Print.

Song, Dong-Ping. *Optimal Control and Optimization of Stochastic Supply Chain Systems.* London: Springer, 2013. Print.

Sonmez, John Z. *Soft Skills: The Software Developer's Life Manual.* Shelter Island: Manning, 2015. Print.

Sottilare, R., Graesser, A., Hu, X., and Holden, H. (Eds.). (2013). *Design Recommendations for Intelligent Tutoring Systems: Volume 1 - Learner Modeling.* Orlando, FL: U.S. Army Research Laboratory. ISBN 978-0-9893923-0-3. Available at: https://gift-tutoring.org/documents/42

Sozański, Krzysztof. *Digital Signal Processing in Power Electronics Control Circuits.* New York: Springer, 2013. Print.

Spence, Ewan. "New Android Malware Strikes at Millions of Smartphones." *Forbes.* Forbes.com, 4 Feb. 2015. Web. 11 Mar. 2016.

"Spyware." *Secure Purdue.* Purdue U, 2010. Web. 11 Mar 2016.

St. Germain, H. James de. "Debugging Programs." *University of Utah.* U of Utah, n.d. Web. 31 Jan. 2016.

"State Laws Related to Internet Privacy." *National Conference of State Legislatures.* NCSL, 5 Jan. 2016. Web. 28 Mar. 2016.

Stallings, William, and Lawrie Brown. *Computer Security: Principles and Practice.* 3rd ed. Boston: Pearson, 2015. Print.

Stallings, William. *Operating Systems: Internals and Design Principles.* Boston: Pearson, 2015. Print.

Stanley, Jay. "'Drones' vs 'UAVs'—What's behind a Name?" *ACLU.* ACLU, 20 May 2013. Web. 27 Jan. 2016.

Stanton, Jeffrey, and Kathryn R. Slam. *The Visible Employee: Using Workplace Monitoring and Surveillance to Protect Information Assets without Compromising*

Employee Privacy or Trust. Medford: Information Today, 2006. Print.

Stein, N. L., and S.W. Raudenbush, eds. *Developmental Cognitive Science Goes to School.* New York: Routledge, 2011. Print

Steiner, Craig (1990) *The 8051/8052 Microcontroller: Architecture, Assembly Language and Hardware Interfacing* Boca Raton, FL: Universal Publishers. Print.

Sterling, Leon, ed. *The Practice of Prolog.* Boston, MA: MIT Press,1990. Print.

Stevens, Tim. "Fitbit Review." *Engadget.* AOL, 15 Oct. 2009. Web. 28 Feb. 2016.

"Storage vs. Memory." *Computer Desktop Encyclopedia.* Computer Lang., 1981–2016. Web. 10 Mar. 2016.

Stokes, Jon. *Inside the Machine: An Illustrated Introduction to Microprocessors and Computer Architecture.* San Francisco: No Starch, 2015. Print.

Stokes, Jon. "RAM Guide Part I: DRAM and SDRAM Basics." *Ars Technica.* Condé Nast, 18 July 2000. Web. 10 Mar. 2016.

Stonebank, M. "UNIX Introduction." *University of Surrey.* U of Surrey, 2000. Web. 28 Feb. 2016.

Streib, James T. (2011) *Guide to Assembly Language. A Concise Introduction* New York, NY: Springer. Print.

Streib, James T., and Takako Soma. *Guide to Java: A Concise Introduction to Programming.* New York: Springer, 2014. Print.

Stroustrup, Bjarne. *The C++ Programming Language* 4th ed. Upper Saddle River, NJ: Addison-Wesley, 2013. Print.

Stuart, Allison. "File Formats Explained: PDF, PNG and More." *99Designs.* 99Designs, 21 May 2015. Web. 11 Feb. 2016.

Sun, Jiming, Vincent Zimmer, Marc Jones, and Stefan Reinauer. *Embedded Firmware Solutions: Development Best Practices for the Internet of Things.* Berkeley: ApressOpen, 2015. Print.

Tabak, Filiz, and William Smith. "Privacy and Electronic Monitoring in the Workplace: A Model of Managerial Cognition and Relational Trust Development." *Employee Responsibilities and Rights Journal* 17.3 (2005): 173–89. Print.

Tabini, Marco. "Hidden Magic: A Look at the Secret Operating System inside the iPhone." *MacWorld.* IDG Consumer & SMB, 20 Dec. 2013. Web. 9 Mar. 2016.

Takahashi, Dean. "The App Economy Could Double to $101 Billion by 2020." *VB.* Venture Beat, 10 Feb. 2016. Web. 11 Mar. 2016.

Talbot, James, and Justin McLean. *Learning Android Application Programming: A Hands- On Guide to Building Android Applications.* Upper Saddle River: Addison, 2014. Print.

Tan, Li, and Jean Jiang. *Digital Signal Processing: Fundamentals and Applications.* 2nd ed. Boston: Academic, 2013. Print.

Tanenbaum, Andrew S., and Herbert Bos. *Modern Operating Systems.* 4th ed. New York: Pearson, 2014. Print.

Tarantola, Andrew. "Why Frame Rate Matters." *Gizmodo.* Gizmodo, 14 Jan. 2015. Web. 11 Mar. 2016.

Tatnall, Arthur, and Bill Davey, eds. *Reflections on the History of Computers in Education: Early Use of Computers and Teaching about Computing in Schools.* Heidelberg: Springer, 2014. Print.

Taylor, Ben. "Cloud Storage vs. External Hard Drives: Which Really Offers the Best Bang for your Buck?" *PCWorld.* IDG Consumer and SMB, 10 July 2014. Web. 7 Mar. 2016.

Taylor, Harriet. "How Your Home Will Know What You Need Before You Do." *CNBC.* CNBC, Jan 6 2016. Web. 11 Mar. 2016.

Taylor, Richard N., Nenad Medvidović, and Eric M. Dashofy. *Software Architecture: Foundations, Theory, and Practice.* Hoboken: Wiley, 2010. Print

"The Digital Millennium Copyright Act of 1998." *Copyright.* US Copyright Office, 28 Oct 1998. Web. 23 Jan 2016.

"The History of the Integrated Circuit." *Nobelprize.org.* Nobel Media, 2014. Web. 31 Mar. 2016.

"The Mind-Blowing Possibilities of Quantum Computing." *TechRadar.* Future, 17 Jan. 2010. Web. 26 Mar. 2016.

"The Printed World." *Economist.* Economist Newspaper, 10 Feb. 2011. Web. 6 Jan. 2016.

"The Unicode Standard: A Technical Introduction." *Unicode.org.* Unicode, 25 June 2015. Web. 3 Mar. 2016.

Thieman, William J., and Michael A. Palladino. *Introduction to Biotechnology.* 3rd ed. San Francisco: Benjamin, 2012. Print.

"Timeline of Computer History: 1963." *Computer History Museum.* Computer History Museum, 1 May 2015. Web. 23 Feb. 2016.

Toal, Ray. "Algorithms and Data Structures." *Ray Toal*. Loyola Marymount U, n.d. Web. 19 Jan. 2016.

Tomei, Lawrence A., ed. *Encyclopedia of Information Technology Curriculum Integration*. 2 vols. Hershey: Information Science Reference, 2008. Print.

Tomlinson, Brian, ed. *Applied Linguistics and Materials Development*. New York: Bloomsbury, 2013. Print.

Tooley, Mike. *Electronic Circuits: Fundamentals and Applications*. 4th ed. New York: Routledge, 2015. Print.

Toomey, Warren. "The Strange Birth and Long Life of Unix." *IEEE Spectrum*. IEEE, 28 Nov. 2011. Web. 28. Feb. 2016.

Topol, Eric J. *The Creative Destruction of Medicine: How the Digital Revolution Will Create Better Health Care*. New York: Basic, 2013. Print.

Tosoni, Simone, Matteo Tarantino, and Chiara Giaccardi, eds. *Media and the City: Urbanism, Technology and Communication*. Newcastle upon Tyne: Cambridge Scholars, 2013. Print.

Touretzky, David S. *Common LISP. A Gentle Introduction to Symbolic Computation* Mineola, NY: Dover Publications. Print.

Treder, Marcin. "Wireframes vs. Prototypes: What's the Difference?" *Six Revisions*. Jacob Gube, 11 Apr. 2014. Web. 4 Mar. 2016.

Tripathy, B. K., and D. P. Acharjya, eds. *Global Trends in Intelligent Computing Research and Development*. Hershey: Information Science Reference, 2014. Print.

Tucker, Allen B., Ralph Morelli, and Chamindra de Silva. *Software Development: An Open Source Approach*. Boca Raton: CRC, 2011. Print.

Tucker, Allen B, Teofilo F. Gonzalez, and Jorge L. Diaz-Herrera. *Computing Handbook*. Boca Raton, FL: CRC Press, 2014. Print.

Tulchinsky, Theodore H., Elena Varavikova, Joan D. Bickford, and Jonathan E. Fielding. *The New Public Health*. New York: Academic, 2014. Print.

Turing, Alan M. "Computing Machinery and Intelligence." *Mind* 59.236 (1950): 433–60. Web. 23 Dec. 2015.

"Turing Test." *Encyclopædia Britannica*. Encyclopædia Britannica, 23 Sept. 2013. Web. 18 Dec. 2015.

"Types of Wireless Networks." *Commotion*. Open Technology Inst., n.d. Web. 3 Mar. 2016.

"Unicode 8.0.0." *Unicode.org*. Unicode, 17 June 2015. Web. 3 Mar. 2016.

"Unmanned Aircraft Systems (UAS) Frequently Asked Questions." *Federal Aviation Administration*. US Dept. of Transportation, 18 Dec. 2015. Web. 11 Feb. 2016.

"Unretouched by Human Hand." *Economist*. Economist Newspaper, 12 Dec. 2002. Web. 14 Mar. 2016.

"USA Patriot Act." *Electronic Privacy Information Center*. EPIC, 31 May 2015. Web. 28 Mar. 2016.

"User Interface Design Basics." *Usability*. US Dept. of Health and Human Services, 2 Feb. 2016. Web. 2 Feb. 2016.

Vacca, John R. *Computer and Information Security Handbook*. Amsterdam: Kaufmann, 2013. Print.

Vacca, John, ed. *Network and System Security*. 2nd ed. Waltham: Elsevier, 2014. Print.

Van den Broek, Egon L. "Beyond Biometrics." *Procedia Computer Science* 1.1 (2010): 2511–19. Print.

Van Roy, Peter. *Concepts, Techniques, and Models of Computer Programming*. Cambridge: MIT P, 2004. Print.

Van Schuppen, Jan H., and Tiziano Villa, eds. *Coordination Control of Distributed Systems*. Cham: Springer, 2015. Print

Vella, Matt. "Nest CEO Tony Fadell on the Future of the Smart Home." *Time*. Time, 26 June 2014. Web. 12 Mar. 2016.

Vetterli, Martin, Jelena Kovacevic, and Vivek K. Goyal. *Foundations of Signal Processing*. Cambridge: Cambridge UP, 2014. Print.

Vick, Paul (2004) *The Visual BASIC .NET Programming Language*. Boston, MA: Addison-Wesley. Print.

Wachter, Robert M. *The Digital Doctor: Hope, Hype, and Harm at the Dawn of Medicine's Computer Age*. New York: McGraw, 2015. Print.

Wagner, Carl. "Choice, Chance, and Inference." *Math.UTK.edu*. U of Tennessee, Knoxville, 2015. Web. 10 Feb. 2016.

Walker, Henry M. *The Tao of Computing*. 2nd ed. Boca Raton: CRC, 2013. Print.

Wang, Jason. "HIPAA Compliance: What Every Developer Should Know." *Information Week*. UBM, 11 July 2014. Web. 26 Feb. 2016.

Wang, John, ed. *Optimizing, Innovating, and Capitalizing on Information Systems for Operations*. Hershey: Business Science Reference, 2013. Print.

Warren, Tom. "Windows Turns 30: A Visual History." *Verge*. Vox Media, 19 Nov. 2015. Web. 26 Feb. 2016.

Watanabe, Shinji, and Jen-Tzung Chien. *Bayesian Speech and Language Processing*. Cambridge University Press, 2015.

Watt, David A. *Programming Language Design Concepts*. West Sussex: Wiley, 2004. Print.

Wei, Hung-Yu, Jarogniew Rykowski, and Sudhir Dixit. *WiFi, WiMAX, and LTE Multi- Hop Mesh Networks: Basic Communication Protocols and Application Areas*. Hoboken: Wiley, 2013. Print.

Weisfeld, Matt. *The Object-Oriented Thought Process*. 4th ed. Upper Saddle River: Addison, 2013. Print.

Weller, Nathan. "A Look into the Future of Web Design: Where Will We Be in 20 Years?" *Elegant Themes Blog*. Elegant Themes, 9 May 2015. Web. 22 Feb. 2016.

Wells, Chris. *The Civic Organization and the Digital Citizen: Communicating Engagement in a Networked Age*. New York: Oxford UP, 2015. Print.

Wexelblat, Richard L. *History of Programming Languages*. New York, NY: Academic Press, 1981. Print.

"What Is a Driver?" *Microsoft Developer Network*. Microsoft, n.d. Web. 10 Mar. 2016.

"What Is E-waste?" *Step: Solving the E-waste Problem*. United Nations U/Step Initiative, 2016. Web. 9 Feb. 2016.

"What Is Personally Identifiable Information (PII)?" *U Health*. U of Miami Health System, n.d. Web. 28 Feb. 2016.

"What Is Pharmacogenomics?" *Genetics Home Reference*. Natl. Insts. of Health, 21 Dec. 2015. Web. 23 Dec. 2015.

"What Is the CompTIA A+ Certification?" *Knowledge Base*. Indiana U, 15 Jan. 2015. Web. 31 Jan. 2016.

"What Is Unicode?" *Unicode.org*. Unicode, 1 Dec. 2015. Web. 3 Mar. 2016.

"What Is Unix?" *Knowledge Base*. Indiana U, 2015. Web. 28 Feb. 2016.

"What Went Wrong? Finding and Fixing Errors through Debugging." *Microsoft Developer Network Library*. Microsoft, 2016. Web. 31 Jan. 2016.

"What You Can Do." *Green Computing*. U of California, Berkeley, n.d. Web. 28 Feb. 2016.

"When Healthcare and Computer Science Collide." *Health Informatics and Health Information Management*. Pearson/U of Illinois at Chicago, n.d. Web. 23 Dec. 2015.

"Why Do You Need to Test Software?" *BBC Academy*. BBC, 23 Feb. 2015. Web. 7 Feb. 2016.

White, Ron. *How Computers Work: The Evolution of Technology*. Illus. Tim Downs. 10th ed. Indianapolis: Que, 2015. Print.

Wieber, Pierre-Brice, Russ Tedrake, and Scott Kuindersma. "Modeling and Control of Legged Robots." *Handbook of Robotics*. Ed. Bruno Siciliano and Oussama Khatib. 2nd ed. N.p.: Springer, n.d. (forthcoming). *Scott Kuindersma—Harvard University*. Web. 6 Jan. 2016

Williams, Brad, David Damstra, and Hal Stern. *Professional WordPress: Design and Development*. 3rd ed. Indianapolis: Wiley, 2015. Print.

Williams, Rhiannon. "Apple iOS: A Brief History." *Telegraph*. Telegraph Media Group, 17 Sept. 2015. Web. 25 Feb. 2016.

Williams, Rhiannon, "iOS 9: Should You Upgrade?" *Telegraph*. Telegraph Media Group, 16 Sept. 2015. Web. 25 Feb. 2016.

Williams, Richard N. *Internet Security Made Easy: Take Control of Your Computer*. London: Flame Tree, 2015. Print.

Wills, Craig E. "Process Synchronization and Interprocess Communication." *Computing Handbook: Computer Science and Software Engineering*. Ed. Teofilo Gonzalez and Jorge Díaz-Herrera. 3rd ed. Boca Raton: CRC, 2014. 52-1–21. Print.

Wilson, Peter. *The Circuit Designer's Companion*. 3rd ed. Waltham: Newnes, 2012. Print.

Winder, Catherine, and Zahra Dowlatabadi. *Producing Animation*. Waltham: Focal, 2011. Print.

"Wireless History Timeline." *Wireless History Foundation*. Wireless Hist. Foundation, 2016. Web. 22 Feb. 2016.

Wolf, Marilyn. *High Performance Embedded Computing: Architectures, Applications, and Methodologies*. Amsterdam: Elsevier, 2014. Print.

Wood, David. *Interface Design: An Introduction to Visual Communication in UI Design*. New York: Fairchild, 2014. Print.

Wood, Lamont. "The 8080 Chip at 40: What's Next for the Mighty Microprocessor?" *Computerworld*. Computerworld, 8 Jan. 2015. Web. 12 Mar. 2016.

Woods, Dan. "Why Adopting the Declarative Programming Practices Will Improve Your Return from Technology." *Forbes*. Forbes.com, 17 Apr. 2013. Web. 2 Mar. 2016.

Worstall, Tim. "Is Unix Now the Most Successful Operating System of All Time?" *Forbes*. Forbes.com, 7 May 2013. Web. 7 Mar. 2016.

Xue, Su. *Data-Driven Image Editing for Perceptual Effectiveness*. New Haven: Yale U, 2013. Print.

Yakowitz, Will. "When Monitoring Your Employees Goes Horribly Wrong." *Inc.* Mansueto Ventures, 14 Feb. 2016. Web. 21 Feb. 2016.

Yamamoto, Jazon. *The Black Art of Multiplatform Game Programming*. Boston: Cengage, 2015. Print.

Yerby, Jonathan. "Legal and Ethical Issues of Employee Monitoring." *Online Journal of Applied Knowledge Management* 1.2 (2013): 44–54. Web. 21 Jan. 2016.

Yu, F. Richard, Xi Zhang, and Victor C. M. Leung, eds. *Green Communications and Networking*. Boca Raton: CRC, 2013. Print.

Yu, P.Y, and M Cardona. *Fundamentals of Semiconductors: Physics and Materials Properties*. Berlin: Springer, 2010. Print.

Zeller, Andreas. *Why Programs Fail: A Guide to Systematic Debugging*. Burlington: Kaufmann, 2009. Print.

Zeng, An-Ping, ed. *Fundamentals and Application of New Bioproduction Systems*. Berlin: Springer, 2013. Print.

Zetter, Kim. "California Now Has the Nation's Best Digital Privacy Law." *Wired*. Condé Nast, 8 Oct. 2015. Web. 28 Mar. 2016.

Zhong, Jian-Jiang, ed. *Future Trends in Biotechnology*. Berlin: Springer, 2013. Print.

Zhou, Weichang, and Anne Kantardjieff, eds. *Mammalian Cell Cultures for Biologics Manufacturing*. Berlin: Springer, 2014. Print

INDEX